MOLECULAR BIOLOGY
(Principles and Practices)

MOLECULAR BIOLOGY
(Principles and Practices)

By

DR. PRIYANKA SIWACH
Lecturer
Department of Biotechnology
Guru Jambheshwar University
of Science & Technology
Hisar (Haryana)

DR. NAMITA SINGH
Lecturer
Department of Biotechnology
Guru Jambheshwar University
of Science & Technology
Hisar (Haryana)

UNIVERSITY SCIENCE PRESS
(An Imprint of Laxmi Publications Pvt. Ltd.)

BANGALORE • CHENNAI • COCHIN • GUWAHATI • HYDERABAD
JALANDHAR • KOLKATA • LUCKNOW • MUMBAI • RANCHI
NEW DELHI • BOSTON, USA

Published by :
UNIVERSITY SCIENCE PRESS
(An Imprint of Laxmi Publications Pvt. Ltd.)
113, Golden House, Daryaganj,
New Delhi-110002

Phone : 011-43 53 25 00
Fax : 011-43 53 25 28

www.laxmipublications.com
info@laxmipublications.com

First Edition : 2007
Reprint : 2009

OFFICES

✆ **Bangalore**	080-26 75 69 30	✆ **Chennai**	044-24 34 47 26
✆ **Cochin**	0484-237 70 04, 405 13 03	✆ **Guwahati**	0361-251 36 69, 251 38 81
✆ **Hyderabad**	040-24 65 23 33	✆ **Jalandhar**	0181-222 12 72
✆ **Kolkata**	033-22 27 43 84	✆ **Lucknow**	0522-220 99 16
✆ **Mumbai**	022-24 91 54 15, 24 92 78 69	✆ **Ranchi**	0651-220 44 64

UMB–9399–200–MOLECULAR BIOLOGY **C—17452/09/04**

Typeset at : ABRO Enterprises, Delhi. **Printed at : Repro Knowledgecast Limited, Thane**

Contents

Preface

Molecular Biology is a distinct discipline dealing with understanding the biological processes at molecular level. It is the result of coalescence of different disciplines like cell biology, genetics and biochemistry and has applications as divergent as plant and animal breeding, archaeology, diagnostic tests for a wide range of inherited conditions, diagnosis and treatment of chronic illnesses, desired manipulation of genome, decoding and exploiting the genetic information in the benefit of mankind, improving nutritional content of food crops and many more. In fact molecular biology techniques have played a decisive role in establishing modern biology as a qualitative and quantitative science. These techniques are based on a thorough understanding of principles of the biological processes as well as principles of modern instrumentation. Many-a-time students regard Molecular biology techniques merely as a collection of empirical instructions for performing qualitative and quantitative analysis. They follow instructions mechanically without understanding the theoretical considerations and principles that underlie various procedures. The aim of this book is to provide the readers an information capsule about some of the popular molecular biology techniques.

The book predominantly centers on a molecular biology laboratory which is routinely handled by undergraduates, post graduates as well as research students in the field of biotechnology. The first unit exclusively deals with the setting up of the laboratory, basic practices done routinely and safety guidelines about different biohazards. We hope this unit will be of immense help to the newcomers who are unaware of the dark side of the fascinating and so popular techniques of molecular biology. The second unit is focused on the various types of calculations involved while performing an experiment. This will guide the students how to prepare different solutions of required concentration and different buffers at different pH. The following units deal with the important techniques which are routinely carried out by students in their practical classroom as well as in their research assignments. Every technique is discussed in detail by giving a theoretical base in the starting and then discussing about principles, requirements as well as the methodology of the process. Some additional information which can make the experiment a success is given in every unit under the heading of 'Remember'. This additional information must always be remembered while performing these experiments.

In preparing this book the major emphasis has been given on clear and simple presentation in a language that is readily understandable by an average student. We are quite aware of the fact molecular biology is a very fast growing area and we really can not claim to include everything that the reader would look for. We have started looking forward to the possibility of modified versions by updating the content as well as incorporating the criticism and suggestions from the scientific community who use this book. We owe every responsibility for any shortcomings.

In the end we like to acknowledge our family members for their support and patience. Dr Priyanka acknowledges the blessings and love of Nani ji, Mummy, Papa, Raj Dadi ji, and Ran Singh Dada ji. She also acknowledges the love and cheerful moments given by her brother Maj Amit, Sister-in-law Poonam and the support given by siseter-in-law Aruna. She can not forget to acknowledge her adorable kids Chaitanya and Chitrakshi whose presence has given fragrance to the life and made her family to blossom like a flower. In the end she acknowledges all the support, patience, guidance and above all the motivation given by her beloved husband Dr. Manoj. There were times when her morale was low enough to drop this assignment; it was only because of his constant motivation and love that she could lead towards completion of the book.

Dr Namita acknowledges all the love and affection given by her Papa, Mummy, Brother Dr. R. K. Singh, Sister, Mrs. Saraswati, Dr. Santosh, Dr. C.K. Singh, Seema, Brother In law Dr. K.C. banger, Dr. Jagjinder Singh and Dr. Birender laura.

We acknowledge the sincere efforts of Prof. R.K. Singh, for helping us in preparing the final design of the book as well as in selection of the right publisher for this manuscript. It was only because of his personal interest that book is published in such a short span.

In the end, we extend our sincere thanks to all our fellow educators and respected teachers (Prof. P.S. Bisen, Prof. V.K. Chowdhury, Prof R.K. Jain, Prof. D.P. Singh).

—AUTHORS

ABBREVIATIONS USED

AFLP	Amplified Fragment Length Polymorphism
CAPS	cleaved Amplified Polymorphic sequence
DarT	DNA qarray Technology
DNA	Deoxyribonuclic acid
EDTA	Ethylene diamine tetra-acetic acid
Kb	Kilobase
ISSR	Inter Simple sequence repeats
MAS	Marker Assistant selection
Min	Minute
MITEs	Miniature inverted reapeat transposable elements
PAGE	Polyacrylamide Gel electrophoresis
PCR	Polymerase chain Reaction
QTL	Quantitative Trait loci
rDNA	recombinant DNA
RAPD	Random Amplified polymorphic DNA
RFLP	Restriction Fragment length polymorphism
RNA	Ribonucleic acid
SNP	Single Nucleotide polymorphism
SSR	Simple Sequence Repeats
Tris	2-amino-2-hydroxymethylpropane-1,3-diol.

Unit I MOLECULAR BIOLOGY LABORATORY—AN INTRODUCTION

Part 1 GENERAL LABORATORY FACILITIES

Contents

- Basic Lay-out
- Personal Protective Equipment
- Research Equipments

BASIC LAY-OUT

A standard molecular biology laboratory should be developed keeping in view the following points:

1. Lockable doors should be there for facilities that house restricted agents.
2. If developing a new laboratory, consider locating away from public areas.
3. Each laboratory should contain a sink for hand washing, Foot, knee, or automatically operated sinks are recommended.
4. Laboratory furniture must be capable of supporting anticipated loading and uses.
5. Spaces between benches, cabinets, and equipment should be accessible to easy cleaning.
6. Chairs and other furniture used in laboratory work should be covered with a non-fabric material that can be easily decontaminated.
7. Install biological safety cabinets in such a manner that fluctuations of the room supply and exhaust air do not cause the biological safety cabinets to operate outside their parameters for containment. Locate biological safety cabinets away from doors,

from windows that can be opened, from heavily traveled laboratory areas, and from other potentially disruptive equipment so as to maintain the required parameters.

8. All exit and entry laboratory doors should be equipped with Air curtain.
9. An eyewash station must be readily available.
10. Illumination should be adequate for all activities, avoiding reflections and glare that could impede vision.
11. Though there are no specific ventilation requirements, planning of new facilities should consider mechanical ventilation systems that provide an inward flow of air without recirculation to spaces outside of the laboratory.
12. Proper Waste disposal facilities should be available for Bio hazard waste.
13. There should be adequate provision for following:
    (*a*) Gas, Water and electricity supplies
    (*b*) Distilled Water unit
    (*c*) A continuous supply of chemicals and glassware
    (*d*) Compressed air and vacuum lines
    (*e*) Pipette washer
    (*f*) Drying and draining racks
    (*g*) Storage space for chemicals, glassware, nutrient media
    (*h*) Water heater
    (*i*) Water bath
    (*j*) Acid proof baths for cleaning glassware
    (*k*) Automatic dishwasher
    (*l*) Glass automizer
    (*m*) Dispensing devices
    (*n*) Markers, labels and vita film
    (*o*) Micropipettes
    (*p*) Research equipments
    (*q*) Personal protective equipment

PERSONAL PROTECTIVE EQUIPMENT

Personal protective equipment, or PPE, is designed to protect researcher from injuries or illnesses resulting from contact with chemical, radiological, physical, electrical, mechanical, or other hazards. PPE includes a variety of devices and garments such as safety glasses, goggles, gloves, earplugs, coverwalls, and respirators. Using PPE is often essential, but it should not be used as a substitute for engineering, work practice, and/or administrative controls to prevent exposure to hazards. Engineering controls involve physically changing the work environment. An example of an engineering control would be a chemical fume hood. Administrative controls involve changing how or when lab workers do their jobs, such as scheduling work and rotating workers to reduce exposures. Work practice controls involve training workers to perform tasks in ways that reduce their exposure to hazards. The personal protective equipments include:

(*a*) **Face and eye protection:** Safety glasses with side shields are required for work with hazardous chemicals. Ordinary prescription glasses are not adequate protection. Contact lenses can be worn safely if appropriate eye and face protection is also worn. Although safety glasses can provide satisfactory protection from injury from flying particles, they do not fit tightly against the face and offer little protection against splashes or sprays of chemicals. Splash goggles should be worn if there is a potential for splash in any operation involving chemicals. Full face shields with splash goggles should be worn when handling large quantities of chemicals, explosive, or highly hazardous chemicals. If work in the lab could involve exposure to lasers, ultraviolet light, infrared light, or intense visible light, specialized eye protection should be worn.

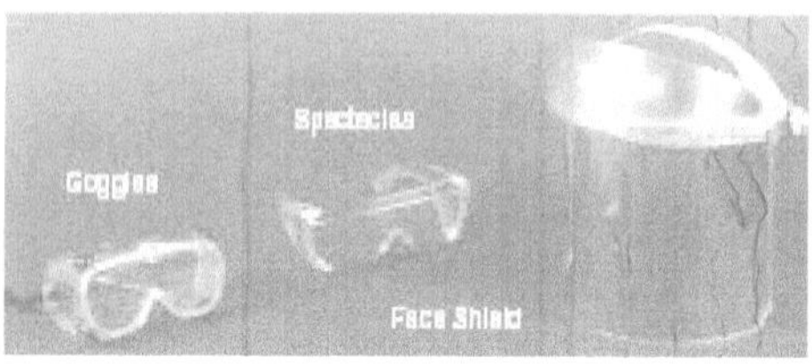

FIGURE 1.1

FIGURE 1.2

(*b*) **Hand protection:** Gloves appropriate to the hazard should be used. It is important that the hands and any skin that is likely to be exposed to hazardous chemicals receive special attention. Proper protective gloves should be worn when handling hazardous chemicals, toxic materials, materials of unknown toxicity, corrosive materials, rough or sharp-edged objects, and very hot or cold objects. Before the gloves are used, it is important that they be inspected for defects. The degradation and permeation characteristics of the glove material selected must be appropriate for protection from from the hazardous chemicals being handled. Glove selection guides (available from most glove manufacturers) should be consulted. Disposable latex gloves are very permeable to most chemicals. They are designed for use with biological hazards, and they should not be used for chemical protection. If latex gloves are used for biological hazards, be aware of latex allergy symptoms: skin rash, inflammation, respiratory irritation, and in rare cases, shock. Be sure to let your supervisor know if you have latex allergy. Gloves should be inspected before use, frequently during use, and replaced immediately if they are contaminated or torn. The use of double gloving may be appropriate in situations involving chemicals of high or multiple hazards. Hands should be washed after removing gloves. **Always remove gloves before leaving lab.**

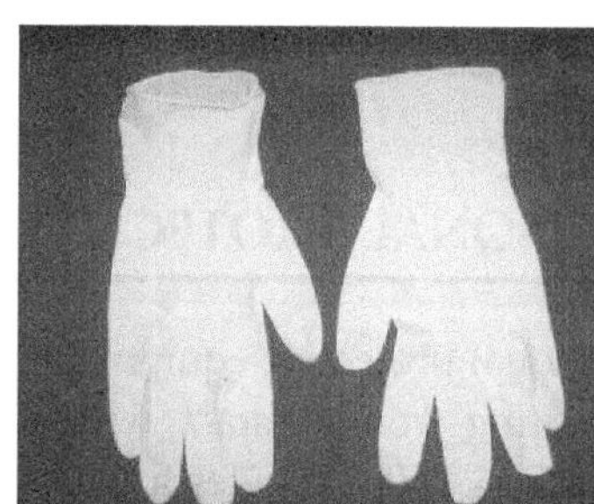

FIGURE 1.3

(*c*) **Clothing:** Wear clothing that protects your skin. Wear long pants - no shorts or short skirts. Wear a lab coat for further protection. Lab coats should be worn with the front fastened and the sleeves completely down, not rolled up. The coat sleeves keep splashes and aerosols from contacting your forearm and wrist. Have a plastic or rubber apron available for working with strong caustics or corrosives. Shoes should completely cover your feet; do not wear open toed shoes, sandals or—sleeper.

(*d*) **Respirator:** The primary method for the protection of laboratory personnel from airborne contaminants should be to minimize the amount of such materials entering the laboratory air. Engineering controls such as chemical hoods, biosafety cabinets, and local exhaust ventilation shall be used to contain and exhaust hazardous emissions. There should be very few instances in a laboratory when respiratory protection is necessary. Respirators may only be considered when engineering controls are not feasible or are inadequate to reduce exposures to acceptable levels.

(*e*) **Dust masks:** There are times when lab staff may wish to use respiratory protection, even when exposures are below the regulatory exposure limit, to provide an additional level of comfort and protection. For example, dust masks are often worn as a precaution when weighing toxic powders, even though there is no quantifiable exposure. However, if a respirator is used improperly **(dust masks are not appropriate for volatile chemicals)** or not kept clean, the respirator itself can become a hazard to the worker.

RESEARCH EQUIPMENT

Vacuum Pump

Bangalore Genie:

- Use for Gel/sample drying.
- Use with vacuum filtration unit.

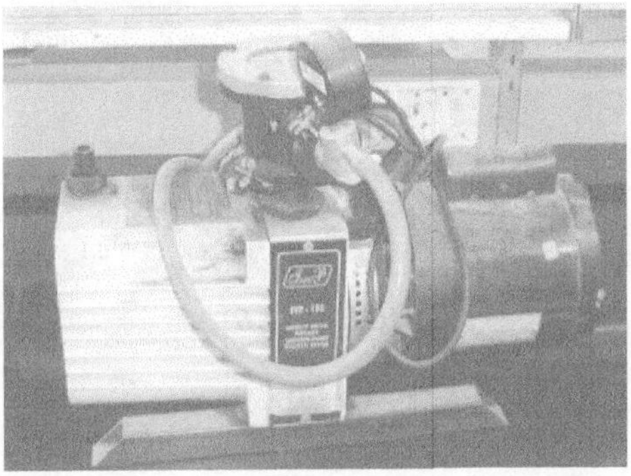

FIGURE 1.4

Inverted Microscope

Leica DMIL, Germany:

- Used For enumeration of cell.
- Inverted Microscope combined with a Visual Display Unit. Used for the observation of culture in tissue culture flasks and Petri-plates.
- Commonly used for counting the cell on haemocytometer.

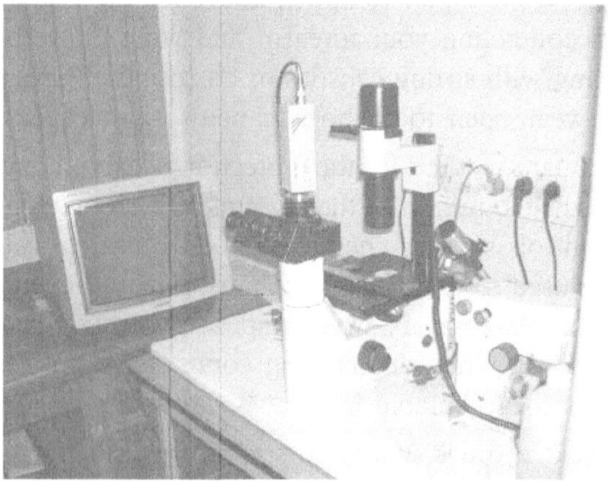

FIGURE 1.5

Compound Light Microscope

ZEISS – 459306, NIKON® – ECLIPSE – E400:

- Compound Light microscope; have very high resolution power used for chromosomal observation on the glass slide.
- Used in the observation of stained Chromosome with Giemsa stain.
- Used for observation of microbial cells.

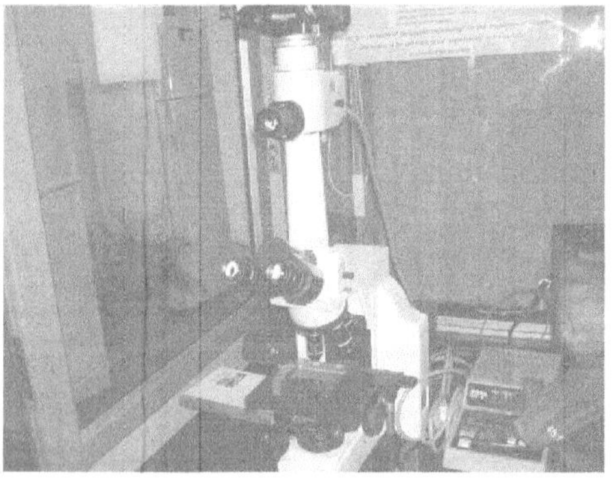

FIGURE 1.6

Centrifuge

REMI, INDIA:

R8C Remi® Laboratory Centrifuge:

- Used for centrifugation of Microbial/Plant/Animal tissue and cell for the formation of pellet during harvesting and reseeding procedure.
- Cells pellets generally formed at 1000 rpm for 10 min. in this centrifuge.
- Refrigerated centrifuge was basically used in the RNA isolation purpose and Separation of temperature sensitive cell organelles/Bio molecules etc. with the help of centrifugal force.

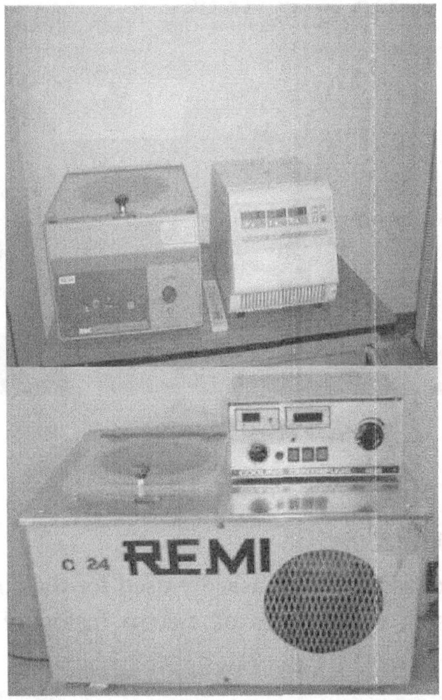

FIGURE 1.7

Hybridization Oven

Amersham pharmacia Biotech,UK:

- Use for maintenance of temperature during hybridization.

FIGURE 1.8

Sequencing Gel Apparatus Along Power Supply

Biorad, UK:

- Use for polyacralamide gel electrophoresis for sequencing.

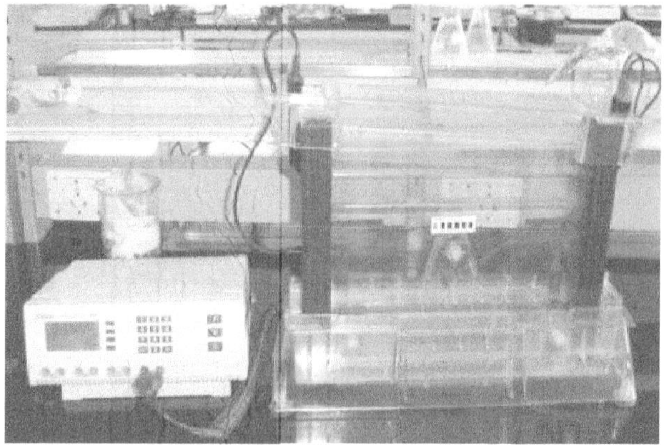

FIGURE 1.9

Electro Cell Manipulator^{tm}

Model BTX - ECM® 2001/830:

- Electro cell manipulator is a device basically used for the supply of high voltage current for few micro-seconds in the process of electro fusion of oocyte-somatic cell during micromanipulation.

FIGURE 1.10

- Electro cell can regulate the electric current upto (2500 volt) for a short duration (15-30 μ sec), this electric stimulation is enough for the fusion of enucleated oocyte-somatic cell couplets.
- Metallic micro electrodes attached with this device are generally used for electro fusion.

Micro Manipulator

Narishinge®, JAPAN:

- Micromanipulator has two Microtool holders and control of x, y and z movements in μm.
- The holders are used to fix Microtool like holding pipette, cutting needle and transfer needle.

- Instrument is basically used for the micromanipulation of oocytes and the transfer of somatic cell into enucleated oocyte.

FIGURE 1.11

Incubators

I. CO_2 Incubator—(Galaxy S, RS Biotech™)

- It provides the optimum temperature (37°C), controlled humidity conditions and 5% CO_2, required for the proper growth of the animal cells.
- A constant electricity supply regulator attached with it for uninterrupted power supply.

FIGURE 1.12

II. CO_2 Incubator—(Bio Center 2001™)

- Used for—Trypsin treatment for the removal of skin fibroblast cells from the collagen monolayer.
- Thawing of FBS, cryovials and maintaining the temperature of normal saline at 37°C for washing of animal tissue.

FIGURE 1.13

BOD Incubator

Caltor, India

- Use for the growth of incubation of microorganism.
- Generally used for the storage of autoclaved and other sterilized material at the fix temperature and other conditions.

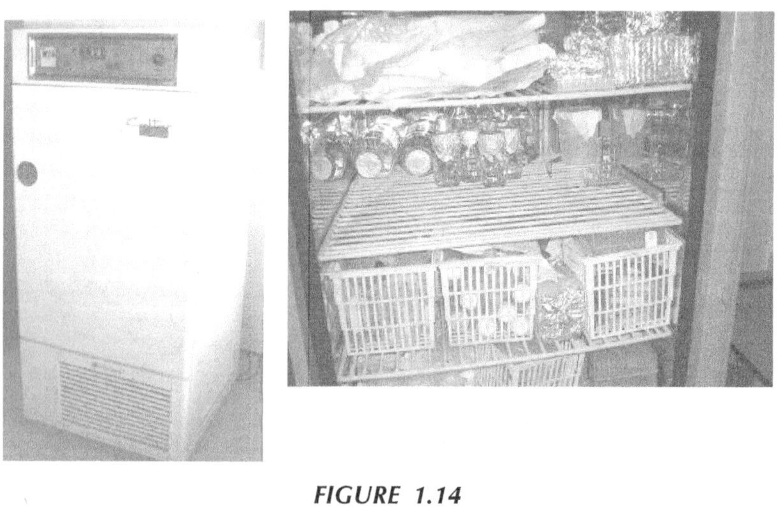

FIGURE 1.14

Laminar Air Flow

- For aseptic manipulations

  Horizontal laminar air flow Manufactured By—WEBCON Instruments Pvt. Ltd. (Horizontal)

  Consists—Main Line System, Filtration System, Illuminations System, UV System, Virus Burn out System.

Consisting Stereozoom Microscope—OLYMPUS—SZX9 (Commonly Used for micro object Searching).

Made By - Toshiba® (INDIA) Laminar Air Flow Assembled With—Motor, UV (Ultra Violet), FV (Front Wave)

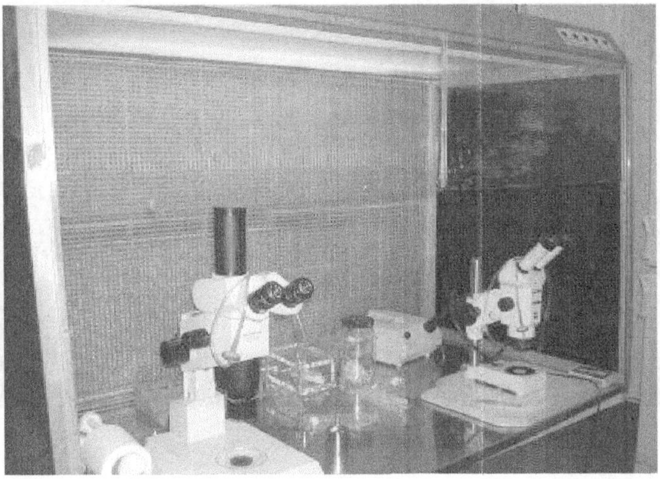

FIGURE 1.15

Vertical Leminar Air Flow

NSW, India

- For aseptic manupulation

FIGURE 1.16

MILI-Q® ASSEMBLY—Double Distilled Water Assembly

Consists 1. Quantum TM (Ultrapure Lonex Cartridge)

- It used for the double distillation of water.

2. MILLIPAK Sterile filter (.22μm)

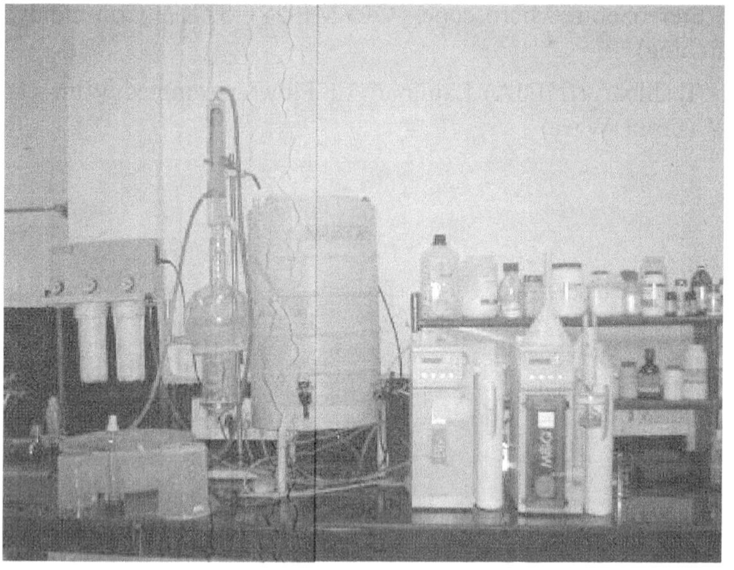

FIGURE 1.17

Ethylene Oxide Gas Sterilizer

- Used for items those are not exposable to the temperature for autoclaving or heat-sterilization like plastic tips, and eppendorf tubes and centrifuge tubes etc.
- Complete sterilization of the surface of heat-labile items required approximately two-three weeks.

FIGURE 1.18

Cryofreezer

CONTROL UNIT AND FREEZING CHAMBER:

(COURTSEY SANYO^{TM} Ultra Low Cryofreezer Consists CRYO-CYL, 35LP Container of Liquid Nitrogen)

FIGURE 1.19

FIGURE 1.20

- It is wide-necked storage vessels cryofreezer, has maximum storage capacity of ampules.
- The cooling rate is regulated by injecting liquid Nitrogen into the chamber.
- At –150°C temperature of cryofreezer, biochemical reactions of the cells cease.
- Preserved the cell lines for their in future use.
- Close-up of a freezing chamber in which cryo vials are placed in cryo boxes.
- In cryo vials 10% DMSO used as a cryo protectant.
- Cryo vials from deepfreeze transfer to the Cryo boxes after two hrs to prevent cryo injuries.

Liquid Nitrogen Freezer

IBP INDIA, CRYOCAN™, TA-55, Capacity 51.5Lit., 20 Lit.

Narrow-necked freezer with storage on canes in canisters.

- In narrow-necked freezer sample stored in square drawers.
- Samples preserved for infinite period in these narrow-necked nitrogen freezers at −196°C.
- For liquid Nitrogen storage
- For Biological sample storage in Liquid nitrogen

FIGURE 1.21

Spin Columns

SDR Molecular A Division of SDR Clinical Technology Sydney, Australia

- Protein purification
- Peptide purification
- DNA purification
- Small molecule, carbohydrate removal
- Radiolabel removal
- Nick translation
- Affinity separation
- Salt removal
- Buffer exchange

FIGURE 1.22

Dialysers

- Purification of biomolecules
- Exchange of buffers
- Removal of detergents
- Sample concentration

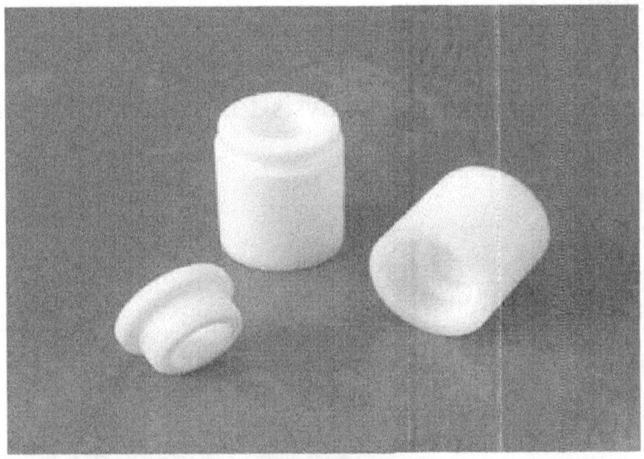

FIGURE 1.23

- HPLC, HPCE
- Removal of excess radiolabel
- Post-PCR clean-up
- GC, GC-MS, NMR

Power Supplies for Gel Electrophoresis

SDR Molecular A Division of SDR Clinical Technology Sydney, Australia

Use for the constant Power supply during the Electrophoresis.

FIGURE 1.24

Semi-dry Blotting Units

SDR Molecular A Division of SDR Clinical Technology Sydney, Australia

FIGURE 1.25

Gel Documentation System

Syngene, UK

- Use for the documentation and gel analysis for protein, DNA and RNA.

FIGURE 1.26

UV Transilluminators

SDR Molecular A Division of SDR Clinical Technology Sydney, Australia

- Use for Quality and Quantity visualization of Gel.

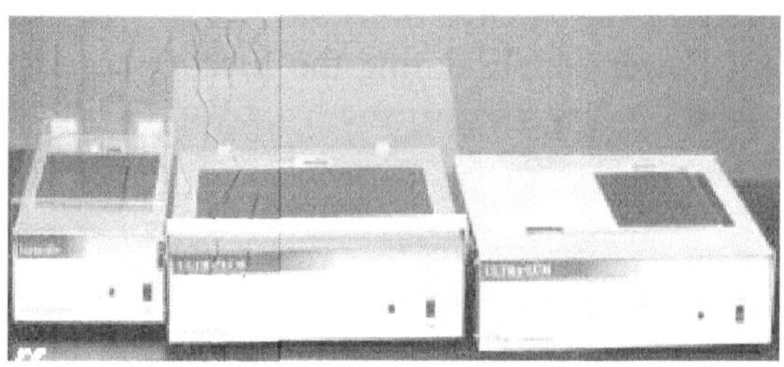

FIGURE 1.27

Mini Horizontal Gel Units

SDR Molecular A Division of SDR Clinical Technology Sydney, Australia

Designed for routine rapid electrophoresis, and ideal for plasmid preps and PCR reactions.

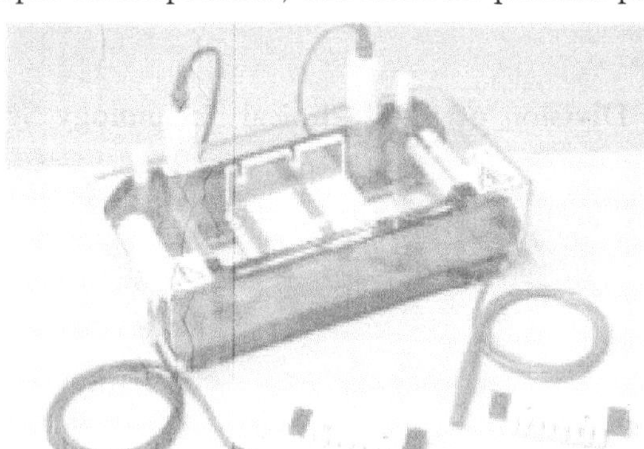

FIGURE 1.28

Cooling Centrifuge

Sigma, USA

- Use for the separation of biomolecules/Cell etc.

FIGURE 1.29

Spectrophotometer

ECIL, India

- Use for spectrophotomitric studies such as Concentration determination of Bio molecules with the help of UV/VIS light.

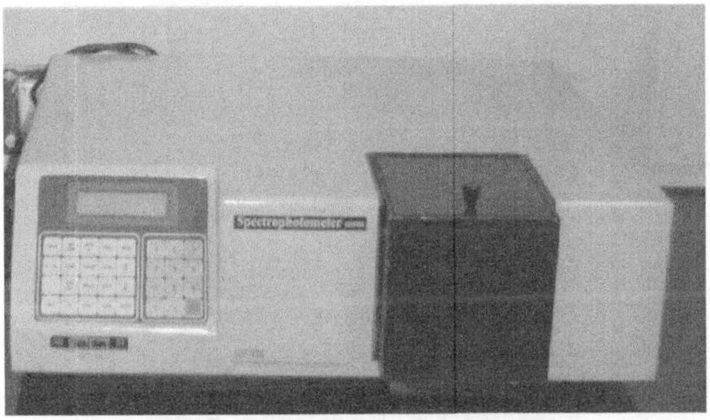

FIGURE 1.30

NIR UV Spectrophotometer

Carry, Finland

FIGURE 1.31

Horizontal Laminar Air flow

Toshiba, India

- For aseptic manipulations

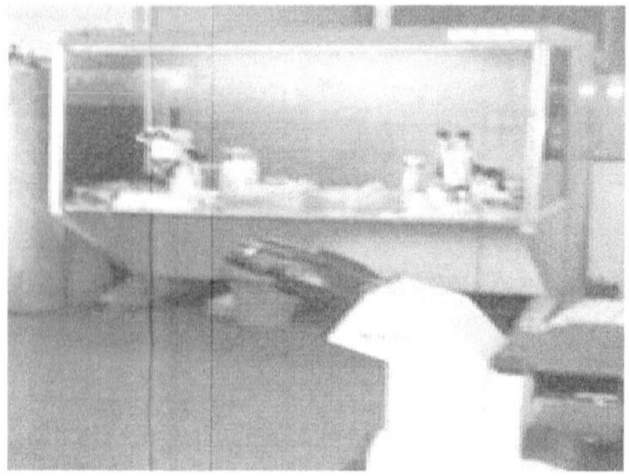

FIGURE 1.32

Rotator Evaporator

Perfit, India

- Use for sample preparations for quick drying.

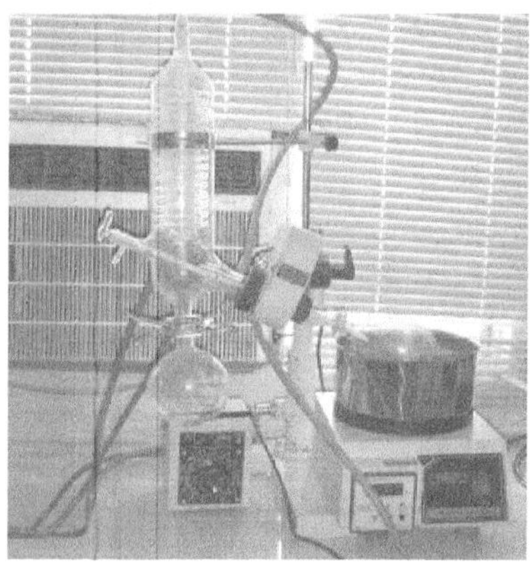

FIGURE 1.33

Laboratory Centrifuge

REMI, India

FIGURE 1.34

Fermentor

Courtesy B. Braun Biotech International

- For bulk culturing of Microbial/Animal/Plant cell.

FIGURE 1.35

Glass Distillation Unit

JSGW, India

- For the preparation of Distilled water for making reagents.

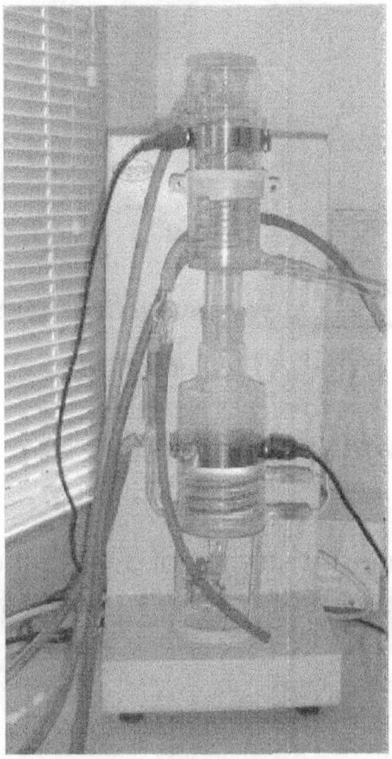

FIGURE 1.36

Programmable Thermalcycler

MJ Research and Strata gene, USA

- Use to carryout PCR reactions
- To carryout real time PCR

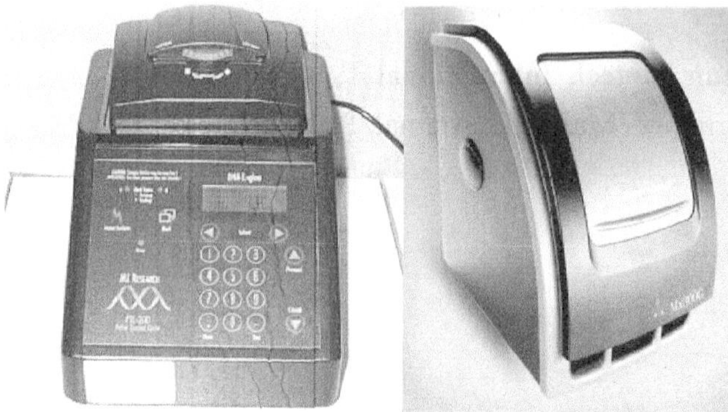

FIGURE 1.37

Electronic Balance and pH meter

A&D Japan and Systronic, India

- Use for setting of pH for different solutions and buffers etc.
- For Weighing of salt/buffers/samples.

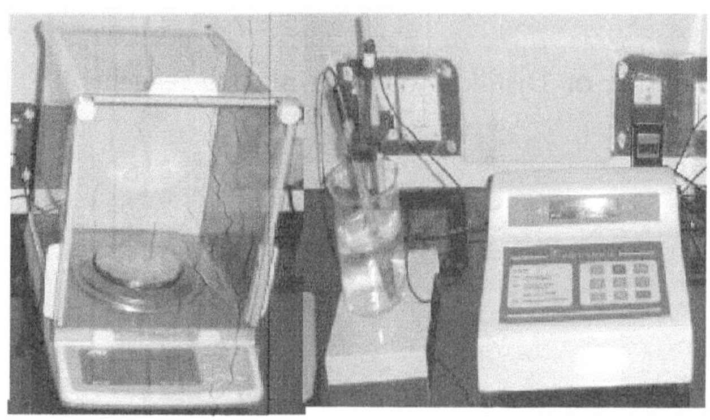

FIGURE 1.38

Horizontal electrophoresis unit with power supply

Biorad, UK

- Use for the Quality and quantity studies of DNA and RNA.

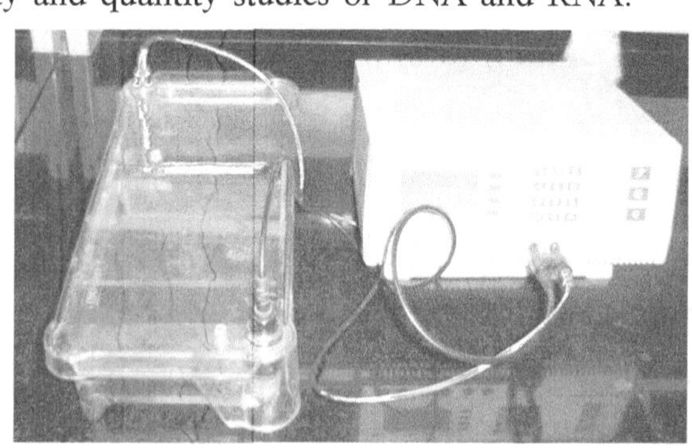

FIGURE 1.39

Vertical Protein Gel Electrophoresis

Biotad UK

- Use for the Quality and quantity studies of protein

FIGURE 1.40

Colony counter

VSI Electronics Pvt. Ltd. India

- For Counting of Bacterial colony in the samples. (Plate count method)

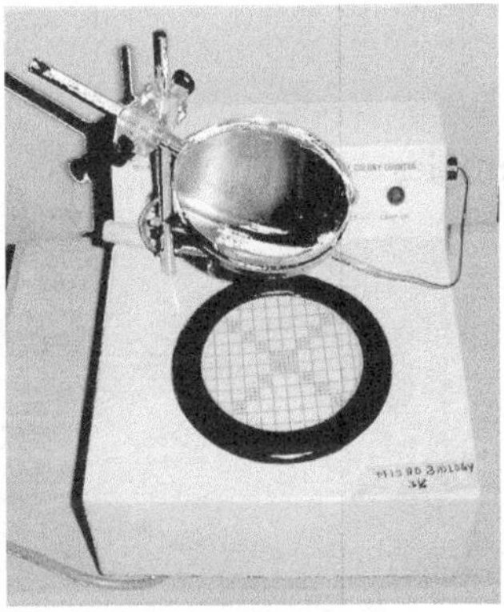

FIGURE 1.41

Water bath

Scientific and industrial corporation Ajadpur new Delhi, India

- For maintained of reaction temperature

FIGURE 1.42

Hot air oven

JSGW, India

- For drying of glass ware/lab ware

FIGURE 1.43

Spinwin

Tarsons, India

- For quick spinning of samples

FIGURE 1.44

Transluminator

Bangalore Genie India

For visualization of PCR Product

FIGURE 1.45

Electroporation Unit

Biorad, UK

- For insertion of Fourine gene.

FIGURE 1.46

Cooling water bath

Pharmacia Biotech

- Use for maintenance of temperature in the experimental system during experiment.

FIGURE 1.47

Fuming Hood

Sonar, India

- Protective equipment for manipulation of Biohazards material.

FIGURE 1.48

Gene Gun

Biorad, UK

- Foreign gene insertion.

FIGURE 1.49

Autoclave

JSW India, Atlantus, India

- Sterilization of media and other lab wears.

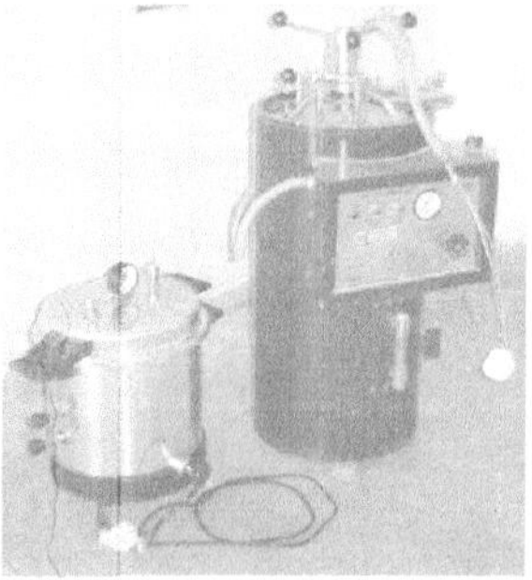

FIGURE 1.50

❑❑❑

Part 2 GENERAL LABORATORY PRACTICES

Contents

CLEANING AND WASHING OF GLASSWARE

New Glassware

Prepare 1% solution of a good glass cleaner such as teepol/liquid detergent. Place new glassware like test tubes, Petri-dishes, pipettes and flasks into the solution and heat to boil. Remove the glassware and rinse thoroughly with warm tap water. Allow to dry before use after rinsing with double distilled water.

New Microscope Slides

Immerse new slides in alcohol containing hydrochloric acid (3%) for several hours. Remove the slides and rinse thoroughly with tap water and finally in distilled water. Wipe the slides dry with soft cotton cloth. While wiping hold the slides by their edges to prevent recontamination with grease. Keep the slides in a covered jar till use.

New Cover Slips

Cover slips are also cleaned free of grease by the method described for the new microscope slides.

Used Glassware

The used glassware and containers should be discarded in 3% lysol or any other disinfectant. All the glassware must be sterilized in an autoclave prior to washing. This must be followed regardless of whether the glassware is contaminated with pathogenic or saprophytic organisms. After sterilization empty the glassware of its contents. Put melted agar into dustbin provided for the purpose. Do not put melted agar into the sink as it will solidify on cooling and shall block the drain pipes. Now wash the glassware like new glassware described earlier.

For molecular/biochemical work immerse the washed glassware in dichromate sulfuric acid solution **(Annexure I)** overnight, remove, wash and rinse in hot water for 5-6 times. Then rinse with Distilled water.

Pipette

Special procedures are required for cleaning glass pipettes. Immediately after use, pipette should be dip in a 0.5% detergent solution in the vertical position if necessary keep immersed in chromic acid overnight, then transfer into pipette washer. Wash continually with tap water to remove any trace of soap, then finally wash with distilled water before drying into the oven.

Used Slides and Cover Slips

Place slides and cover slips in a pan containing 1% solution of a good cleaner. Heat till boiling and keep for a few minutes. Remove the material, rinse thoroughly in warm tap water and finally in distilled water. Wipe dry with a soft clean cloth using the same precautions as given under new slides and cover slips. Store them in a covered jar.

STERILIZATION TECHNIQUES

For molecular biology study with pure cultures, the glassware and culture media must be free from any other kind of organism at the time of use. Sterilization is the process which makes the objects free from all liceing organism. The sterile condition must be maintained indefinitely by protecting from atmospheric contamination by methods like plugging the test tubes, flasks, bottles etc. in aluminium/stainless steel/copper containers before sterilization. A complete destruction of organisms can be achieved by exposure to lethal physical and chemical agents. Solutions of vitamins, amino acids and other growth factors, which are destroyed by heat treatment, are sterilized by filtration. There are different methods of sterilization.

1. Heat Sterilization

The most commonly used form of physical sterilization is heat. The heat used may be dry heat as in case of hot air oven or wet heat as in autoclave. Moist heat kills the organisms by denaturing the proteins while dry heat kills organisms mainly by oxidation of cell components. Two commonly used methods of heat sterilization are:

(*a*) **Hot air oven:** It is a double walled insulated cabinet equipped with a heating element and for circulating hot air at a constant temperature. It also contains a temperature regulator and thermostat to adjust & maintain the temperature. Time temperature combination of 160°C for 1½ to 2 hours or 180°C for 30 minutes is sufficient for effective sterilization.

(*b*) **Autoclave:** The autoclave is a double jacketed cylindrical metal vessel which can withstand a steam pressure of 50-60 psi. There is heating element in the base to generate steam. It has control panel for adjustment of pressure and thermometer to check the temperature inside the autoclave. It is also equipped with a safety valve to prevent explosion because of excessive steam pressure and a steam outlet to evacuate air from the chamber. A stand or resting plate is also provided to keep the material to sterilize.

(*c*) **Principle:** Water boils at 100°C at atmospheric pressure. If the atmospheric pressure is increased the boiling temperature of water will also increase. Thus if the steam pressure inside the closed vessel increases to 15 psi (2 atm), the temperature rises to 121.6°C. All the organisms including heat resistant bacterial spores are killed if kept for 15-20 minutes. All liquids, which are not heat sensitive and articles that are likely to be charred by dry heat can be sterilized in an autoclave. To avoid incomplete sterilization following precaution should take

- Incomplete displacement of air from the chamber must be avoided.
- Steam must have access to the material to be sterilized.
- After sterilization, the steam release valve should be opened a little so that steam may escape slowly.
- All the flasks, tubes etc. should be plugged or bottles by screw-capped.

Table 2.1 Outlines of the properties of various heat decontamination methods. For everyday laboratory purposes, autoclaving is the preferred method, unless the item can not withstand the heat and/or moisture of autoclaving.

Table 2.1. Heat decontamination methods

Method Principle/Condition	*Advantages*	*Disadvantages*	*Uses*
DRY HEAT: Thermal inactivation: destroys by oxidation	Non-corrosive, simple design and principle	Less effective than moist heat; requires longer times and/or higher temperatures	Materials that are damaged by, or are impenetrable to, moist heat
Hot Air Oven: 160-180°C for 2-4 hours	Penetrates water-insoluble materials (*e.g.*, grease and oil) less corrosive to metals and sharp instruments than steam	Slow diffusion, penetration; loading, packing critical to performance; not suitable for reusable plastics	Anhydrous materials, such as oils, greases and powders; laboratory glassware, instruments, closed containers
Red-heat Flame: Oxidation to ashes (burning)	Rapid	Initial contact with flame can produce a viable aerosol; possibility of accidental fire	Inoculating loops, needles

Incineration: Oxidation to ashes (burning) 1-60 mins: temperatures may exceed 1000°C	Reduces volume of waste by up to 95%	Improper use may lead to emission of pathogens in smoke; requires transport of infectious waste; excess plastic (>20%) content reduces combustibility	For decontamination of waste items prior to disposal in landfill
MOIST HEAT: Irreversible coagulation of (microbial) proteins	More rapid and more effective than dry heat		
Pasteurization: Heating to below boiling point (generally 77°C) for up to 30 mins	Can be used on heat sensitive liquids and medical devices; low cost	Not reliably sporicidal	Milk and dairy products; some heat-sensitive medical equipment
Tyndallization (Fractional sterilization): Heating to 80-100°C for 30 mins on 3 successive days, with incubation periods in between	Resistant spores germinate and are killed on the second and third days	Time consuming; not reliably sporicidal	Heat sensitive materials such as bacteriologic media, solutions of chemicals, biological materials
Boiling: Maximum temperature obtainable is approx. 100°C 10-30 mins	Minimal equipment required	Cumbersome; not practical for everyday lab use; not reliably sporicidal	Small instruments and equipment
Autoclaving: Steam under pressure 121°C/15 psi for 15-90 mins (gravity displacement autoclave); 132°C/27 psi for 4-20 mins (pre-vacuum autoclave)	Minimal time required; most dependable sterilant for lab use	Loading and packing critical to performance; shielding dirt must first be removed; maintenance and quality control essential; damages heat-sensitive items	Preparation of sterile glassware, media and instruments; decontamination of reusable supplies and equipment; decontamination of infectious waste

2. Filter Sterilization

Filtration involves the sieving of liquids or gas through a screen like material that has pores small enough to retain micro-organisms. The micro-organisms are retained because of small size of filter pores and partly by adsorption on the walls of pores while passing through filters pore size of the filters (0.1 – 0.22 micron). Initially the filters are made up of high quality ceramics, which are brittle and get logged easily. Filters used today are made up of cellulose, asbestos and acetyl cellulose. Some commonly used filters are:

Seitz filter: Made of asbestos cellulose mixture.

Sintered glass filters: Prepared of fusing fine glass fragments.

Membrane filters: Consists of porous disc of cellulose esters.

Filtration is useful for sterilization of substances that are heat sensitive *e.g.,* vitamins, amino acids, enzyme solution, cell extracts etc.

3. Radiation Sterilization

(*a*) **Ultraviolet sterilization:** Micro-organisms in the air are killed by exposure to UV radiations and thus used for sterilization of large surface area or air space. UV radiation has poor penetration because it is strongly absorbed by media components. However once reached to a cell, it penetrates the cytoplasm and is absorbed by purine, pyrimidine bases of DNA & aromatic amino acids such as tryptophan, phenylalanine and tyrosine etc. Once absorbed by pyrimidine bases, it causes changes in DNA where two adjacent pyrimidine dimer to give an intra stand cyclobutane ring or a pyrimidine (6-4)—pyrimidine. The presence of pyrimidine dimer in DNA strand prevents DNA replication resulting in gap which must be repaired for the survival of cell. If two such dimers are formed in two opposite strands, the DNA molecule will break and cell may die.

(*b*) **Gamma radiation:** Ionizing electromagnetic radiation like x-rays when absorbed by molecule, high energy electrons are emitted by its atoms. The ejected electrons are absorbed by other atoms, creating a chain of ionizations in the irradiated substances. This activity excites chemical groups. In DNA causing the production of highly reactive, short lived chemical radicals. Such radicals may break DNA stands causing mutations or death of cell. β-rays emitted from Co^{60} are of sufficient energy & when intact with DNA molecules in a cell, the phosphodiester backbone is broken when enough breaks are produced; the DNA is unable to replicate or repair those phosphodiester bonds. Consequently the cell dies.

The major use of γ-radiation is the sterilization of plastic wares like disposable Petri plates, tubes or hypodermic syringes. They are also used to prevent spoilage of packaged food without causing any significant change in appearance taste, texture or nutritive value of the product.

4. Chemical Sterilization

The use of chemical agents is done for sterilization of heat labile materials. These chemical agents are highly reactive molecules that modify the molecules inside the cell resulting in destruction of their biological activities. Some of the commonly used chemical sterilizing agents along with advantages and disadvantages are listed in **Table 1.2.2.** Ethylene oxide or propylene oxide are volatile chemicals which boil at low temperature. These chemicals act as alkylating agents and react with sulphahydral and carboxyl group of cell molecules. Ethylene oxide is in liquid form at and below 10°C. It boils at 10.7°C and turned into gaseous form. It is added to solution in liquid form at concentration varying from 0.5 to 1% at 4°C or is used in gaseous form at temperature above its boiling point when used in gaseous form, the treatment is given in a special chamber in which temperature is below 50°C. Thus it is used for sterilization of plastic material.

Ethylene oxide is chemically unstable and decomposes to ethylene glycol that is non-volatile, toxic to human lungs and explosive. Thus it is used only in industries for sterilization of heat labile materials.

Table 2.2. Different chemical sterilizing agents

Name	*Effective Concentrations, Contact times*	*Advantages*	*Disadvantages*	*Some Uses*
Sodium hypochlorite solution (liquid bleach)	100-10,000 ppm (.01-1%) free chlorine 10-60 min (3,000 ppm for broad spectrum)	Broad spectrum; inexpensive; widely available; bactericidal at low temperature.	Toxic, corrosive to skin and metals; efficacy decreases as pH increases; inactivated by organic matter; deteriorates under light and heat: shelf life of dilutions is less than 1 week.	General disinfectant; waste liquids; surface decontamination; emergency spill clean up; instrument disinfection.
Calcium hypochlorite granules, powder, tablets	As for liquid bleach	As for liquid bleach, but more stable	As for liquid bleach above, except shelf life is longer	As for liquid bleach
NaDCC (Sodium dichloroisocyanurate) powder, granules, tablets	As for liquid bleach	More stable than hypochlorites	Toxic; corrosive; inactivated by organic matter	As for liquid bleach
Chloramine-T4 (Sodium tosylchloramide) powder or tablets	As for liquid bleach	More stable, less affected by organic matter than hypochlorites; longer activity than hypochlorites	Deteriorates under humidity, light and heat	As for liquid bleach
Chlorine dioxide	Demand-release of chlorine dioxide in situ	Longer activity than other chlorine compounds; less corrosive, less toxic than other chlorine compounds; effective at pH 6-10	Aqueous solutions decompose under light	Instrument disinfection; gas sterilization of germ-free animal chambers
Iodophors	30-1,000 ppm (.003-0.1%) free iodine 10-30 min	Broad spectrum; germicidal over a wide pH range; generally nonstaining, less toxic and less irritating than aqueous or alcoholic iodine solutions	Not consistently sporicidal; efficacy reduced by organic matter; some iodophor solutions support growth of Pseudomonas 7	Germicidal soaps and antiseptics; surface decontamination; work surface wipedown; instrument disinfection

Alcohols	70-80% ethanol 60-95% isopropanol 10-30 min	Low toxicity; rapid action; low residue; non-corrosive	Rapid evaporation limits contact time; flammable, eye irritant; may damage rubber, plastic, ineffective against bacterial spores	Skin disinfectant (antiseptic); surface decontamination; benchtop, cabinet wipedown
Phenolic compounds	400-50,000 ppm (.04-5%) 10-30 min	Tolerant of organic load; "hard" dilution water leaves an active residue (may be desirable on some surfaces); biodegradable	Pungent odour; corrosive; some forms toxic; not sporicidal; limited activity against viruses; leaves a residual film (undesirable in culture systems); may support growth of bacteria 1	Instruments and equipment disinfection; disinfection of floors and other surfaces; antiseptic soaps and lotions
Quarternary ammonium compounds	500-15,000 PPM (.05-1.5%) 10-30 min	Combined detergent and germicidal activity; stable; working dilutions have low toxicity	Not sporicidal; limited activity against viruses, mycobacteria; most formulations not readily biodegradable; may support growth of bacteria 2	Surface decontamination; equipment wipedown; antiseptic formulations available; floors and walls
Hydrogen peroxide compounds	3-30% aqueous solution 10-60 min 6% for 30 min may kill spores	Rapid action; no residue; low toxicity; environmentally safe	Limited sporicidal activity; corrosive to some metals; potentially explosive at high concentrations; stock solutions irritating to skin and eyes	Surface decontamination; instruments and equipment
Per Acetic Acid (PAA)	.001-.3% aqueous solution gas phase: 2-4% 5-120 min	Broad spectrum; sporicidal at low temperature; can tolerate organic load; rapid action; nontoxic decomposition products; leaves no residue	Pungent odour; corrosive to some metals; shelf life of dilutions is less than 1 week; stock solutions irritating to skin and eyes; stock must be protected from heat & light; gas phase: respiratory irritant; fire hazard above 55°C	instruments and equipment; gas phase sterilization of chambers for germ-free animals

Glutaraldehyde	0.5-2.5% alkalinized aqueous solution 2-30 mins; up to 12 hours to kill all spores	Broad spectrum; does not corrode metal; can tolerate organic load	Broad spectrum; does not corrode metal; can tolerate organic load	Cold sterilant and fixative; surface decontamination; instruments, equipment, glassware
Formalin (37% aqueous formaldehyde)	3-27% formalin (1-10% formaldehyde) in 70-90% alcohol 10-30 min	Broad spectrum; inexpensive; does not corrode metal; can tolerate organic load	Pungent odour; skin, eye and respiratory tract irritant; potential carcinogen (animal studies); may require 24 hrs or more to kill all spores	Cold sterilant and fixative; surface decontamination; instruments and equipment
Formaldehyde (gas)	1-3 hours	As for formalin; effective penetration	As for formalin; flammable; poor penetration of covered surfaces	On site decontamination of biological safety cabinet HEPA filters; enclosed areas
Ethylene Oxide Gas	50-1200 mg/L 1-12 hours	Broad spectrum; no heat or moisture evolved; penetrates packaging materials	Flammable, reactive; toxicess: potential carcinogen and mutagen; some sterilized items may need more than 24 hours for outgassing	Heat or moisture sensitive supplies, instruments and equipment

CALIBRATION OF VOLUMETRIC GLASSWARE

In order to perform any biological analysis, reagent solutions are prepared in such a manner as to facilitate the assignment of a concentration value to the reagent solution which is as accurate as possible. Concentrations are expressed in units of mass/volume: therefore, it is critical that the mass of the reagent used, as well the final volume, be known as accurately as possible. Glassware that has been precisely manufactured to contain or deliver exact volumes of liquids are given the title "volumetric". Even with the additional accuracy provided by volumetric glassware, there is no assurance that the volume marked on the outside of the vessel is the same as the volume it actually contains or delivers, and the analyst should test them for accuracy. In this experiment, certain pieces of volumetric glassware will be tested, or calibrated, to determine as accurately as possible the volume of liquid that it contains.

The results of these calibrations should be used in all subsequent experiments in which the vessels are employed, ensuring the highest possible accuracy.

Volumetric glassware is calibrated by weighing the amount of pure H_2O it contains at a given temperature. When calibrating glassware, it is important to realize that:

(1) the density of H_2O as well as the volume of glass containers vary with temperature,

(2) the buoyancy of air effects the observed weight of low density objects.

The variation of H_2O density and vessel volume can be controlled by simply letting the vessel and liquid equilibrate to room temperature prior to calibration. Common errors in this process are using either hot or cold water to fill the vessel, or placing the vessel in a oven to dry and using it before it has equilibrated to room temperature. To avoid these errors, draw the water into a clean container (not plastic), cover, and let stand 20-30 minutes prior to use. If the vessel to be used must be dry (which is not very often), rinse it with alcohol or acetone instead of oven drying. Purging with air is not advisable since oil originating in the compressors frequently contaminates air lines.

To attain the highest possible accuracy, the effects of air buoyancy must be taken into account. To illustrate, when an empty vessel is weighed, it is actually not empty, but contains air. When the vessel is filled, the liquid placed in it displaces the air, the net effect of which is that the weighed object appears lighter than it actually is. To correct for air buoyancy, the following relation is used:

$$M = \frac{m'(1 - da/dw)}{1 - da/d}$$

where

m = true mass

m' = observed mass of object in air

da = density of air (0.0012 g/ml)

dw = density of the calibration weights (8.0000 g/ml)

d = density of weighed object (0.9978 g/ml for H_2O)

The effects of air buoyancy on the observed mass of an object decreases as the density of the object weighed increases and, in practice, it is only necessary to consider it when weighing low density liquids. Use care and good technique when using the balances. Always keep in mind that the analysis starts here: an inaccurate mass obtained for standards or reagents used to prepare solutions will carry through the entire analysis, no matter how careful the actual analysis part is carried out. Pick a scale in the balance room, and use the same one each time to decrease variability. Keep the balances clean using the brushes provided, and remember to close the glass doors to decrease air currents around the pan. Gently place and remove objects from the pan, and check the level bubble often. If the scale is hard to tare or the readings particularity unstable, inform the instructor. You should also be aware that gloves are available for you to use in the lab when working with reagents.

Procedure

All glassware must be clean prior to use. A check to see if glassware is clean is to notice if H_2O smoothly covers it when drained out: if patches or large droplets remain, the vessel is contaminated and is in need of cleaning. If a detergent or cleaning solution is used, be sure to rinse the vessel several times with first tap, and finally distilled H_2O to remove any remaining cleaning solution. Draw the distilled water to be used in the calibration procedure into a clean container (not plastic), cover with a watch glass, and let stand 20-30 minutes prior to use if it is not already at ambient temperature.

Calibration of Transfer Pipette

(*a*) Clean and dry a 50 ml beaker. Carefully weigh the beaker, recording the weight to the nearest 0.1 mg. Fill the transfer pipet to be calibrated with H_2O, and drain the contents into the weighed beaker. The protocol for using transfer pipets is as follows:

(*b*) fill the pipet past the graduation mark by applying suction (from a bulb) Wipe off any obvious droplets that adhere to the exterior of the pipet. Drain until the bottom of the meniscus reaches the graduation mark. Use your finger to control the flow of solution from the pipet by holding over the top of the pipet after removing the bulb.

(*c*) place the pipet tip into the receiving vessel and drain. When the liquid is 1/2 to 3/4 drained, touch the pipet tip to the wall of the receiving vessel and maintain contact until all the liquid has drained. after draining, break the contact with the vessel wall, and then touch the pipet tip against the vessel wall again to remove the last drop of liquid. Do not blow out the remaining liquid. Weigh the beaker and water. In this manner, calibrate the 2.00, 5.00, 10.00, and 25.00 ml transfer pipets. Repeat the calibration at least three times for each pipet. Replicate measurements should agree to within 0.01 g or 0.02 g (0.02 g is ok if having trouble).

Calibration of a Buret

(*a*) Place the buret in a suitable stand and fill with distilled H_2O.

(*b*) Drain to the zero mark and let stand for about 5 minutes. Check the tip for air bubbles, which, if present, must be removed. If the meniscus changes position during the 5 minute waiting period, the stopcock is loose and needs to be refitted.

(*c*) If no bubbles or leakage is noted, drain the buret at a reasonable speed (about 6 seconds per ml) to the 10.00 ml mark into a weighed flask (prepared as above). Be sure to catch any remaining liquid that adheres to the buret tip by touching it to the vessel wall.

(*d*) Determine the weight of H_2O delivered.

(*e*) Refill the buret, drain to the zero mark, and repeat the process **at least** a further two times. Drain the buret as close as possible to the 10.00 ml mark each time.

(*f*) In this manner, calibrate the buret up to 50.00 ml (0-10 ml, 0-20 ml, 0-30 ml, 0-40 ml, and 0-50 ml). Replicate measurements should again agree to within 0.02 g Always remember to return to the zero mark for each measurement.

Calibration of Pipets

From the weight of H_2O delivered and it's density, calculate the actual volume of liquid that the pipet delivers. Be sure to correct the observed weight of H_2O in air for buoyancy effects. The corrected volume should be reported as the mean of the determinations and the pipet marked for identification, since this correction will be applied in all subsequent work that employs the pipet.

Calibration of the Buret

For each 10.00 ml interval, calculate a correction for the volume delivered by first correcting the weight of H_2O observed for buoyancy effects, and then transform the weight of H_2O delivered to volume delivered via the density. Subtract the "theoretical" volume delivered from the actual volume delivered as determined by the calibration procedure. Report your results as a table of buret interval against correction value for that interval. Use the data to prepare a plot (see figure 1) of the buret interval (abscissa) versus the correction (ordinate). Place this plot to the inside cover of your lab notebook, and use the corrections when calculating the results of subsequent titrations.

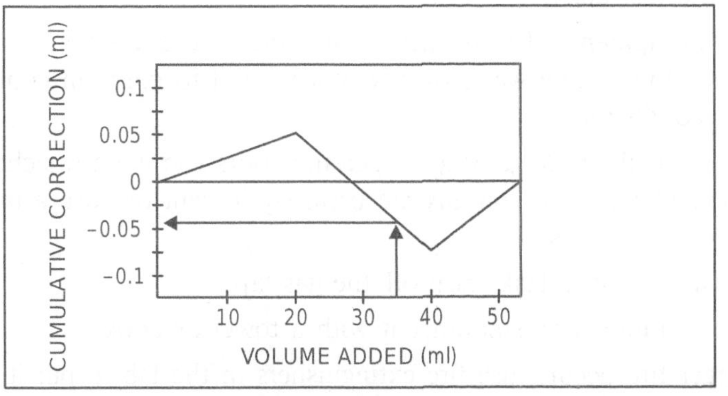

FIGURE 2.1 **A hypothetical buret calibration graph. Corrections are made as in the linerising from 35 ml titrant delivered.**

Calibration of Micropipette

The accuracy and success of any experiment depends on the accuracy of tools use during experiment. Micropipettes is one such tool for the calibration weight several volumes taken with micropipette on the electronic microbalance to determine that the micropipette in drawing the correct volume of water, 1 ml of water correspond to 1 g. Use an adjustable micro pipetting device in a following way:

- Set the digital micrometer to the desired volume using the adjustment nobe.
- Attach a new disposable tip to the shaft of the pipette.
- Press on firmly with a slight twisting motion.
- Depress the plunger to the first positive stop, immerse the disposable tip into the sample liquid to a depth of 2 to 4 mm and allow the push button to return slowly to the up position and wait 1 to 2 seconds.
- To dispense sample, place the tip end against the side wall of the receiving vessel and depress the plunger to the second stop to achieve final blow-cut. With draw the device from the vessels carcfully with the tip sliding along the inside wall of the vessel. Allow the plunger to return to the up position. Discard the tip by depressing the tip ejector button. Special care is needed during the drowning organic solvents.
- Allows keep micropipette at natural stop position.

Always store micropipette at high position.

GENERAL INSTRUCTIONS WHILE WORKING IN THE LABORATORY

Following guidelines explain the hazardous nature of Molecular laboratory work and direct the reader to a wide range of information available on laboratory safety measure :

1. **Handling of alcohol:** Both ethyl as well as methyl alcohol is highly inflammable. Keep containers of alcohol covered and away from open flame.
2. **Breakage of glassware:** In case a glass objects break, do not pick the pieces of glass with bare hands. Sweep the pieces with a broom and dust pan and put into the waste containers marked for this purpose. However be careful if thermometer breaks and report to the instructor.
3. **Electrical Equipments:** Ensure that your hands are dry when you handle electrical connectors. Don't allow water or any other liquid to come in contact with electrical cords and conductor.
4. **Fire:** If you smell a leakage of gas, open the doors and windows check all the gas taps and tighten. Don't switch on any electrical equipment or burn a matchstick until the room is free of the gas.
   - If gas burns from a leak, turn off the gas tap.
   - In case of minor fire, smoother it with a towel or book.
   - If a larger fire occurs, use fire extinguishers in the lab to put it out.
   - In cases of large fire involving the lab itself, evacuate the room after turning all gas burners off & unplugging the electrical equipment.
5. **Accidents:** Following things should be done:
   - Report all accidents to your instructor immediately.
   - For spills in or near eyes, use eyewash.
   - For larger spills on the body, use shower.
   - For heat burns, chill the affected part with ice immediately.
   - In case of cut or abrasion, use bandage.
6. **Chemical safety:**
   - A Material Safety Data Sheet (MSDS) should be available for every chemical use in lab. Read these and follow the recommendations for safe use and disposal of the material.
   - Use the chemical as soon as possible. If you need to store it, take proper care of all the instructions mentioned by the manufacturer in this context.
   - Confirm again the details on the label before using the chemical, implementing all necessary safety measures. Further precautions will be required for a reagent without detailed information.
   - Be sure to wear protective clothing while dealing with chemical/reagents.
   - Take appropriate protective measures considering the characteristics of the reagent with reference to the related regulations, papers and MSDS.
   - Reagents are manufactured on the assumption that they are rapidly used after sale. However when you keep some portion, the same or greater attention as that of the first usage, should be given.
   - The use of reagents/chemical operation should be implemented under the guidance and supervision of an expert.

- After using, waste materials, unused chemicals and old reagents should be properly treated conforming to corresponding rules and regulations **(Annexure II).**

7. **Cryogenic Safety:** See Part 3, Unit I
8. **Working on fume hoods and laboratory ventilation equipment:**
   - Make sure that the exhaust blower is operating and air is entering into the fume hood or ventilation equipment. Check the magnehelic gauge or audible/visual (AV) alarm. The magnehelic gauge should register 0.1 inches of water or slightly higher. The AV alarm should not indicate any problems.
   - A qualitative fume hood check can also be accomplished by taking a small strip of Kim-Wipe and place it near the opening. If the fume hood is operating properly the tissue should be drawn inward. If the strip of tissue is not drawn inward, please notify Health & Safety officer of your institute.
   - Do not place your face inside a unit. Keep your hands out as much as possible.
   - Do not store chemicals or equipment inside your fume hood if they are not needed for the current experiment or process.
   - Clean up spills as soon as possible. Periodically clean the interior areas of your fume hood. Use a neutral cleaner such as Simple Green and dispose of contaminated wipes and paper towels as solid chemical waste.
   - The hood or ventilation equipment should be in continuous operation if any volatile, toxic or corrosive materials are stored or used in the unit.
   - Avoid blocking off the baffle exhaust slots located in back of the fume hood. Elevate large equipment 2 inches off the surface or base of the fume hood.
   - Be aware of other room ventilation factors that may interfere with the unit's operation. Opened doors, windows, blocked exhaust ports or heating and air conditioning vents can negatively impact fume hood/ventilation equipment performance.
   - Avoid cross drafts and disruptive air currents in front of the fume hood.
   - Use your fume hood sash as a safety shield when working in a fume hood. Work with the sash at the proper level as indicated by the certification sticker placed on the fume hood.
   - Never work in a fume hood with a broken sash. Never attempt to lower a fume hood sash that has been broken
   - Prepare a plan of action in case of an emergency, such as a power failure. Your plan should include what actions are to be taken to make the operation and materials in the unit safe.
   - Know where your fire extinguisher is located. Verify that it is approved for the materials being used.
   - Periodically check for proper airflow through the unit by checking the magnehelic, AV.
9. **Preparation of reagents:**
   - Store chemicals/solvents at appropriate temperatures as mentioned on the label.

- The highest purity chemicals and double distilled water should be used for preparation of reagents and solutions. (Weight of the substance, brand, batch No. of the substance, type of water used, amount of alkali or acid used to dissolve should be recorded then and there).
- Label all reagents. The label should contain the name of the solution/reagent, concentration (% or Molar), date of preparation of the reagent and initials of the person who prepared. Unlabelled solutions/reagents should be discarded.
- Follow proper guidelines while preparing a reagent for use *e.g.,* Phenol must always be equilibrated with buffer when used in molecular biology experiments **(Annexure VIII).**
- Store the reagents in appropriate bottles and at appropriate temperatures. (Light sensitive reagents and solutions should be stored in brown/dark bottles).
- Keep record of the reagent preparation and experimental details correctly and completely. Store all the data sheets supplied along with fine chemicals and enzymes. Note down the date of expiry, Batch No and Catalogue No for labile substances.

10. Experiments and record keeping:

- Make sure the laboratory and laboratory benches are clean (all spillover of culture, corrosive solvents should be cleaned immediately and appropriately).
- Make sure that the balance is cleaned after your weighing.
- Make sure that the glassware, plastic wares, media, etc., are ready for the experiment (if needed sterilize them previous day itself).
- Make sure that you have studied and understood the principle and methodology of each experiment. Discuss the experiment with your guide in advance. Keep appropriate controls. Copy the protocols of the experiment in the record note book. Attach print outs of instruction sheets and supporting materials to record.
- Make sure that the glassware, plastic wares, media, etc., are ready for the experiment (if needed sterilize them previous day itself).
- Label all plates, tubes, cultures etc., correctly before starting the experiment
- Make sure that the equipments are in working condition and available for the experiment. If it requires prior permission and reservation in advance.
- Make sure that you know the proper operations of the equipments correctly and if not assistance is available.
- Standard graphs, radioactive counting and enzyme assays should be done in triplicates.
- For standard graphs, plot all the points and draw the line, which passes through maximum number of points.
- Use a record notebook of permanent binding. Number all pages before using the record notebook. Use permanent ink in writing the record. Don't rely on memory and enter all data directly into the record book then and there. Avoid abbreviations, code names and code numbers in recording the data. Start new experiment on a new page. Every experiment should have date, title and objective(s). All calculations

should be done step by step and recorded neatly. Do not overwrite on mistakes. Photographs, printouts, chromatographs, etc. should be labeled to indicate what they are supposed to represent and then paste them directly on to the notebook as soon as the experiment is over. Do not record data on loose sheets. Do not take the record book outside the laboratory. Relationship between two variables in experiments can be best represented using graphs rather than in tables. AJI graphs should be titled; dated and fixed in the record book on the same day.

11. **Others:**
    - Never mouth-pipette any solution (acids, phenols, peroxides, organic solvents, culture of organisms, etc.).
    - Wear a laboratory coat (use gloves when necessary).
    - Do not eat, drink or smoke inside the laboratory.
    - Aseptic techniques should be followed rigorously at all times.
    - All microbial cultures are potential biohazards so dispose off them after autoclaving or after treating with 10% formalin for 30 min.
    - Dispose of all the wastes properly. Radioactive wastes should be stored separately.
    - No chatting, gossiping and loud discussions should be done inside the laboratory as it affects your concentration on the work.
    - Do not store food materials and drinks in the refrigerator where other chemicals and solvents are stored. (Consumption of such stored food items may lead to cancer).
    - While disposing any type of waste (lab-trash, solid waste, lab sharps), proper guidelines should be followed **(Annexure II).**

❑❑❑

Part 3 LABORATORY SAFETY GUIDELINES AND HEALTH HAZARDS

Contents

INTRODUCTION

Safety consideration is of paramount concerns during the preparation and execution of molecular laboratory experiments. Laboratory safety guidelines are based on identifying the sources of hazard and by assessing the possible risks associated. This chapter gives an overview of common hazards encountered in the molecular biology laboratory and some measures to minimize or eliminate them. Depending upon the nature of experimental work in molecular biology laboratory Chemical, Physical and Biological hazards, three principal areas are identified.

Physical Hazards: This includes Fire, pressure and ionization and non ionization radiation.

Chemical Hazards: Chemicals are considered highly hazardous for many reasons. They may cause cancer, birth defects, induce genetic damage, cause miscarriage, or otherwise interfere with the reproductive process. Or they may be a cholinesterase inhibitor, cyanide, or other highly toxic chemical that, after a comparatively small exposure, can lead to serious injury or even death. Working with compounds like these generally necessitates implementation of additional safety precautions. Below are definitions of the classes of chemicals that are considered highly hazardous and descriptions of the safety precautions to consider when designing experiments involving them. Depending on the amount and manner in which such substances are used, lab-specific written safe operating procedures may be required.

Biological Biohazards: Biohazards are infectious agents or biologically derived infectious material, which presents or may present a risk to the well being of a person or the community.

PHYSICAL HAZARDS AND BIOSAFETY

Working in a Molecular Laboratory requires the use of various types of equipment. Be familiar with the equipment before using it. User should know how the equipment works, how it operates, what are the safe Guards, its maintenance requirements. Before beginning any experiment, be sure that each piece of equipment required is in proper working order. If equipment does not operate properly or is in need of repair, immediately take it out of service so that it will not be used by other persons. Tag the equipment to indicate the deficiency and notify the laboratory supervisor. Always prepare a checklist to check that all equipments:

- have adequate controls and safeguards.
- are installed in a safe location with adequate ventilation, if required.
- are being used only for its designated purpose. When special adaptation is required, it should be done only with the assistance and direction of Biomedical Engineering or Physical Plant.
- are not used in flammable or combustible atmospheres. Most pieces of laboratory equipment can be a potential source of ignition.

Some of the commonly used equipments with safety guidelines are discussed below:

1. Refrigerators

Laboratory refrigerators are appropriate equipment for the storage of materials that must be kept cold to preserve their function. These include:

- biological specimens and preparations.
- chemicals which are heat sensitive.
- drugs.

Misuse:

- Routine storage of chemicals.
- Mixed storage of substances used for living systems with toxic and/or hazardous chemicals.
- Food storage, which is **NEVER ACCEPTABLE** in laboratory refrigerators. Lunches requiring refrigeration may only be stored in other refrigerators that are identified as acceptable for the storage of food and beverages. Such refrigerators must be located outside the laboratory.

Flammable material storage: Flammable chemicals **cannot** be stored in conventional refrigerators. Electrical sparks from a conventional refrigerator can ignite the flammable vapors that build up inside. Unless a cold room is ventilated and has a fire suppression sprinkler system, do not store flammable liquids there. Two kinds of refrigerators are approved for storage of flammables:

Flammable Materials Refrigerator: These have no spark sources within the refrigerator cabinet. There are, however, spark sources outside the refrigerator cabinet from switches, motors, relays, etc. These spark sources can ignite flammable vapors present outside of the refrigerator. A bottle of flammable liquid dropped and broken near one of these refrigerators can easily be ignited by sparks.

Explosion-proof Refrigerators: These refrigerators have all spark sources completely sealed inside and are safe for flammable atmosphere both within and outside of the refrigerator cabinet.

Conventional refrigerators in laboratories and cold rooms that are not safe for flammable storage must be labeled **"NO STORAGE OF FLAMMABLES".**

Labeling of Storage Containers: All materials stored in laboratory refrigerators must be kept in closed containers and labeled with the specific contents. If a refrigerator is used by multiple researchers, the individual refrigerator shelves should denote the name of the researcher or the individual materials should be labeled with the name of the responsible person. Seals should be vapor-tight and spill-proof in case the container is tipped over. Glass stoppers, aluminium foil caps and similar closures are not acceptable.

Location of refrigerator in the Laboratory: Refrigerators should be:

- Placed against fire-resistant walls.
- Equipped with heavy-duty electrical cords
- Protected by a separate circuit breaker.

Defrosting of Refrigerator/Freezer: It is necessary to defrost a laboratory refrigerator. One should exercise caution to prevent injury and/or contamination from items in the freezer or refrigerator during this process. If the unit is used to store biological, radioactive or hazardous materials, make sure that the inside of the unit is decontaminated before defrosting. After decontamination, place the unit with a sign indicating that the unit was cleaned/disinfected by (name) on (date). This will show anyone who may need to use the freezer/refrigerator or make repairs to it that it is safe to use or repair.

2. Heating Devices

Electrical devices that supply heat for reactions or separations are commonly used in laboratories. The use of Bunsen burners and other open flames is highly discouraged. If they are absolutely necessary, they must never be left on and unattended.

Electrically heated devices include: hotplates, heating mantles, oil baths, air baths, hot-tube furnaces, hot-air guns, ovens.

Hazards: If used improperly, these devices can cause:

- electrical hazards
- fire hazards
- burns to the user

Baths: If baths are required to be hot at the start of each day, they should be equipped with timers to turn them on and off at suitable hours.

Never use flammable or combustible solvents in a heated bath unless housed in a chemical hood.

Safe Practices: Before using any heating device:

- check to see if the unit has an automatic shutoff in case of overheating.
- note the condition of electrical cords and have them replaced as required
- make sure it has been maintained as required by the manufacturer.
- before leaving an area for any extended period of time, check to see that all heating units in use without automatic shut-off have been turned off.

Open Flames: Bunsen burners, alcohols burners, or other open flames may not be used inside biosafety cabinets. When deemed absolutely necessary, touch-plate microburners equipped with a pilot light to provide a flame on demand may be used. Internal cabinet air disturbance and heat buildup will be minimized. The burner must be turned off when work is completed. Small electric 'furnaces' are available for decontaminating bacteriological loops and needles and are preferable to open flame inside the BSC. Disposable loops can also be used.

3. Vacuum Systems

The following general precautions should be observed when using vacuum systems.

- Every laboratory vacuum pump must have a belt guard in place when it is in operation
- The power cord and switch, if any, must be free of observable defects.
- Use a trap on the suction line to prevent liquids from being drawn into the pump.
- If vapors are being drawn through the pump, a cold trap should be inserted in the suction line to prevent contamination of the pump oil.
- Place a pan under the pump to catch any oil drips.
- Vacuum lines leading from an experimental procedure must always be equipped with traps to prevent contamination of vacuum equipment or house lines. The output of each pump should be vented to a properly functioning fume hood. Do not discharge into an enclosed space, such as a cabinet, as this may cause an explosion.

Glassware: Glassware used for vacuum distillations or other uses at reduced pressure must be properly chosen for its ability to withstand the external pressure of the atmosphere.

- Only round-bottom vessels may be subjected to vacuum unless specially designed, such as Erlenmeyer-type filtration flasks.
- Each vessel must be carefully inspected for defects such as scratches or cracks prior to use.

Implosion: Implosion occurs when atmospheric pressure propels pieces inward creating small fragments that are then propelled outward with considerable force. Because all vacuum equipment is subject to failure by implosion, vacuum operations must be enclosed by blast shielding or conducted in a chemical hood with lowered sash and blast shielding.

Dewar Flask: Dewar vessels have a vacuum between the walls and some types can be dangerous when they fail.

- Glass types can propel glass into the eyes and should be wrapped from top to bottom with cloth tape such as electrician's friction tape. (Mylar tape can be used if transparency is needed.)
- Large Dewars encased in metal and stainless steel vacuum containers do not require wrapping.

Desiccators: Glass desiccators are often subjected to partial vacuum due to cooling of the contents. They have inherent strains due to glass thickness and the relatively flat surface of the top and bottom. It is strongly recommended that you either:

- obtain the available desiccators guard made of perforated metal; or
- use a molded plastic desiccators, which is spherical and has high tensile strength.

4. Centrifuges

- Do not attempt to operate a centrifuge until you have received instruction in its specific operation.
- Read the operation manual.
- Ask an experienced colleague to demonstrate procedures. Ask questions about hazards and emergency action.
- When operating a centrifuge, you are responsible for the condition of the machine and rotors both during and at the end of your procedure. This means:
- proper loading.
- controlling speed to safe levels.
- safe stopping.
- removal of materials.
- cleanup.

Ultra Centrifuge: Ultra centrifuge rotors require special, cleaning procedures to prevent scratching of surfaces, which can lead to stress points and possible rotor failure during a run.

Potential Problem to Watch: An unbalanced load can cause damage to seals or other parts. Keep the lid closed during operations. Shut down and stop the rotor if you observe anything abnormal, such as noise or vibration. To avoid broken tubes examine tubes for signs of stress when loading the rotor and discard tubes that look suspicious.

Selection criteria: In selecting a centrifuge, carefully consider location, type, and use. Other considerations include:

- balance capability each time centrifuge is used.
- adequate shielding against accidental "flyaways".
- suction cups or heel brakes to prevent "walking".
- accessibility of parts, particularly for rotor removal.
- lid equipped with disconnect switch which shuts off rotor if the lid is opened.
- safeguard for handling flammables and pathogens. This may include positive exhaust ventilation or a safe location.
- positive locking of head.

- electrical grounding.
- locations where vibration will not cause bottles or equipment to fall off shelves.
- knowledge of chamber entry during a power failure.

5. Autoclaves

Do not attempt to operate an autoclave until you have received instruction in its specific operation.

– Read the operation manual.
– Ask an experienced colleague to demonstrate procedures. Ask questions about hazards and emergency action.

Preparations:

- Inspect drain strainer daily—clean when blocked.
- Fill liquid containers no more than 50–75% full.
- Loosen caps or use vented closures on bottles. (NEVER tighten caps on non-vented bottles).
- Leave space between items to allow steam to circulate.
- NEVER use plastic bags, which are impervious to steam, unless the top of the bag is loosened to allow steam penetration.
- Autoclave reusable syringes and needles in a pan of disinfectant.
- Do NOT autoclave cellulose nitrate media due to decomposition hazard.
- Place the load in secondary containment. Special plastic bins (polypropylene) or stainless steel bins can be purchased for this purpose. Secondary containment will serve to contain the broken vessels or ruptured bags that sometimes result from routine autoclaving.
- Where a load to be decontaminated is comprised of largely dry materials in an autoclave bag, the addition of water to the bag may facilitate steam penetration, although it is important that caution be used so as not to create aerosols of infectious microorganisms.

Removing the load: Following precaution must take during unloading:

- Check that chamber pressure is zero prior to removing the load.
- Allow load to cool before opening the chamber. If the chamber must be opened before it has cooled, wear a lab coat; face protection (face shield), heat-insulating gloves, and closed toe shoes to avoid steam exposures.
- Stand behind the door and open it slowly. BEWARE OF A RUSH OF STEAM.
- After slow exhaust cycle, open the door and allow liquids to cool before removing.

Potential hazards:

- Potential for generation of scalding jets of steam.
- Handling of superheated solids and liquids.
- Liquids can boil over if pressure is reduced too quickly or if removed too soon after cycle.

- Cellulose nitrate media, such as centrifuge tubes and nitrocellulose filters, can be highly reactive and can decompose explosively under the high heat and pressure conditions present in autoclaves.

6. Fume Hoods and Laboratory Ventilation

One of the primary safety devices in a laboratory is a chemical fume hood. A well-designed hood, when properly installed and maintained, can offer a substantial degree of protection to the user, provided that it is used appropriately and its limitations are understood. This section covers a number of topics aimed at helping laboratory workers understand the limitations and proper work practices for using fume hoods and other local ventilation devices safely.

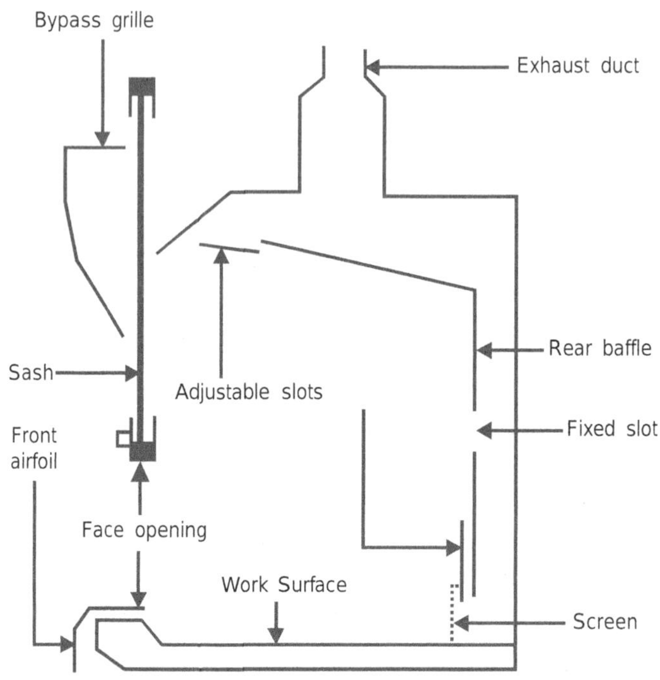

FIGURE 3.1.

Hood face velocity. Based on a number of studies and the recommendations of several laboratory safety guidance documents, the following face velocity criteria are recommended:

- ***Above 150 feet per minute (fpm):*** Unacceptable for laboratory use.
- ***95-125 fpm:*** Provides adequate control of inhalation exposure to most hazardous substances, including radioactive materials and particularly hazardous substances.
- ***80-95 and 125-150 fpm:*** Adequate for manipulation of laboratory quantities of hazardous materials except radioactive materials and particularly hazardous substances.
- ***Below 80 fpm:*** Use approved by EHS on case by case basis, based on activities, placement of hood, smoke tests, etc. **60-80 fpm:** Acceptable for chemical storage.

CHEMICAL HAZARDS AND BIOSAFETY

All the reputed supplier of laboratory chemical provide risk and safety information with those products that present a potential hazard. The label will carry symbol indicating the hazard category **(Annexure III).**

General Rules for Handling Chemicals

The first step in designing a safe operating procedure is to become familiar with the potential health and physical hazards of the chemicals involved. Once appraised of the hazards, the researcher must then select appropriate control strategies to assure safe use and disposal of the chemicals. Engineering controls such as chemical hoods are required for processes releasing airborne contaminants. If work practice and engineering controls do not reduce the hazard to a safe level, then personal protective equipment may be required. Lab workers must all plan for the possibility of accidental spills and releases.

The following general guidelines are to be used when handling any chemical. All personnel are responsible for knowing these general rules, plus any other rules that apply to the specific chemical with which they are working. databases and websites, are available to find additional health and safety information on a wide variety of chemicals.

- know the locations of safety showers, and eyewash stations, as well as other emergency equipment. Know details of their operation.
- conduct procedures that involve hazardous volatile chemicals or that may result in the production of aerosols or dangerous gases in a properly functioning chemical hood.
- be alert to unsafe conditions, and call attention to them so that corrections can be made.
- consider any unlabeled chemical solution hazardous until it is identified.
- discard chemicals that have changed in color or appearance using approved disposal procedures.
- allow only authorized personnel in the laboratory.
- use mechanical devices for all pipeting procedures; never use mouth suction.
- wash hands frequently—always before leaving the laboratory and prior to eating, smoking, applying cosmetics, etc.
- remove gloves before leaving the laboratory. Do **NOT** wear gloves out of the lab into elevators or while typing on lab computers.
- protect the face, skin, and eyes at all times by wearing appropriate protective clothing and equipment to avoid direct contact with the chemical (*i.e.,* chemical goggles, gloves, apron or lab coat, etc.). Remove these items before leaving the laboratory. Do **NOT** wear lab coats or other potentially contaminated protective equipment out of the lab into elevators, during lunch breaks, or launder lab coats at home.

Depending on the level of hazard, chemicals can be categorized in different heads like :

Highly Hazardous Chemicals

It includes selected carcinogens, reproductive toxins, and substances with a high degree of acute toxicity. General laboratory safety practices and procedures are usually sufficient for operations involving hazardous chemicals of mild to moderate risk. When highly hazardous chemicals are involved, however, it may be necessary to take additional steps when dealing with highly hazardous materials.

Standards Operating Procedures (SOPS)

When working with highly hazardous chemicals it is necessary to develop lab-specific written standard operating procedures (SOP's) relevant to safety and health considerations. The purpose of the SOP's is to outline the risks associated with the highly hazardous chemicals in use as well as to describe the steps that lab staff will take to mitigate those risks. SOP's for highly hazardous chemicals can be substance specific or procedure specific, depending on the needs of a particular laboratory.

1. Confine operations involving highly hazardous materials to a designated work area in the laboratory. This designated area can be the entire laboratory, an area of the Laboratory, or a device such as a chemical fume hood. Warning signs are used to clearly indicate which areas are designated and the nature of the hazard. Use of designated areas need not be restricted to highly hazardous materials, as long as all lab staff are aware of the nature of the substances being used, and the special precautions, laboratory skill and safety discipline required to work in the area.
2. Limit access to laboratories where highly hazardous chemicals are in use to appropriately trained and authorized personnel. Depending on the materials and the circumstances of use, access control can be achieved by prohibiting unauthorized visitors and/or physical barriers (such as closing laboratory doors while highly hazardous chemicals are in use or storing highly toxic chemicals in locked cabinets).
3. Procedures involving highly toxic chemicals that can generate dust, vapors, or aerosols must be conducted in a hood, glove box, or other suitable containment device. These devices should be checked for acceptable operation prior to conducting experiments with highly hazardous chemicals. If experiments are to be ongoing over a significant period of time frequent checks of hood function or the installation of a flow-sensing device with an audible or visual indicator of performance should be considered.
4. Experiments conducted with highly hazardous chemicals may need to be carried out in work areas designed to contain accidental releases. Hood trays and other types of secondary containers should be used to contain inadvertent spills, and careful technique must be observed to minimize the potential for spills and releases.
5. Proper waste disposal practices should be use for the disposal of the toxic material **(Annexure II).**

   Solid Waste:

   - Solid waste in the labs must be stored in either approved, labeled containers, or in the fume hood in plexiglas containers which are labeled as "caution radioactive material".
   - All waste must have all radiation symbols defaced before it is put into the proper waste container.
   - All waste must be in clear plastic bags, and different nuclides must be stored separately.
   - Each container must have a Radioactive Waste Card attached with the Authorized User name, date, nuclide, chemical form, and amount of nuclide.

Liquid Waste:

- All liquid waste will be considered either bulk liquid or liquid scintillation vials.
- Bulk liquid must be contained in tightly sealed glass or plastic bottles that have no leaks or cracks.
- Liquid scintillation vials must be packaged in easily movable containers.
- Any liquid requiring a pick-up must have a Radioactive Waste Card attached with the authorized user name, date, nuclide, chemical form, and amount of nuclide.
- Liquid scintillation fluid that is also a hazardous waste must be marked as "Hazardous Waste".
- pH of all liquids must be determined and documented on the Radioactive Waste Card.

Animal Carcasses:

- Handled on a case-by-case basis. Contact the Radiation Safety Office of your institute prior to the use of animals.
- Any other liquid waste that is also a Hazardous Waste must be marked as Hazardous Waste as well as radioactive.

Biological/Infectious Waste:

- Must be collected per Infectious Waste Protocols.
- Radioactive Waste Card attached.
- Custodial Staff should not pick up Biological/Infectious Waste mixed with Radioactive Waste.

6. Chemicals like ethidium bromide which is a potent carcinogen must be inactivated before disposal **(Annexure IV).**
7. Equipment used for the handling of highly hazardous chemicals may need to be suitably isolated from the general laboratory environment. Laboratory vacuum pumps used with these chemicals should be protected by high-efficiency scrubbers or HEPA filters and vented into an exhaust hood. Decontamination, when appropriate, should be conducted in a designated hood.

Some of the commonly used hazardous chemicals are discussed below:

Sodium Pyrophosphate

It is an irritant and may be harmful by inhalation, ingesation or skin absorption. Wear appropriate gloves and safety glasses. Don't breath the dust.

SDS

It is toxic and irritant and poses a risk of sever damage to the eyes. It may be harmful by inhalation, ingesation, or skin absorption. Wear appropriate gloves and safety goggles. Don't breath the dust.

Phenyl methyl sulfonyl fluoride (PMSF)

It is a highly toxic cholinesterase inhibitor. It is extremely destructive to the mucous membrane of the respiratory tract, eyes and skin. It may be fatal by inhalation, ingesation or skin absorption. Wear appropriate gloves and safety glasses and always use in a chemical fume hood. In case of contact, immediately flush eyes or skin with copious amount of water and discard contaminated clothing.

$NaH_2PO_4/Na_2HPO_4/Na_3PO_4$

It is an irritant the eyes and skin. It may be harmful by inhalation, ingestion or skin absorption. Wear appropriate gloves and safety goggles. Don't breath the dust.

Solvents

These should be handled with great care. Wear appropriate gloves and a face protector and use in a chemical hood.

Streptomycin

It is toxic and a suspected carcinogen and mutagen. It may cause allergic reactions. It may be harmful by inhalation, ingesation or skin absorption. Wear appropriate gloves and safety glasses.

Silver nitrate

It is strongly oxidizing agent and should be handled with care. It may be harmful by inhalation, ingesation or skin absorption. Avoid contact with skin. Wear appropriate gloves and safety glasses. It can cause explosions upon contact with other oxidizing materials.

Sodium hydroxide

Sodium hydroxide and solutions containing NaOH are highly toxic and caustic and should be handled with great care. Wear appropriate gloves and a face mask. All others concentrated bases should be handled in a similar manner.

Acetic acid (conc.)

It must be handled with great care. It may be harmful by inhalation, ingesation, or skin absorption. Wear appropriate gloves and goggles and use in a chemical fume hood.

Acrylamide (unpolymerised)

It is a potent neuro-toxin and is absorbed through the skin (the effect are cumulative). Avoid breathing the dust. Wear appropriate gloves and a face mask when weighing powered acrylamide and methylene bis acrylamide, use in chemical fume hood. Polyacrylamide is considered to be non toxic but it should be handled with care because it might contain small quantities of unpolymerised acrylamide.

Coomassic brilliant blue

It may be harmful by inhalation, ingesation or skin absorption. Wear appropriate gloves and safety glasses.

Citric acid

It is irritant and may be harmful by inhalation, ingesation or skin absorption. It poses a risk of serious damage to the eyes. Wear appropriate gloves and safety goggles. Don't breath the dust.

Chloramphenicol

It may be harmful by inhalation, ingesation or skin absorption and is a carcinogen. Wear appropriate gloves and safety glasses. Use in a chemical fume hood.

Cycloheximide

It may be fetal if inhaled, ingested or absorbed through the skin. Wear appropriate gloves and safety glasses. Use in a chemical fume hood.

Nitro blue tetrazolium chloride (NBT)

It is hazardous, handle with care.

Lysozyme

It is caustic to mucous membranes. Wear appropriate gloves and safety glasses.

Iodoacetamide (C_2H_4INO)

It can ankylate the amino group in proteins and can therefore cause problems if the antigen is being purified for amino acid sequencing. It is toxic and harmful by inhalation, ingesation or skin absorption. Wear appropriate gloves and safety glasses and use only in a chemical fume hood. Don't breath the dust.

Chloroform ($CHCl_3$)

It is irritating to the skin, eyes, mucous membranes and respiratory tract. It is a carcinogen and may damage the lever and kidney. It is also volatile. Avoid breathing the vapor. Wear appropriate gloves and safety glasses and always use in a chemical fume hood.

Actinomycin D

It is a teratogen and a carcinogen. It is highly toxic and may be fatal if inhaled, ingested or absorbed through the skin, it may also cause irritation. Avoid breathing the dust. Wear appropriate gloves and safety glasses and always use in a chemical fume hood. Solutions of actinomycine D are light sensitive.

Bacterial strain shipping

The Department of Health Education and Welfare (HEW) has classified various bacteria into different categories with regard to shipping requirements Sanderson and Zeigler, Method in Enzymology 204:248-264(1991). Nonpathogenic strains of *E.coli* (such as K12) and *B. subtilis* are in class 1 and are considered to present no or minimal hazard under normal shipping conditions. However, *Salmonella, Haemophilus* and certain strains of Streptomyces and *Pseudomonas* are in class 2. Class 2 bacteria are "Agents of ordinary potential hazard: agent which produces disease of varying degrees of severity…. But which are contained by ordinary laboratory techniques". For detailed regulations regarding the packaging and shipping of class 2 strains, see Sanderson and Ziegler (Methods enzymology: 248-264)(1991) or the instruction brochure by Alexander and Brandon (Packaging and Shipping of Biological materials at ATCC (1986) available from the American Type Culture Collection, Rockville Maryland).

Bisacrylamide

It is potent neuro toxic and is absorbed through the skin (the effect are cumulative). Avoid breathing the dust. Wear gloves and a face mask when weighing powdered acrylamide and methylne bis acrylamide.

3, 3'-Diamino benzidine tetra hydro chloride (DAB)

It is a carcinogen. Handle with extreme care. Avoid breathing vapors. Wear appropriate gloves and safety glasses and use in a chemical fume hood.

Dimethyl sulfoxide (DMSO)

It may be harmful by inhalation or skin absorption. Wear appropriate gloves and safety glasses. Use in a chemical fume hood. DMSO is also combustible. Store in a tightly closed container. Keep away from heat sparks and open flame.

Dithiothreitol (DTT)

It is a strong reducing agent that emits a foul odor. It may be harmful by Inhalation, ingesation or skin absorption, when working with the solid form or highly concentrated stocks. Wear appropriate gloves and safety glasses and use in a chemical fume hood.

Glutaraldehyde

It is toxic. It is readily absorbed through the skin and is irritating or destructive to the skin, eyes, mucous membranes and upper respiratory tract. Wear appropriate gloves and safety glasses. Always use in a chemical fume hood.

Glycine

It may be harmful by inhalation, ingesation or skin absorption. Wear gloves and safety. Avoid breathing the dust.

Hydrogen peroxide (H_2O_2)

It is corrosive, toxic to the skin. It may be harmful by inhalation, ingesation and skin absorption. Wear appropriate gloves and safety glasses and use only in a chemical fume hood.

Isopropyl-β-D-thiogalacto pyranoside (IPTG)

It is harmful by inhalation, ingesation or skin absorption. Wear appropriate gloves and safety glasses.

β-Mercaptoethanol (2-mercaptoethanol), $HOCH_2CH_2SH$

It may be fetal if inhaled or absorbed through the skin and is harmful if ingested. High concentrations are extremely destructive to the mucous membrane, upper respiratory tract, skin and eyes. B-mercaptoethanol has very foul odor. Wear appropriate gloves and safety glasses. Always use in a chemical fume hood.

Proteinase-K

It is an irritant and may be harmful by inhalation, ingesation or skin absorption. Wear appropriate gloves and safety glasses.

Radioactive substances

When planning an experiment that involves the use of radioactivity, include the physsico-chemical properties of the isotopes (half life, emission type and energy), the chemical form of the radioactivity, its radioactive concentration (specific activity), total amount and its chemical concentration. Order and use only as much as really needed. Always wear appropriate gloves, lab coat and safety goggles when handled radioactive material. X-ray and gamma rays are electromagnetic wavers by radioactive materials. They may be emitted isotopically from the source or may be focused into a beam. Their potential danger depends upon the time period exposure, the intensity experienced and wavelength used. Be sure that appropriate shielding is usually of lead or other similar material. The thickness of shielding is determined by the energy of the-ray or gamma rays. The appropriates guidelines follows when use and dispos radioactive materials.

Always monitor thoroughly after using radio isotopes. A convenient calculator to perform routine radio activity calculations can be found at http://www.graphad.com/calculators/radcalc.cfm.

Highly Reactive Chemicals

Such chemicals are characterized with following properties:

These are normally unstable and readily undergoes violent change without detonating; reacts violently with water; forms potentially explosive mixtures with water; when mixed with water, generates toxic gases, vapors, or chemicals in a quantity sufficient to present a danger to human health or the environment; cyanide or sulfide-bearing waste which, when exposed to pH conditions between 2 and 12.5, can generate toxic gases, vapors, or chemicals in quantity sufficient to present a danger to human health or the environment; capable of detonating or exploding when subjected to a strong initiating source or if heated under confinement; and readily capable of detonation or explosive decomposition or reaction at standard temperature and pressure.

Types of reactive chemical:

Chemicals that exhibit the characteristic of reactivity can be grouped as follows :

Pyrophoric Chemicals: Readily oxidized, will ignite spontaneously in air. Examples of chemicals that are considered to be pyrophoric include such diverse substances as silane, diborane, lithium hydride, white phosphorus, etc.

Water-reactive chemicals: React violently with water. Examples of chemicals that are considered water-reactive include the alkali metals, aluminium chloride, phosphorus pentachloride, phosphorus pentoxide, all hydrides, etc.

Oxidizing materials: React vigorously at ambient temperatures when in contact with reducing materials or may evolve oxygen at room temperature under slight heating. Examples of chemicals that are considered oxidizers include chlorates, permanganates, nitrates, peroxides, etc.

Reducing materials: React vigorously at ambient temperatures when in contact with oxidizing materials. They are compounds, which remove oxygen from other compounds or lower valence states, often with vigorous evolution of heat. Examples of chemicals that are considered reducers include all hydrides, all acetylides, alcohols, etc.

Shock-sensitive materials: Explosive at standard temperature or if heated under confinement. Highly reactive chemicals with explosive properties require special storage, handling, and disposal procedures. Examples of chemicals that are considered shock-sensitive include picrates, azides, polynitro compounds, peroxides, etc.

Heat sensitive chemical: These highly reactive chemicals with explosive properties require special storage handling and disposal procedures.

Peroxides and peroxide forming chemicals: Organic peroxides are a special class of compounds with unusual stability problems that make them among the more hazardous substances handled in the laboratory. Specific chemicals have a strong tendency to form peroxides on exposure to air. Their presence as a contaminant in a reagent or solvent may change the course of a reaction. The peroxides as a class are low-power explosives, hazardous because of their great sensitivity to shock, sparks, heat, or friction. Some laboratory chemicals that are capable of forming explosive levels of peroxides include: diethyl ether, methyl isobutyl ketone, 2-propanol, tetrahydrofuran, dioxanes, acetaldehyde.

Perchloric acid: Perchloric acid is a very strong oxidizing agent that can undergo vigorous, self-sustained decomposition or explosive reactions when catalyzed or exposed to heat. The majority of accidents involving perchloric acid are due to its mixture with organic material or to the formation of the anhydrous acid, which is extremely unstable and explodes upon contact with wood, paper, or organic solvents. Cold perchloric acid, at less than 70% $HClO_4$ by weight, is a less powerful oxidizing agent. As the percentage or the temperature increases, the oxidizing power of the perchloric acid also increases.

Combustible materials, such as wood or oil, contaminated with perchloric acid are highly flammable and dangerous. Such materials may explode spontaneously, upon impact, or when in contact with heat or flame.

Other perchlorates are also sensitive to heat and shock, such as heavy metal and organic perchlorates, hydrazine perchlorate, and ammonium perchlorate.

Precautions for safe use of perchloric acid:

- Whenever possible purchase the 60%-by-weight $HClO_4$ grade.
- Perchloric acid should be stored on a non-metal surface away from: combustible materials; organic solvents; metal hydrides; alcohols; sulfuric acid; and acetic anhydride.
- Inspect monthly for signs of discoloration that indicate development of the anhydrous form of perchloric acid.
- Wear goggles and rubber gloves whenever perchloric acid is handled.
- To avoid a violent reaction in procedures requiring perchloric acid for wet digestion, pretreat the sample with nitric acid to destroy oxidizable material.
- Procedures involving heated perchloric acid require use of a special chemical hood, termed a perchloric acid hood. This type of hood:
- is constructed of stainless steel
- has a duct wash down system, and
- has no exposed organic coatings or sealing compounds
- Procedures involving heated perchloric acid are prohibited.

Other explosive chemicals: In general, other compounds containing the following functional groups tend to be heat- and shock-sensitive:

- acetylide; diazo; nitroso; and inorganic nitrates.
- Hydrogen and chlorine gases react explosively in the presence of light.
- Hydrazine can explode in contact with iron rust.

Safety Guidelines for Unidentified Chemical Waste

Chemicals that cannot be identified should be considered unknown hazardous waste. Unknown wastes cannot be legally transported or disposed.

If you find unknown hazardous waste, please adhere to the following guidelines:

1. Contact immediately lab staff.
2. Find out as much as you can about how the waste was generated.
3. Keep the material in your laboratory or work area and try to remove waste material from the work place/laboratory.

Please DO NOT

- pour unknown chemicals down the sink
- mix unknown chemicals with any other chemicals for consolidation
- bring unknown chemicals to a regular waste pickup unless instructed by lab staff/ supervisor.
- abandon unknown chemicals in the work area

To avoid generating unknown hazardous waste by doing the following:

- Label all chemicals in the laboratory in a meaningful way.
- Dispose of spent materials and chemicals use promptly.
- Before moving out of a work area or leaving the laboratory, go through the laboratory or work area with your supervisor or the new occupant to determine which chemicals need disposal and to identify anything that is not properly labeled.
- When relocating from one work area to another, do not leave any chemicals behind unless specific arrangements have been made with the new occupant.

Safety Guidelines for Cryogenic Material

A cryogenic material is any substance that must be cooled to a temperature of –130°C or lower to change from a gas to a liquid. Cryogenics have several distinguishing characteristics:

- They are extremely cold (–120 to –270°C)
- Their primary cooling mechanism is vaporization (latent heat), and
- They have an extremely high expansion ratio (averaging 800: 1) when their physical state changes from liquid to vapor/gas.

Because of these special characteristics, cryogenic materials must be handled with care. The most common cryogen used is **liquid nitrogen.** Additional cryogens commonly used are: helium; hydrogen; argon; oxygen; and methane.

General precaution: Personnel shall be thoroughly instructed and trained in the nature of the hazards associated with cryogenics and how to avoid those hazards.

- Any person using cryogenics should have a thorough knowledge of: procedures; operation of equipment; safety devices; properties of materials used; and use of personal protective equipment.
- Equipment and systems should be kept scrupulously clean.
- Mixing of gases or fluids should be strictly controlled to prevent the formation of flammable or explosive mixtures. Extreme care should be taken to avoid contamination of a fuel with an oxidant or the contamination of an oxidant with a fuel.
- Proper consideration should be given to the properties of the gas involved when venting storage containers and lines. Venting of large storage vessels should always be done outdoors to prevent an accumulation of flammable, toxic, or inert gas in the work area. Smaller, lab sized containers can be vented into a chemical hood system.

Storage containers: Cryogenic fluids are usually stored in properly insulated containers designed to minimize the loss of product due to boil-off.

NOTE *Boil-off gases can freeze the skin or eyes faster than liquid or metal contact.*

- A Dewar flask is the most common container for cryogenic fluids. It is a double-walled, evacuated container made of metal or glass, with a vacuum between the walls.
- Larger quantities of cryogenic fluid require double-walled metal containers of evacuated construction.
- Exposed glass should be taped to minimize the flying glass hazard if the container should break or implode.
- Liquids should be transferred from the metal Dewar vessels with special transfer tubes or pumps designed for that particular application.

Hazard factors: Cryogens present many hazards. All may be present concurrently and must be considered when introducing a cryogenic system:

- ultra cold temperatures;
- flammability;
- high pressure gas, resulting in over-pressurization of containers and pressure vessels;
- Displacement of oxygen/asphyxiation.

Cryogenic burns: All cryogenics can cause cold burns or frostbite when in contact with human skin.

- Do not overfill containers
- Never make direct contact with cryogenic liquids, un insulated cryogenic pipes or equipment.
- Use tongs
- Stay out of the path of boil-off gases.
- Wear suitable personal protective equipment when handling (goggles, appropriate gloves, lab coat etc.)
- Transfer or pour cryogens slowly to minimize boiling and splashing.
- Use a phase separator or special filling funnel when filling a dewar or transferring cryogens.
- Ensure all secondary containers are secured when filling.
- Ensure that all dewars are positioned so the pressure relief valves and rupture disks vent paths are directed away from personnel, critical equipment or designated work areas.
- Inspect and maintain cryogenic equipment, and remove equipment from service when it does not meet manufacturer's operating specifications.

Flammability: Fire and explosion are hazards associated with cryogenics.

The source and hazards are:

- Hydrogen, methane, and acetylene, where the gases themselves are flammable.
- Oxygen. Its presence will increase the flammability of ordinary combustibles. Keep all organic materials and other flammable substances away from contact with oxygen.
- Liquified inert gases. Liquid nitrogen and helium can condense oxygen from the atmosphere, causing oxygen entrapment in unsuspected areas.
- Extremely cold surfaces can condense oxygen from the atmosphere.

High pressure gas: The hazard from high-pressure gas is always present when cryogenic fluids are used or stored.

- Since liquified gases are usually stored at or near their boiling point, there is always some gas present in the container.
- The large expansion ratio from liquid to gas causes a buildup of high pressure due to evaporation of the liquid.
- Ensure pressure relief devices are utilized and are sized for maximum possible back pressure. Inspect at regular intervals for leakage, frosting and dirt accumulation.
- Ensure an oxygen alarm is present in the work area when appropriate
- Periodically test and calibrate the oxygen alarm.

NOTE *(If an oxygen monitor alarm goes off while you are in the work area, evacuate the room immediately. If it goes off in another room, do not enter the room.)*

- Use natural ventilation and chemical hoods where appropriate.
- Avoid transporting containers in enclosed spaces.

Materials: Materials for cryogenic service must be carefully selected because of the drastic changes in the properties of materials when they are exposed to extremely low temperatures. Suitable materials are:

- Metals such as: stainless steel, copper, brass, bronze, monel, and aluminium. Proper soldering of metals is important.
- Non-metals such as: Dacron, Teflon, and Kel-F.

  When dealing with non-metals, chemical reactivity between the fluid or gas and the storage containers and equipment must be carefully evaluated.

Safety Guidelines for Corrosive Chemicals

The term "corrosive" refers to substances that rapidly attack skin, eyes, and other living tissue. Such chemicals commonly include strong acids or bases, dehydrating agents, and oxidizing agents.

Examples of corrosive chemicals are:

- Acids: Sulfuric, Nitric, Hydrofluoric (HF) Acid.
- Bases: Sodium Hydroxide, Ammonia.
- Dehydrating Agents: Sulfuric acid, Sodium hydroxide, Phosphorus pentoxide, Calcium oxide.
- Oxidizing Agents: Picric acid, Chromic acid, Perchloric acid, Peroxides, Nitrates, Nitrites.

Personal exposure hazards: Strong acids and bases may cause serious damage to the skin and eyes. Inhaling the vapors of corrosive chemicals can cause severe bronchial irritation or pulmonary edema.

Requirements for protection and emergence facilities:

Eyewash/Shower: If the eyes or body of personnel may be exposed to corrosive material, suitable facilities for quick drenching or flushing of the eyes and body shall be provided located within 10 seconds walking time from the location of a hazard. Emergency eyewash/ shower facilities must be clearly visible and unobstructed so that the unit is immediately available in case of a splash to the eyes/body. Lab personnel should be familiar with the location of this equipment so they can locate it readily in an emergency situation. The Physical Plant shall test the units monthly.

First aid: Immediately remove any contaminated clothing.

- Wash the affected area with copious amounts of water; and
- Get first aid or medical help.

Storage guidelines: To properly store corrosive chemicals, follow these guidelines.

- Strong oxidizers such as perchloric acid present fire and explosion hazards when in contact with organics. Store in glass containers in a metal tray, away from organic, flammable, dehydrating, or reducing agents.
- Separate acids from bases.
- Separate acids from flammables.
- Store all corrosives on a low shelf near floor level avoid storage on the floor.

Protective clothing Guidelines: When using corrosive chemicals, follow this rule.

Always wear rubber apron, gloves, and safety goggles, or a face shield.

Accident prevention guidelines: Always Remember: When diluting an acid, add the acid slowly to the water to reduce the reactive effect **(Always remember AAA (Always add Acid to the water))**

Do not:

- Mix concentrated acids and bases together; or
- Use corks or rubber stoppers with strong oxidizing agents.

BIOLOGICAL HAZARDS AND BIOSAFETY

Biohazards are infectious agents or biologically derived infectious material, which presents or may present a risk to the well being of a person or the community.

The Concept of "Biohazard"

Today, however, the content of this word has changed to a much greater degree, and been enlarged to what is related to laypeople. It has changed in the following ways.

First, harmful effects caused by accidents or careless handling of pathogenic microorganisms (or pathogens) can spread not only to researchers in the laboratories but to people and the environment outside laboratories or laboratory facilities. In other words, the "biohazard" has become the concept that communities around laboratory facilities should pay sufficient attention to.

Second, as a result of the rapid development and expansion of biotechnology, there has emerged a new situation where it is no longer sufficient to regard only pathogenic microorganisms existing in nature as causes of biohazards. That is, there arises a situation where we are forced to be concerned about the occurrence of biohazards in the process of implementing biotechnology and their emergence through the accidental creation of by-products different from what researchers intend to produce. Now there have been gradually accumulated noteworthy experimental results showing that biotechnology itself including genetic engineering essentially as its main part may produce an unpredictable danger, causing results different from what researchers aimed to produce. It follows from this argument that we need to include the danger produced by today's biotechnology in the contemporary concept of "biohazard."

Categories of Biohazards

Human, animal and plant pathogens; Bacteria, including those with drug resistance, Plasmids, Fungi, Viruses, including oncogenic viruses, Prions, Parasites, Rickettsia, Chlamydiae. All human blood, blood products, tissues and certain body fluids; Cultured cells (all human or certain animal) and potentially infectious agents these cells may contain, Allergens, Toxins (bacterial, fungal, plant, etc.),Certain recombinant products, Clinical, necropsy and surgical specimens (tissues, fluids, etc.) Infected animals and animal tissues , Recombinant DNA, Human gene transfer, Human recombinant vaccine trials, Transgenic animals, Transgenic plants, Biomolecules with disease-causing potential, Oncogenes.

Some of the above quoted biological agents are discussed in **Annexure V.** Different biological agents are classified into different risk group **(Annexure VI)** which help in designing the safety guidelines while handling them.

Exposure to biohazards. People working in molecular biology laboratory are at risk of biological biohazards because of constant exposure. Depending on the exposure level, we can classify them into following five categories:

1. The first is that workers in medical, health-care and biotechnological laboratory facilities and visitors to them become infected with pathogenic microorganisms or genetically modified microorganisms or other products as listed.
2. The second is that biohazardous materials contained in exhaust air, waste water and waste from bio facilities infect residents living around the facilities and contaminate the environment.
3. The third is that the biohazardous materials existing latently in the natural environment (or settling in animals other than human beings) proliferate and disperse due to the change of environmental conditions, causing infection damage to people, animals or plants in a wide area and disrupting ecosystem.
4. The fourth is that biohazardous materials existing latently in ingested foods and drinks cause infection, poisoning, allergy, cancer, etc. This corresponds with the concept of so-called "food poisoning."
5. The fifth is that on occasions the administration of biological, vaccines, antitoxic sera, blood component preparations, antibiotics, etc. causes a fatal side reaction—headache, fever, asthma, convulsion, paralysis, etc. This is the disaster which is generally called "vaccination accident".

Bio-safety Related with Biological Hazards

Bio-safety is concerned with the containment methods required when managing parasites, infectious agents and infected or potentially infected animals, tissues or other materials. The purpose is to reduce exposure of laboratory workers, other persons and the outside environment to potentially hazardous agents.

Laboratory Safety Protocols for Infectious Material:

Basic requirements for a laboratory using infectious materials are:

Table 3.1. People at risk of biological hazards in molecular laboratory

Procedure	*Persons at Risk*	*Mode of Transmission*
Collection of Animal blood sample	Animal Sample collecting worker	Contaminated needles, contaminated hands or gloves of Laboratory worker Skin punctures by needle or broken specimen container Contamination of hands by blood
Transfer of specimens (within laboratory)	Laboratory personnel Transport worker	Contaminated exterior of specimen container Broken container Spill or splash of specimen
Serology, Virology, Microbiology, Biotechnology and other clinical laboratories	Laboratory personnel	Skin puncture or contamination of skin or mucous membrane Contaminated exterior of specimen container Contaminated work surface Spill or splash of specimen Broken specimen container Perforated gloves
Cleaning and maintenance	Laboratory personnel Support staff	Skin puncture or skin contamination Spills or splashes Contaminated work surface
Waste disposal	Laboratory personnel Support staff Transport worker	Contact with contaminated waste Puncture wounds and cuts
Shipment of specimens (to other centers)	Transport worker Postal worker	Broken or leaking specimen containers and packages

- Ensure that all laboratory personnel, including service and custodial staff and visitors, understand the chemical and biological dangers associated with the lab. Affix biohazard signs to Laboratory Information Cards on doors outside laboratories where biohazardous material is handled or stored. Post the spill response protocol in a visible location in the laboratory.

- Restrict laboratory access and keep doors locked when the laboratory is unattended. Keep the facility clean and free of clutter. Ensure that emergency safety devices (*e.g.,* fire extinguishers, eyewashes, etc.) are easily accessible and in working order.
- Ensure that all personnel, students and visitors adhere to policies for eye and face protection and for protective clothing Remove lab coats or gowns and gloves before leaving the laboratory; never wear lab clothing in eating facilities.
- Avoid eating, drinking, smoking, storage of food and food utensils, application of cosmetics or lip balm and insertion or removal of contact lenses in the laboratory.
- Restrain long hair. Avoid wearing loose clothing or jewellry, shorts and open-toed shoes or sandals.
- Observe "Universal Precautions" when collecting, processing, storing, shipping or transporting human blood and body fluids; *i.e.,* handle such specimens as if infected with a bloodborne pathogen such as hepatitis B or C or human immunodeficiency virus (HIV).
- Carry out procedures so as to minimize risks of splashes, spills and generation of aerosols.
- Refrain from pipetting by mouth.
- Use hypodermic needles only when absolutely necessary. Do not bend, break, shear or recap used needles.
- Wash hands after handling infectious material (even when gloves have been worn) and before leaving the laboratory.
- Decontaminate all contaminated materials before disposal or reuse.
- Decontaminate laboratory surfaces following any spill of biohazardous materials and at the end of each workday.
- Report all spills and accidents/incidents.

Risks Assessment

The term "biotechnology" describes a variety of techniques for manipulation of cells; biotechnology has long been used for purposes such as selective breeding of animals and food production (bread, yogurt, beer). More recently, in vitro incorporation of segments of genetic material from one cell into another ("recombinant DNA technology") has resulted in altered organisms that can manufacture products such as vaccines, hormones, interferons and enzymes. Genetically engineered organisms are used for treatment of waste and spills, and plants can be made resistant to cold, disease, pests and drought.

However, biotechnology carries with it the potential for harm. A genetically altered organism may be directly pathogenic or toxic or, if released into the environment, crowd out beneficial organisms, transfer undesirable genetic traits to wild species or mutate into a pathogenic form.

The risks associated with recombinant DNA technology are to be assessed before implementing it. In India different Biosafety committees have been constituted for this purpose **(Annexure VII).**

Containment

The term "containment" is used in describing measures used to provide a barrier between the infectious organism(s) being handled and the worker (and, ultimately, the community at large). Containment is achieved through two ways:

Physical Containment

Four Biosafety levels are described in Physical Containment. These biosafety levels consist of combinations of laboratory practices and techniques, safety equipment, and laboratory facilities appropriate for the operations performed and are based on the potential hazards imposed by the agents used and for the laboratory function and activity. Biosafety Level 4 provides the most stringent containment conditions, Biosafety Level 1 the least stringent.

The objective of physical containment is to confine organisms containing recombinant DNA molecules and to reduce the potential for exposure of the laboratory worker, persons outside of the laboratory, and the environment to organisms containing recombinant DNA molecules. Physical containment is achieved through the use of laboratory practices, containment equipment, and special laboratory design. Emphasis is placed on primary means of physical containment which are provided by laboratory practices and containment equipment. Special laboratory design provides a secondary means of protection against the accidental release of organisms outside the laboratory or to the environment. Special laboratory design is used primarily in facilities in which experiments of moderate to high potential hazard are performed.

Combinations of laboratory practices, containment equipment, and special laboratory design can be made to achieve different levels of physical containment.

Bio-safety requires a well-designed and functional laboratory. The laboratory is not necessarily separated from the general traffic patterns in the building. Work is generally conducted on open bench tops using standard microbiological practices. Special containment equipment or facility design is neither required nor generally used. Laboratory personnel have specific training in the procedures conducted in the laboratory and are supervised by a scientist with general training in microbiology or a related science.

Bio-safety Level-1 is suitable for experiments involving:

1. Microorganisms of minimal or no biohazard potential under normal handling conditions. Such agents are designated Risk Group (WHO).
2. Recombinant DNA activity requiring Bio safety containment including animal studies which involve the construction of most transgenic species.
3. Non-recombinant cell or tissue culture studies that do not involve infectious plant or animal virus.
4. Management of animal populations that are free of zoonotic organisms or are not part of a study which involves organisms or chemical substances which may require higher containment considerations.

Biological Containment

Experiments involving recombinant DNA lend themselves to a second containment mechanism, namely, the application of highly specific biological barriers. Natural barriers exist that limit either: (*i*) the infectivity of a vector or vehicle (plasmid or virus) for specific hosts, or (*ii*) its dissemination and survival in the environment. Vectors, which provide the means for recombinant DNA and/or host cell replication, can be genetically designed to decrease, by many orders of magnitude, the probability of dissemination of recombinant DNA outside the laboratory In consideration of biological containment, the vector (plasmid, organelle, or virus) for the recombinant DNA and the host (bacterial, plant, or animal cell) in which the vector is propagated in the laboratory will be considered together. Any combination of vector and host which is to provide biological containment shall be chosen or constructed so that the following types of "escape" are minimized: (*i*) survival of the vector in its host outside the laboratory, and (*ii*) transmission of the vector from the propagation host to other non-laboratory hosts.

Combination of Physical and Biological Containment

Since these three means of containment are complementary, different levels of containment can be established that apply various combinations of the physical and biological barriers along with a constant use of standard practices. Categories of containment are considered separately in order that such combinations can be conveniently expressed in the WHO Guidelines.

Physical containment conditions within laboratories may not always be appropriate for all organisms because of their physical size, the number of organisms needed for an experiment, or the particular growth requirements of the organism. Likewise, biological containment for microorganisms may not be appropriate for all organisms, particularly higher eukaryotic organisms. However, significant information exists about the design of research facilities and experimental procedures that are applicable to organisms containing recombinant DNA that is either integrated into the genome or into microorganisms associated with the higher organism as a symbiont, pathogen, or other relationship. This information describes facilities for physical containment of organisms used in non-traditional laboratory settings and special practices for limiting or excluding the unwanted establishment, transfer of genetic information, and dissemination of organisms beyond the intended location, based on both physical and biological containment principles. Research conducted in accordance with these conditions effectively confines the organism.

Table 3.2. Possible alternate combinations of physical and biological containment safeguards

Classification of Physical and Biological Containment	*Alternative Physical Containment*			*Alternate Biological Containment*
	Lab Facilities	*Lab Practices*	*Lab Equipment*	
BL3/HV2	BL3	BL3	BL3	HV2
	BL3	BL3	BL4	HV1
BL3/HV1	BL3	BL3	BL3	HV1
	BL3	BL3	BL2	HV2
BL4/HV1	BL4	BL4	BL4	HV1
	BL4	BL4	BL3	HV2

BL - Biosafety Level

HV - Host-Vector System

Biological Safety Cabinets

Biological safety cabinets reduce the risk of airborne infection by reducing the escape of aerosolized infectious agents into the laboratory environment. In addition to protecting workers, some biological safety cabinets protect the work inside the cabinet from airborne contamination (product protection). Biological safety cabinets minimize contact between the operator and pathogens through the use of directional airflow, HEPA filtration of supply and/or exhaust air, and, in some cases, a physical barrier such as a plastic or glass shield.

1. **Hepa filters:** HEPA (High Efficiency Particulate Air) filters are an essential component of the biological safety cabinet, and have particle removal efficiencies of 99.97% or better for 0.3 micron diameter particles. This size particle is used as the basis for filter definition because it is considered the most difficult to remove. Thus, a filter that can trap 0.3-micron diameter particles can easily eliminate other sizes.

   HEPA filters consist of continuous sheets of glass fiber paper pleated over rigid corrugated separators and mounted in a wooden or metal frame **(Figure 3.2).** The filter medium is delicate and should never be touched. As well, the gaskets used to seal the filter frame to the cabinet must not be disturbed; thus the biological safety cabinet should not be moved without subsequently being tested and certified.

   While HEPA filters remove particulates from an air stream, they are not effective at collecting chemical gases or vapors. Thus, it is inadvisable to use recirculating Class II cabinets with agents which have significant amounts of hazardous volatile or radioactive components. Although Class III and 100% exhaust Class II cabinets can be used in experiments which involve use of chemicals of moderate toxicity, it should be remembered that these cabinets are not explosion-proof. Use of flammable or explosive products is to be avoided unless the cabinet has been specifically designed for their use.

2. **Classes of Biological Safety Cabinets (Figure 3.1).** HEPA-filtered clean air is directed across the work surface toward the front, blowing any generated aerosols into the operator's face. This type of cabinet provides product protection only and should be used solely for procedures such as assembly of sterile apparatus or computer parts.

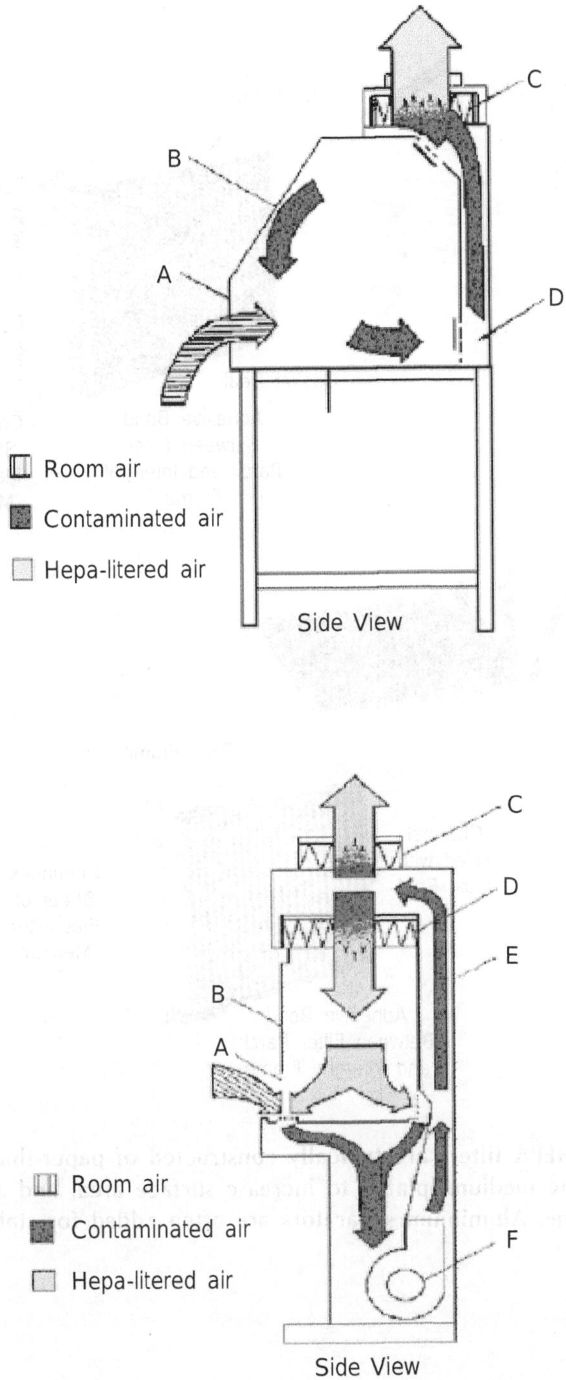

FIGURE 3.2. **Schematic diagram of biological safety cabinet (vertical (*a*) and Horizontal (*b*)).**

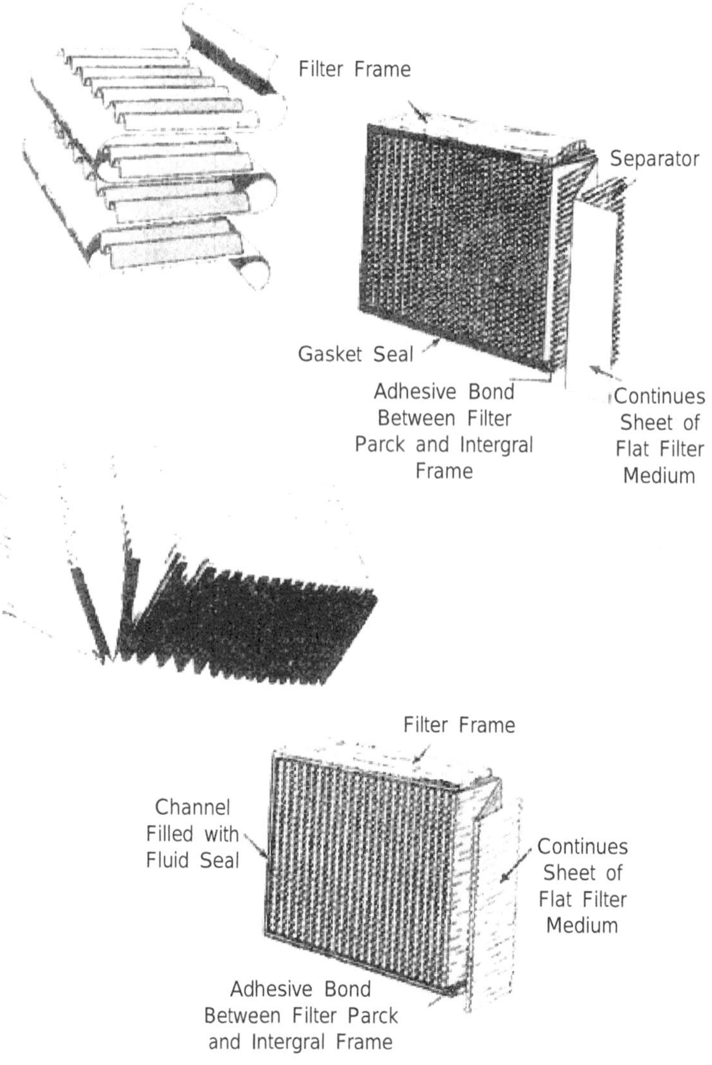

FIGURE 3.3. **HEPA filters are typically constructed of paper-thin sheets of borosilicate medium, plated to increase surface area, and affixed to a frame. Aluminium separators are often added for stability.**

Unit II CALCULATIONS USED IN MOLECULAR BIOLOGY

PART 1 CALCULATIONS USED IN MOLECULAR BIOLOGY

Part 1

CALCULATIONS USED IN MOLECULAR BIOLOGY

Contents

UNITS AND MEASUREMENTS CALCULATIONS

Units

The units used in this book SI units Syste'm International d'Units based on the metric system. These were approved in 1960 by the General Conference of Weights and Measures and being adopted by scientific laboratories throughout the world. They a coherent system of unit so that if two unit quantities are multiplied or divided, then the answer is the unit of the resultant quantity. In this way the number of multiplies and submultiples of units now in use will be reduced.

Basic Units

There are seven basic units on which all others are based and these are set out in following Table 1.1:

Table 1.1. Basic SI units

Physical Quantity	*Name*	*Symbol*
Length	Metre	m
Mass	Kilogram	kg
Time	Second	s
Amount of substance	Mole	mol
Thermodynamic temperature	Kelvin	k
Electric current	Ampere	A
Luminous intensity	Candela	cd

Derived Units

In addition to those above, there also a number of derived SI units obtained by appropriate combination of these basic units. For convenience, these derived units are given in Table 1.2:

Table 1.2. The special names and symbols of some derived SI units

Physical Quantity	*Name*	*Symbol*	*Units*
Frequency	Hertz	Hz	s^{-1}
Force	Newton	N	Kg m s^{-2}
Pressure	Pascal	Pa	N m^{-2}
Energy or work	Joule	J	N m
Power	Watt	Wa	J s^{-1}
Electric charge	Coulomb	C	A s
Electric capacitance	Farad	F	A s V^{-1}
Potential difference	Volt	V	W A^{-1}
Resistance	Ohm	Ω	V A^{-1}
Conductance	Siemens	S	mho m^{-1}
Radioactivity	Becquerel	Bq	S^{-1}
Absorbed dose of radiation	Gray	Gy	J kg^{-1}
Customary temperature	Degree Celsius	°C	°C = K – 273.15

None of these units takes the plural form so that 10 volt is written 10 V not 10 Vs and 2 meters is written 2m not 2ms.

Prefixes

Sometimes units may be to large or too small and in this case, in order to avoid writing too many zeroes; a prefix is placed before the symbol of the unit (Table 1.3). The recommended multiples of fractions of a unit change by 1000 each time. Thus 0.000012 mol is written 12 μ mol and 25400 m is written 25.4 km. the combination of prefix and symbol is regarded as a new symbol so that mm^3 means $(10^{-3}\ m)^3$ not $10^{-3}\ m^3$. There is therefore no space, point or full stop between the prefix and the symbol. However, a space is left between symbol in derived units and for the sake of clarity a point above the line is often used. For example:

$$ms = \text{millisecond } (\textit{i.e.,})\ 10^{-3}\ s,$$

where as

$$m\,.\,s = \text{metre} \times \text{second } (\textit{i.e.,})\ m \times s$$

Table 1.3. Common prefixes for the SI units likely to be used in molecular biology experiments work

Multiples			*Fractions*		
Factor	*Prefix*	*Symbol*	*Factor*	*Prefix*	*Symbol*
10^6	Mega	M	10^{-3}	milli	m
10^3	Kilo	K	10^{-6}	micro	μ
			10^{-9}	nano	n
			10^{-12}	pico	p

Table 1.4. Some universal constants

Constant	*Symbol*	*Value*	*Units*
Avogadro's number	N_A	6.023×10^{23}	$kmol^{-1}$
Gas constant	R	8.314×10^3	$J\ kmol^{-1}\ K$
Boltzmann constant	k	1.381×10^{-23}	$J\ K^{-1}$
Charge on the electron	e	1.602×10^{-19}	C
Faraday	F	9.649×10^{-7}	$C\ kmol^{-1}$
Planck's constant	h	6.626×10^{-34}	$J\ s^{-1}$
Gravitational constant	g	9.81	$m\ s^{-2}$
Velocity of light in a vacuum	c	2.998×10^8	$m\ s^{-1}$

Physical Constants

Some of the universal which the biologist is likely to encounter are given above in Table 1.4 together with SI units.

Units Used in Conjunction with SI

There are some units used in Laboratory work which are unlikely to be replaced altogether because of their convenience and it is probable that they will continue to be used in conjunction with SI units for some time.

Unit for Volume

The SI unit for volume is, the cubic meter (m^3) but this is large and the litre is accepted as the unit of volume in Laboratory work.

The liter is exactly equal to one cubic decimeter (1 decimeter = 10^{-1} m = dm) so that:

1000 litres = 1 cubic metre = m^3

1 litre (l) = 1 d m^3 = 10^{-3} m^3

1 millimeter (ml) = 1 cm^3 = 10^{-6} m^3

1 microlitre (μl) = 1 mm^3 = 10^{-9} m^3

Unit for Mass

The gram used as an elementary and it association with prefixes (μg, mg) adopted for the basic unit of mass now known as a kilogram.

1 g = 1000 mg (milligram) = 10^{-3} Kg

1 mg = 1000 μg (microgram) = 10^{-3} g

1 μg = 1000 ng (nanogram) = 10^{-6} g

1 ng = 1000 pg (picogram) = 10^{-9} g

1 pg = 1000 fg (fentogram) = 10^{-12} g

Part per million (ppm) = μg per gram or mg per kg = 10^{-6}

Part per billion (ppb) = ng per gram or μg per kg = 10^{-9}

Unit for Time

The basic SI unit for time is the second but the common units of time (*e.g.,* minute, hour, and year).

MAKING PERCENT SOLUTIONS

Success of any experiment depends how accurately you perform the experiments. Good solution making is an important skill for the molecular biology experiments. This chapter deals with the calculations involve in the solution making for molecular biology experiments.

Common Definitions

pH: pH is used to quantify the acidity of the solution. The p indicated that you take the negative log of what follows, which is H, signifying the concentration of protons. pH is therefore a measure of the concentration of protons, or H^+, ions in the solution. As the number of protons in solution increases, the magnitude of the pH decreases : so the lower the pH of a solution, the more acidic it is.

Reagents: A compound or the solution use to prepare reaction mixture.

Solute: The dissolved phase of a solution.

Solvent: The dispersing phase of a solution. The substrate in which a solute is dissolved.

Solution: A homogeneous mixture of two or more substances; one or more solute dissolve in the solvent. The solutes can be solid, liquid, or gase; solvent can be liquid or gases.

Concept and Calculations of Concentration

Concept of Mole

In Molecular Biology studies we frequently deal with number of particles such as molecules, atoms, ions that are involved in biochemical processes.

As we know that one gram atom (a g atom) of an element contains 6.023×10^{23} atoms (Avogadro's number, N).

Avogadro's number is equal to the carbon atoms present in exactly twelve grams of C^{12} (Carbon isotope). By one g atom of an element, we mean it atomic weight expressed in g.

For example, the atomic weight of Ca is 40.08, hence one g of atom Ca means 40.08 grams of Ca and this would contain 6.023×10^{23} Ca atoms or 0.1 gm atom *i.e.,* 40.08/10 = 4.008 gm. The Ca would contain $6.023 \times 10^{23}/10 = 6.023 \times 10^{22}$ Ca atoms. The number of g atoms of an element present in a given sample is calculated by:

$$\text{Number of g atom} = \frac{\text{weight of element in gm}}{\text{atomic weight of an element}}$$

It is also known that one g molecular weight of any substance contains 6.023×10^{23} molecules of that substance. For example a mole of oxygen is equivalent to 6.023×10^{23} molecules of oxygen. A mole of water (H_2O) is equivalent to 6.023×10^{23} molecules of water.

The number of moles in a substance is calculated as

$$\text{No. of moles} = \frac{\text{wt. of the substance}}{\text{mol. wt. of the substance}}$$

Suppose in a given volume of a solution 4 g of NaCl is present than the number of moles of NaCl in the solution will be 4/58.453 = 0.0684 mole, where 58.453 is the molecular wt. of the NaCl.

The term mole, refers to a fixed number *i.e.,* 6.023×10^{23}. It does not refer to any fixed amount of weight of a substance. The correct use of the term mole is, one mole of hydrogen atoms and not one mole of hydrogen. However, we say a mole of hydrogen meaning 2 g of hydrogen (2 is molecular weight of hydrogen). For example, one mole of electrons means 6.023×10^{23} electrons.

Gram Formula Weight/Molecular Weight

The gram formula weight/molecular weight of a substance is equivalent to sum of its atomic weights. For example, the gram molecular weight of NaOH (sodium hydroxide) is 40.00, the atomic weight of Na (22.99 g) + atomic weight of oxygen (16.00) + atomic weight of Hydrogen (1.01).

Problem 1. *One gram of SO_4^{2-} ions contains how much of SO_4^{2-} ions ?*

Solution: One gram ion weight is the formula weight of the ion expressed in gram.

The formula weight of SO_4^{2-} ion is 32 + 64 = 96, hence one gram ion of SO_4^{2-} means 96 grams of Sulphate ions. A gram ion weight is also written briefly as a gram ion.

Problem 2. *What is the molecular weight of NaCl (Sodium Chloride).*

Solution:

$$\begin{aligned} \text{Atomic weight of Na} &= \ 22.99 \\ \text{Atomic weight of Cl} &= +\ 35.45 \\ \hline \text{Molecular weight of NaCl} &= \ 58.44 \end{aligned}$$

Molar and Formular Solutions

If one g mole of a substance is present in a litre of solution it is called a **molar solution.** If a litre of solution contains 40 g of NaOH *i.e.,* one g of NaOH, the solution is called a molar solution or the molarity of solution will be unit.

$$\text{Molarity of solution} = \frac{\text{No. of g mole}}{\text{Volume of solution in litre}}$$

If 3 g mole of a substance are present in 10 litres of its solution, the molarity of the solution will be 3/10 = 0.3 or the solution will be said to be 0.3 molar or 0.3 M.

If the formular weight of the substance is considered, then the term formality is used. For example, the formula weight of NaCl is 58.44, hence if 58.44 g NaCl is present in a litre of solution, it will be called a formal solution. If 58.44 g NaCl is present in 10 litres of solution, the formality of the solution would be 1/10 = 0.1 Thus,

$$\text{Formality of a solution} = \frac{\text{No. of g formula weight}}{\text{Volume of solution in litre}}$$

Problem 3. *How are 100 ml of 0.2 M NaCl prepared ?*

Solution: Molecular weight of NaCl is 58.44.

First calculate how much grams are required for 1 L of 0.2 M solution. This can be done by setting up a ratio stating "58.44 g is to 1 M as x grams is to 0.2 M". //This relationship, expressed mathematically, can be written as follows and we can solve for x:

$$\frac{58.44}{1\text{ M}} = \frac{x\text{ g}}{0.2\text{ M}}$$

Because units on both sides of the equation are equivalent, (If we were to multiply one side by the other, all terms would cancel) we will disregard them. Multiplying both sides of the equation by 0.2 gives

$$\frac{0.2}{1} \times \frac{58.44}{1} = \frac{x}{0.3} \times \frac{0.2}{1}$$

$$11.68 = x$$

Therefore, to prepare 1 L of 0.2 M NaCl, 11.688 grams of NaCl are required.

Another ratio now needs to be calculated to determine how many grams of NaCl are needed if 100 ml of a 0.2 M NaCl solution are being prepared. It can be expressed verbally as "11.688 grams is to 1000 ml as x grams is to 100 ml" or calculated by,

$$\frac{11.688}{1000\text{ ml}} = \frac{x}{100\text{ ml}}$$

Multiply both sides of the equation by 100 to cancel out the denominator on the right side of the equation and to isolate x.

$$11.688 \times \frac{100}{1000} = x$$

$$1.1688 = x$$

Therefore, to prepare 100 ml of 0.2 M NaCl solution, 1.1688 grams of NaCl are dissolved in distilled water to a final volume of 100 ml.

Problem 4. *How many moles of NaOH are present in 10 ml of a 0.25 M solution ?*

Solution: 0.25 M solution of NaOH contains 0.25 moles of NaOH/L.

We want to know how many moles are in 10 ml of a 0.25M solution of NaOH, we need to use conversion factor to convert litres to millileters. Equation as follows:

x mol = 10 ml × 1 L/1000 ml × 0.25 mol/L

x mol = 10 × 1 × 0.25/1000 = 0.0025 mol

Therefore, 10 ml of 0.25 M NaOH contains 0.0025 moles of NaOH.

How to Dilute Molar Solutions

Most of the time it is convenient prepare concentrated stock solution of commonly used reagents (Buffers, Reaction mixtures).

A specific volume of a stock solution at a particular concentration can be added to buffer or reagent mixture so that it contains that component at same concentration less than that in the stock.

More concentrated solutions being used to prepare a lower concentrated solution. A dilution of stock solution is being perform. There are several methods that can be used to calculate the concentration of diluted reagents.

Mostly used method is use of following approaches.

1. **By using equation: $C_1V_1 = C_2V_2$ where,**

   C_1 = Initial concentration of the stock solution.

   V_1 = Amount of stock solution taken to perform the dilution.

   C_2 = Concentration of the diluted sample and

   V_2 = Final total volume of the diluted sample.

   Example: If you were asked how many μl of 10% SDS should be used to make 5 ml of 4% SDS.

   The $C_1V_1 = C_2V_2$ equation could be used this can be done as follows:

   $$(10\%) \times V_1 = 4 \times 5000 \ \mu l$$

   $$V_1 = 4\% \times 5000 \ \mu l/10\% = 2000 \ \mu l$$

   The % units cancel since they are in both the numerator and the denominator of equation, leaving μl as the remaining unit. Therefore, you would need 2000 μl of 10% SDS plus 3000 μl (5000 μl, 2000 μl, 3000 μl) of distilled. water to make 4% SDS solution from a 10% SDS solution.

2. **Dimensional analysis a general approach to solve problems of concentration as follows:**

   Desired concentration = Initial concentration × conversion factor × (unknown volume/ Final volume)

   Using this approach the problem of discovering how many μL of 10% SDS are needed to make 5 ml of 4% SDS is written as follows:

   $$10\% \times 1 \text{ ml}/1000 \ \mu\text{l} \times x \ \mu\text{l/ml} = 4\%$$

   All terms on the left side of the equation will cancel except for % solving for x μl gives the following result:

   $$(10\%)\ x/5000 = 4\%$$
   $$x = 4 \times 5000/10$$
   $$x = 2000 \ \mu\text{l}$$

   Since x is a μl amount, you need 2000 μl of 10%. SDS in a final volume of 5 ml to make 4% SDS. Notice how similar the last step of the solution to this equation is to last step of the equation using the $C_1V_1 = C_2V_2$ approach.

   Making a conversion factor part of the equation obviate the need for performing two separate calculations, as is required when using the $C_1V_1 = C_2V_2$ approach. For this reason, dimensional analysis is the method used for focusing problems of concentration most of the time.

How to Prepare Percent Solution

Many reagents are prepared as a percent of solute (such as salt, NaOH, Cesium chloride, Sucrose, SDS etc.) dissolved in solution.

Depending on the physical state of the solute its concentration can be expressed as % w/v or % v/v. A percentage in weight/volume refers to the weight of solute (in grams) in a total of 100 ml of solution. A % v/v refer to the amount of liquid solute (in ml) in a final volume of 100 ml of solution.

Problem 5. *How can the 30% (w/v) of acrylamide solution prepared?*

Solution: Weight out 30 grams of Acryamide and dissolve in distilled water so that the final volume of the solution with the acrylamide completely dissolved, in 100 ml. It is 30% (w/v) of acrylamide solution.

Problem 6. *How can the 500 ml of 95% (v/v) solution of ethanol prepared?*

Solution: 95% of 500 ml is calculated by multiplying 0.95 (the decimal from the 95%) by 500 ml.

$$0.95 \times 500 = 475$$

Therefore, to prepare 500 ml of 95% ethanol measure 475 ml of 100% ethanol and add 25 ml of distilled water to bring the final volume to 500 ml.

How to Dilute % Solutions

For solving dilution problem involving percentages, express the percent solutions as fractions of 100.

It can be written as an equation in which the concentration of the stock solution write on the left side of the equation and the required final concentration is on right side of the equation. The stock solution volume require to add to the volume of the final mixture should also be expressed as a fraction and this part also put on the left side of the equals sign.

Example: If 25 ml of 20% sugar solution to be prepared from 50% sugar solution stock solution.

It can be written:

$$\frac{50}{100} \times \frac{x \text{ ml}}{25 \text{ ml}} = \frac{20}{100}$$

Where x ml is the required volume of the stock:

Multiply numerator together and denominators together.

$$\frac{50x \text{ ml}}{2500 \text{ ml}} = \frac{20}{100}$$

Multiply both sides of the equation by 2500.

$$\frac{2500}{1} \times \frac{20x}{2500} = \frac{20}{100} \times \frac{2500}{1}$$

$$50\ x = \frac{50{,}000}{100}$$

$$50\ x = 500$$

Divide each side of the equation by 50

$$\frac{50x}{50} = \frac{500}{50}$$

$$x = 10$$

Therefore, to prepare 25 ml of 20% sugar solution using a 50% sugar stock solution, combine 20 ml of 50% sugar stock with 5 ml of distilled water.

How to Convert Molority to Percent

As we discuss molarity is a concentration of grams per 100 ml, it is a simple calculation to convert it to a percent value, an expression of a gram amount in 100 ml.

Problem 7. *Express 1.5 M NaCl as a percent solution.*

Solution: The gram molecular weight of NaCl is 58.44.

First to calculate how many grams of NaCl in 1.5 M NaCl. Solution as follows :

$$\frac{58.44}{1 \text{ M}} = \frac{x \text{ g}}{1.5 \text{ M}}$$

$$\frac{(58.44)\ (1.5)}{1} = x$$

$$87.66 = x$$

Therefore, to prepare a 1.5 M solution of NaCl 87.66 grams of NaCl are dissolved in a total volume of 1 liter. Percent is an expression of concentration in parts per 100.

To set up the state.

"87.66 gms is to 1000 ml as x g is to 100 ml"

$$\frac{87.66g}{1000 \text{ ml}} = \frac{xg}{100 \text{ ml}}$$

$$87.66 \times \frac{100}{1000} = x$$

$$x = 8.766$$

Therefore a 1.5 M NaCl solution contains 8.766 grams of NaCl in 100 ml. Which is equivalent to a 8.76% NaCl solution.

Normality

1 N solution is equivalent to the gram molecular weight of a compound divided by the number of hydrogen ions present in solution (*i.e.,* dissolved in one liter of water).

Example: Gram molecular weight of HCl is 36.46 in solution 1 H^+ ion combine with Cl^- to form HCl. 1 N HCl solution contains $\frac{36.46}{1}$ = 36.46 grams of HCl in 1 liter. A 1N HCl solution.

Therefore, it is equivalent to a 1 M HCl solution.

Normality and molarity are related by the equation.

$$N = nM$$

where N is equal to the number of replaceable H^+ (or Na^+) or OH^- ions per molecule.

MAKING BUFFER SOLUTIONS

Concept of pH and Buffer

Buffers are solutions which could tolerate small changes in pH during the reactions. A Buffer is composed of a weak acid and its salt. All biochemical reactions in cells and tissues eventuate in aqueous medium. These reactions depend upon the rigid regulation of hydrogen ion activity, in *in vivo* condition. In order to understand these biological processes in vitro, it is necessary to *mimic* these conditions in artificial media, where as in hydrogen ion concentration plays a portentous role. The term pH denotes the hydrogen ion concentration or activity of a given solution. The symbol '*p*' denotes the *negative logarithm* while 'H' represents the *hydrogen* ion concentration. It is defined by the following expression

$$pH = \log 1/[H^+] = -\log [H^+]$$

One of the important properties of pure water is its *amphoteric* nature. It serves as a proton donor (acid) and acceptor (base). This is represented by the following equation of ionic equilibrium of water,

$$H_2O = H^+ + OH^-$$
$$H_2O + H^+ = H_3O^+ \text{ (hydronium ions)}$$

In pure water, the degree of ionization at equilibrium is small. For every 10^7 molecules of water, one molecule is in ionized form at 25°C. The ion product of $[H^+]$ $[OH^-]$ in aqueous solution (at 25°C) always equals 1×10^{-14} M. This is based on the following expression:

$$K_w = [H^+]\,[OH^-] = (55.5 \text{ M})\,(1.8 \times 10^{-16} \text{ M}) = 1.0 \text{ x } 10^{-14} \text{ M}^2$$

Where,

(*i*) K_w is ion product of water at 25°C.

(*ii*) 55.5 M is the concentration of water at 25°C (calculated by dividing weight of one liter water in grams by gram molecular weight, *i.e.*, 1000 g ÷ 18.02 = 55.5 M).

(*iii*) $1.8 \text{ X } 10^{-16}$ M is the value of K (equilibrium constant for reversible ionization of water – $Keq = [H^+]\,[OH^-]/[H_2O]$ based on the electrical conductivity measurement of pure water.

Thus at neutral *pH* the concentration of H^+ and OH^- ions are exactly equal. This is arrived by the following equation:

$$K_w = [H^+][OH^-] = [H^+]^2 \qquad \text{Equation 1}$$

Solving the equation I for $[H^+]$ gives

$$[H^+] = \sqrt{K_w} = \sqrt{1.0 \text{ X } 10^{-14} \text{ M}^2}$$
$$(H^+) = (OH^-) = 10^{-7} \text{ M}$$

Equation 1 is useful in calculating hydrogen ion concentration in solution.

A buffer is a solution that can maintain a nearly constant pH when diluted, or when strong acids or bases are added. A buffer solution contains a conjugate weak acid-base pair in approximately equal concentrations. Such a solution works because there are two components present at all times. There is acid present in the solution to neutralize any added strong base. And there is base present to neutralize any added strong acid. This works only for weak conjugate acid-base pair. If they were strong (such as HCl and NaOH) they would instantaneously get neutralized.

Example: Consider a solution containing both acetic acid, CH_3COOH, and acetate ions, CH_3COO^-.

Any strong base that is added to the solution is neutralized by acetic acid:

$$CH_3COOH(aq) + OH^-(aq) \longrightarrow CH_3COO^-(aq) + H_2O(aq)$$

Any strong acid that is added to the solution is neutralized by acetate:

$$CH_3COO^-(aq) + H^+(aq) \longrightarrow CH_3COOH(aq)$$

The amount of strong acid or base that a buffer can neutralize is called the buffer capacity. After the strong base or acid is neutralized, equilibrium will be reestablished.

Since the concentrations of the conjugate acid-base pair in a buffer are usually high, there is very little change in the acid-base pair concentrations as the system establishes equilibrium.

Therefore, the $[H^+]$ concentration can be calculated from the equilibrium expression using the pre-equilibrium concentrations of the conjugate acid-base pair.

In the example above how could we produce the solution containing CH_3COOH and CH_3COO^- ? In general a buffer can be prepared by:

1. Mixing a weak acid and a salt of its conjugate base in solution.
2. Mixing a weak acid and enough strong base to neutralize a portion of the weak acid.
3. Mixing a weak base and enough strong acid to neutralize a portion of the weak base.

Problem 8. *What is the pH of a buffer solution containing 0.100 moles of both CH_3COOH and CH_3COO^- in 0.100 L of water ?*

Solution:

1. The only species present in solution that take part in the equilibrium are acetic acid and acetate ion. The pre-equilibrium concentrations of acetic acid and acetate ion are given as 1.00 M (0.100 moles/0.100 L). In buffers, the concentrations of the weak acid and conjugate base are high, so the $[H^+]$ of water (1.0×10^{-7} M) can always be neglected.
2. The balanced equilibrium reaction is: $CH_3COOH(aq) \rightleftharpoons H^+(aq) + CH_3COO^-(aq)$ and the equilibrium constant expression is:

$$K_a = \frac{[H^+][CH_3COO^-]}{[CH_3COOH]}$$

3. It is seldom necessary to calculate Q in buffer problems. See explanation in step 4.
4. As stated in the introduction above, this step is not necessary in buffer problems. The concentrations of CH_3COOH and CH_3COO^- do not change appreciably as equilibrium is established.
5. We can now calculate the equilibrium concentrations using the equilibrium constant expression:

$$K_a = \frac{[H^+][CH_3COO^-]}{[CH_3COOH]}$$

$$K_a = \frac{(x)(1.00\ M)}{(1.00\ M)}$$

$$x = K_a$$

$$x = 1.8 \times 10^{-5}$$

This problem asked for the pH of the solution.

$$[H^+] = x = 1.8 \times 10^{-5}\ M$$

$$pH = -\log [H^+] = -\log (1.8 \times 10^{-5})$$

$$pH = 4.74$$

Notice that when the concentrations of the weak acid and conjugate base are equal, the pH of the buffer solution equals pK_a.

Preparing a Buffer Solution

Adapted from Robinson, R. A., and Stokes, R. H., **Electrolyte solutions, the measurement and interpretation of conductance, chemical potential, and diffusion in solutions of simple electrolytes,** 2nd ed., rev. London, Butterworths, 1968.

Buffer solutions (pH of the first component and volume of the second component of each of the following 10 buffers is given in Table 2.5.)

Buffer 1-50 mL 0.2 M KCl + volume indicated (in mL) 0.2 M HCl

Buffer 2-100 mL 0.1 M potassium hydrogen phthalate + volume indicated (in mL) 0.1 M HCl.

Buffer 3-100 mL 0.1 M potassium hydrogen phthalate + volume indicated (in mL) 0.1 M NaOH.

Buffer 4-100 mL 0.1 M KH_2PO_4 + volume indicated (in mL) 0.1 M NaOH.

Buffer 5-100 mL 0.1 M tris (hydroxymethyl) aminomethane + volume indicated (in mL) 0.1 M HCl.

Buffer 6-100 mL 0.025 M $Na_2B_4O_7.10H_2O$ (borax) + volume indicated (in mL) 0.1 M HCl.

Buffer 7-100 mL 0.025 M $Na_2B_4O_7.10H_2O$ (borax) + volume indicated (in mL) 0.1 M NaOH.

Buffer 8-100 mL 0.05 M $NaHCO_3$ + volume indicated (in mL) 0.1 M NaOH.

Buffer 9-100 mL 0.05 M Na_2HPO_4 + volume indicated (in mL) 0.1 M NaOH.

Buffer10-50 mL 0.2 M KCl + volume indicated (in mL) 0.2 M NaOH.

Table 1.5. Information for preparing above mentioned buffers

Buffer 1		*Buffer 2*		*Buffer 3*		*Buffer 4*		*Buffer 5*	
pH	*volume added*	*pH*	*volume added*	*pH*	*volume added*	*pH*	*volume added*	*pH*	*volume added*
1.00	134.0	2.20	99.0	4.10	2.6	5.80	7.2	7.00	93.2
1.10	105.6	2.30	91.6	4.20	6.0	5.90	9.2	7.10	91.4
1.20	85.0	2.40	84.4	4.30	9.4	6.00	11.2	7.20	89.4
1.30	67.2	2.50	77.6	4.40	13.2	6.10	13.6	7.30	86.8
1.40	53.2	2.60	70.8	4.50	17.4	6.20	16.2	7.40	84.0
1.50	41.4	2.70	64.2	4.60	22.2	6.30	19.4	7.50	80.6
1.60	32.4	2.80	57.8	4.70	27.2	6.40	23.2	7.60	77.0
1.70	26.0	2.90	51.4	4.80	33.0	6.50	27.8	7.70	73.2
1.80	20.4	3.00	44.6	4.90	38.8	6.60	32.8	7.80	69.0
1.90	16.2	3.10	37.6	5.00	45.2	6.70	38.6	7.90	64.0

2.00	13.0	3.20	31.4	5.10	51.0	6.80	44.8	8.00	58.4
2.10	10.2	3.30	25.8	5.20	57.6	6.90	51.8	8.10	52.4
2.20	7.8	3.40	20.8	5.30	63.2	7.00	58.2	8.20	45.8
		3.50	16.4	5.40	68.2	7.10	64.2	8.30	39.8
		3.60	12.6	5.50	73.2	7.20	69.4	8.40	34.4
		3.70	9.0	5.60	77.6	7.30	74.0	8.50	29.4
		3.80	5.8	5.70	81.2	7.40	78.2	8.60	24.4
		3.90	2.8	5.80	84.6	7.50	82.2	8.70	20.6
		4.00	0.2	5.90	87.4	7.60	85.6	8.80	17.0
						7.70	88.4	8.90	14.0
						7.80	90.6	9.00	11.4
						7.90	92.2		
						8.00	93.4		

Buffer 6		*Buffer 7*		*Buffer 8*		*Buffer 9*		*Buffer 10*	
pH	*volume added*	*pH*	*volume added*	*pH*	*volume added*	*pH*	*volume added*	*pH*	*volume added*
8.00	41.0	9.20	1.8	9.60	10.0	10.90	6.6	12.00	12.0
8.10	39.4	9.30	7.2	9.70	12.4	11.00	8.2	12.10	16.0
8.20	37.6	9.40	12.4	9.80	15.2	11.10	10.2	12.20	20.4
8.30	35.4	9.50	17.6	9.90	18.2	11.20	12.6	12.30	25.6
8.40	33.2	9.60	22.2	10.00	21.4	11.30	15.2	12.40	32.4
8.50	30.4	9.70	26.2	10.10	24.4	11.40	18.2	12.50	40.8
8.60	27.0	9.80	30.0	10.20	27.6	11.50	22.2	12.60	51.2
8.70	23.2	9.90	33.4	10.30	30.4	11.60	27.0	12.70	64.4
8.80	19.2	10.00	36.6	10.40	33.0	11.70	32.4	12.80	82.4
8.90	14.2	10.10	39.0	10.50	35.6	11.80	38.8	12.90	106.0
9.00	9.2	10.20	41.0	10.60	38.2	11.90	46.0	13.00	132.0
9.10	4.0	10.30	42.6	10.70	40.4	12.00	53.8		
		10.40	44.2	10.80	42.4				
		10.50	45.4	10.90	44.0				
		10.60	46.6	11.00	45.4				
		10.70	47.6						
		10.80	48.5						

COMMONLY USE BUFFER SYSTEMS IN MOLECULAR LABORATORY EXPERIMENTS

2 × BES—Buffered Saline

Dissolve 1.07 g of BES (N, N-bis [2-hydroxyethy1]-2-aminoethanesulfonic acid), 1.6g of NaCl, and 0.027 g of Na_2HPO_4 in a total volume of 90 ml of distilled H_2O. Adjust the pH of the solution to 6.96 with HCI at room temperature, and then adjust the volume to 100 ml with distilled H_2O. Sterilize the solutions by passage through a 0.22 μm filter, and store in aliquots at – 20°C.

2 × HEPES—Buffered Saline

Dissolve 1.6 g of NaCl, 0.074 g of KCl, 0.027 g of $Na_2HPO_4.2H_2O$, 0.2 g of dextrose, and 1g of HEPES in a total volume of 90 ml of distilled H_2O. Adjust the pH to 7.05 with 0.5 N NaOH, and then adjust the volume to 100 ml with distilled H_2O. Store in 5-ml aliquots at – 20°C.

HEPES Buffer

For 50 mM HEPES buffer (pH 7.0), mix 25 ml of 200 mM HEPES (52.06g/liter of HEPES Na salt), 11.8 ml of 100 mM NaOH, and distilled H_2O to 100 ml. Sterilize the solution by passage through a 0.22 μm filter.

PIPES Buffer

For 50 mM PIPES buffer (pH 6.8), mix 25 ml of 200 mM PIPES and 20.7 ml of 100 mM NaOH and distilled H_2O to 100 ml.

Potassium (K) Phosphate Buffer

For 50 mM K-phosphate buffer (pH 7.5), mix 50 ml of 100 mM (KH_2PO_4 (13.60g/liter) plus 40.9 ml of 100 mM NaOH diluted to 100 ml with distilled H_2O.

0.1 M Potassium Phosphate Buffer at 20°C

To prepare 1 M stocks, add 136g of monobasic (KH_2PO_4) OR 174.2 g of dibasic (K_2HPO_4) potassium phosphate to l liter of distilled H_2O. Combine the following volume (Table 1.6) of the 1 M stocks to achieve the desired pH. Dilute the combined 1M stock solution to 1000 ml with distilled H_2O.

Table 1.6. Preparation of 0.1 M potassium phosphate buffer

pH	*Volume of 1 M K_2HPO_4 (ml)*	*Volume of 1 M KH_2PO_4 (ml)*
5.8	8.5	91.5
6.0	13.2	86.8
6.2	19.2	80.8
6.4	27.8	72.2

6.6	38.1	61.9
6.8	49.7	50.3
7.0	61.5	38.5
7.4	80.2	19.8
7.2	71.7	28.3
7.6	86.6	13.4
7.8	90.8	9.2
8.0	94.0	6.0

pH is calculated according to the Henderson-Hasselbalch equations:

$$pH = pK + \log \text{(proton acceptor)/proton donor}$$

where $pK' = 6.86$ at 25°C

1 M Potassium Phosphate Buffer (pH 7.6) for Plasmid Broth

Add 69 g of KH_2PO_4 to 300 ml of deionized H_2O. Adjust pH to 7.6 with potassium hydroxide (KOH). Adjust to 500 ml with deionized H_2O. Aliquot to 100-ml bottles and autoclave. Store at room temperature. To use, add 10 ml/liter of medium.

0.1 M Potassium Phosphate Buffer (pH 7.2–7.4)

Add 4.5 g of KH_2PO_4 (anhydrous) and 11.6 g of K_2HPO_4 (anhydrous) to distilled H_2O to make a total volume of 1 liter. The pH of the final solution will be 7.2–7.4. This buffer is stable for 1 month at 4°C.

0.1 Sodium (Na) Acetate Buffer (pH 5.0-5.2)

A. 200 mM Na acetate, 27.22 g/liter $NaOAc.3H_2O$.

B. 200 mN acetic acid, 12.05 ml of glacial acetic acid diluted to 1 liter with distilled H_2O.

For 100 mM Na acetate buffer at pH 5.0, mix 14.8 ml of B, 35.2 ml of A, and add distilled H_2O to 100 ml. For pH 5.2 buffer, mix 10.5 ml of B, 39.5 ml of A, and add distilled H_2O 100 ml.

Sodium (Na) Borate Buffer

A. 200 mM boric acid, 12.4 g/liter H_3BO_3.

B. 50 mM *di*-Na *tetra*-borate 19.05 g/liter $Na_2B_4O_7.10H_2O$

For pH 9, mix 50 ml of A and 59.0 ml of B and dilute to 200 ml with distilled H_2O.

Sodium (Na) Cacodylate Buffer

A. 100 mM Na cacodylate, 21.40 g/liter Na $(CH_3)_2AsO_2.3H_2O$

B. 100 mM HCl, 8.3 ml of concentrated HCl diluted to 1 liter with distilled H_2O.

Mix 50 ml of A plus indicated volumes of HCl (see below) and add distilled H_2O to 100 ml for 50 mM Na cacodylate buffer.

pH	*ml of 0.1 M HCl*
6.0	29.55
6.4	18.75
6.8	9.30
7.0	6.30
7.2	4.15

Sodium Citrate (Na Citrate)

tri-Na citrate *di*-hydrate has a molecular weight of 294.10. Unbuffered Na citrate is used for the hypotonic swelling of nuclei in cytology. Dissolve citric acid *tri*-Na salt *di*-hydrate in distilled water at either 1% or 0.5%, depending on protocol.

0.1 M Citrate Buffer (pH 5.5)

Add 9.6 g of citric acid (anhydrous) and 4.4 g of NaOH pellets to distilled H_2O to a final volume of 500 ml, and adjust pH with 10 N NaOH. Autoclave and store at room temperature.

Or

Mix 4.7 volumes of 0.1 M citric acid with 15.4 volumes of 0.1 M Na_3 citrate.

1 M Sodium Phosphate Buffer (pH 6.0-7.2)

Mixing 1 M NaH_2PO_4 (monobasic) and 1 M Na_2HPO_4 (dibasic) stock solutions in the volumes designated in the table 1.7 below result in 1 liter of 1M sodium phosphate buffer of the desired pH. To prepare the 1 M stock solutions, dissolve 138 g of $NaH_2PO_4.H_2O$ (monobasic; m.w. = 138 g/mole) in sufficient H_2O to make a final volume of 1 liter and dissolve 142 g of Na_2HPO_4 (dibasic, anhydrous; m.w. = 142 g/mole) in sufficient H_2O to make a final volume of 1 liter.

Table 1.7. Preparation of sodium phosphate buffer at pH 6.0-7.2

Volume of 1 M NaH_2PO_4 (ml)	*Volume of 1 M Na_2HPO_4 (ml)*	*Final pH*
877	123	6.0
850	150	6.1
815	185	6.2
775	225	6.3
735	265	6.4
685	315	6.5
625	375	6.6
565	435	6.7

510	490	6.8
450	550	6.9
390	610	7.0
330	670	7.1
280	720	7.2

1 M Sodium Phosphate Buffer (pH 7.4)

Add 3.1 g of $NaH_2PO_4.H_2O$ and 10.9 g of Na_2HPO_4 (anhydrous) to distilled water to make a total volume of 1 liter (Table 1.8). The pH of the final solution will be 7.4. This buffer can be stored for up to 1 month at 4°C.

Table 1.8. Preparation of 1 M sodium phosphate buffer at pH 7.4

pH	*Volume of 1 M Na_2HPO_4 (ml)*	*Volume of 1 M NaH_2PO_4 (ml)*
5.8	7.9	92.1
6.0	12.0	88.0
6.2	17.8	82.2
6.4	25.5	74.5
6.6	35.2	64.8
6.8	46.3	53.7
7.0	57.7	42.3
7.2	68.4	31.6
7.4	77.4	22.6
7.6	84.5	15.5
7.8	89.6	10.4
8.0	93.2	6.8

Dilute the combined 1 M stock solution in 1000 ml with distilled H_2O. pH is calculated according to the Henderson-Hesselbalch equation:

$$pH = pK' + \log \text{(proton acceptor)/proton donor}$$

where $pK' = 6.86$ at 25°C

Phosphate Buffered Saline (PBS) (pH 7.2-7.4)

Historically, this commonly used reagent was sodium phosphate-buffered saline (150 mM NaCl plus 10 mM sodium phosphate). PBS has been modified to suit different applications. D-PBS (8 mM sodium phosphate, 2 mM potassium phosphate, 140 mM NaCl, 2.7 mM KCl (pH 7.4) with or without 0.5 mM $MgCl_2$ and 0.9 mM $CaCl_2$) can be purchased as 1x and 10x stock buffers (Table 1.9). One common preparation of PBS (including potassium) and another (omitting potassium) are provided. This is Ca^{++}/Mg^{++}-free phosphate-buffered saline and in the same as Dulbecco's PBSA.

Table 1.9. Preparation of phosphate buffered saline at pH 7.2-7.4

Component and Final Concentration	*Amount to Add Per 1 Liter*
137 mM NaCl	8g
2.7 mM KCl	200 mg
10 mM Na_2HPO_4O (Dibasic, anhydrous)	1.44 g
2 mM KH_2PO_4O (monobasic, anhydrous) H_2O	240 mg to make 1 liter

Dissolve the components in approximately 0.9 liter of H_2O. Adjust the pH to 7.2–7.4 with HCl and then adjust the final volume to 1 liter with H_2O. Dispense in convenient volumes and sterilize by autoclaving for 20 minutes at 15 psi on liquid cycle. Store at room temperature.

When used for tissue culture, the pH is 7.2. For other uses (*e.g.,* in situ hybridization), the pH is 7.4. PBS for tissue culture can be purchased already made up as a sterile 10x or 1x solution. If Ca^{++} and Mg^{++} are added, the amounts are 0.133 g of $CaCl_2.2H_2O$ and 0.10 g of $MgCl_2.6H_2O$ per liter. This is the same as PBSABC of Dulbecco.

Note that for in situ hybridization, some laboratories may use a different formulation of phosphate-buffered saline, *e.g.,* NaCl (7.6 g/liter), Na_2HPO_4 (3.8 g/liter), and NaH_2PO_4 (0.42 g/liter). The pH will be 7.4 based on the ratio of monobasic and dibasic sodium phosphate. This formulation appears to give results in biochemical studies identical in the one given above. It will not be suitable for tissue culture.

1 × PBS (Without Potassium) (pH 7.2)

Prepare 0.1 M sodium phosphate buffer (pH 7.2) as above. Add NaCl to a final concentration of 0.9%.

NOTE *This buffer can be stored for up to 1 month at 4°C. If PBS is to be stored for longer periods of time, 0.01% sodium azide or thimerosal should be included. If preservatives must be avoided, PBS can be stored as a 10x stock solution (1 M sodium phosphate, 9% NaCl) and diluted to 1x as needed. The salt concentration in 10x PBS is too high for organisms to grow.*

Sorensen's Phosphate Buffer

Solution A

Sodium phosphate, dibasic ($Na_2HPO_4.2H_2O$)	11.876 g
distilled H_2O	1000 ml

Solution B

Potassium phosphate

Monobasic (KH_2PO_4) 9.08 g

distilled H_2O 1000 ml

The desired pH can be obtained by adding enough Solution B to Solution A given below in Table 1.10 to make a total volume of 100 ml:

Table 1.10. Attaining desired pH of sorenson's phosphate buffer

Solution A (ml)	*pH of buffer*
0.6	4.9
2.3	5.3
4.9	5.6
12.1	6.0
26.4	6.4
49.2	6.8
61.2	7.0
67.0	7.1
72.6	7.2
77.7	7.3
81.8	7.4
85.2	7.5
88.5	7.6
93.6	7.8
96.9	8.0

8.5% Saline (for Dilutions)

85 g of NaCl in 1 liter of distilled H_2O.

10x BS (Buffered Saline)

Dissolve 110 g of $Na_2HPO_4.7H_2O$, 30 g of KH_2PO_4, and 8.5 g of NaCl in 100 ml of deionized H_2O. Adjust to 200 ml with H_2O. Autoclave, Before use, dilute 10 ml into 100 ml of sterile H_2O for 1x BS. Store at room temperature.

20x SSC (Saline Sodium Citrate)

Dissolve 175.3 g of NaCl and 88.2 g of sodium citrate in 800 ml of H_2O. Adjust the pH to 7.0 with a few drops of a 10 N solutions of NaOH. Adjust the volume of 1 liter with H_2O. Dispense into aliquots according to need. Sterilize by autoclaving. Dilute to working concentration with distilled H_2O. Store at room temperature.

20x SSPE (Saline, Sodium, Phosphate, EDTA)

Dissolve 175.3 of NaCl, 27.6 g of $NaH_2PO_4.H_2O$, and 7.4 g of EDTA in 800 ml of H_2O. Adjust the pH of 7.4 with NaOH (~6.5 ml of a 10 N solution). Adjust the volume to 1 liter with H_2O. Dispense into aliquots. Sterilize by autoclaving. Store at room temperature.

50 mM Tricine Buffer (pH 7)

For 50 mM tricine buffer (pH 7.0 at 20°C), mix 25 ml of a 200 mM solution of Tris tricine (35.83 g/liter) and 2.7 ml of 100 mM NaOH to 100 ml with distilled H_2O.

1 M Tris-HCl Buffers

Mixing the volumes of concentrated HCl (11.6 N) designated in the table below with 121 g of Tris base (m.w. = 121 g/mole) result in 1 liter of 1 M Tris-Cl buffer of the desired pH at 25°C. To prepare the 1 M buffer, dissolve 121 g of Tris base in approximately 0.9 liter of H_2O, add the appropriate volume of concentrated HCl and adjust the final volume to 1 liter with H_2O (Table 1.11).

Table 1.11. Preparation of Tris-HCl buffer

Volume of concentrated HCl (ml)	*pH*
8.6	9.0
14	8.8
21	8.6
28.5	8.4
38	8.2
46	8.0
56	7.8
66	7.6
71.3	7.4
76	7.2

NOTE *some pH electrodes cannot accurately measure the pH of Tris solutions. Be sure to check the information provided by the electrode manufacturer.*

Tris solutions should be kept in a closed container, because they will absorb carbon dioxide from the air; they should be colorless. Tris-HCl buffers can also be made by mixing Tris base and Tris-HCl in the appropriate quantities, but these must be desiccated before being weighed.

NOTE *Tris has a significant temperature coefficient. As a temperature of the solution decreases from 25°C to 5°C, the pH increases an average of 0.03 pH units per degree centigrade. Conversely, as the temperature increases from 25°C to 37°C, the pH decreases an average of 0.025 pH units per degree centigrade. Slight effects on pH based on the total Tris concentration have also been noted. Most molecular biology protocols do not account for these effects.*

50 mM Tris Buffers

Tris buffers (0.05 M) of the desired pH can be made by mixing 50 ml of 0.1 M Tris base with the indicated volume of 0.1 HCl and then adjusting the volume of the mixture to 100 ml with H_2O (Table 1.12).

Table 1.12. Dilution of 0.1 N HCl to obtain desired pH of 50 mM Tris buffers

Desird pH (25°C)	*Volume of 0.1 N HCl/100 ml*
7.10	45.7
7.20	44.7
7.30	43.4
7.40	42.0
7.50	40.3
7.60	38.5
7.70	36.6
7.80	34.5
7.90	32.0
8.00	29.2
8.10	26.2
8.20	22.9
8.30	19.9
8.40	17.2
8.50	14.7
8.60	12.4
8.70	10.3
8.80	8.5
8.90	7.0

Tris-buffered Saline (TBS)

TBS is a commonly used buffer, and is used as an alternative biological substrate to PBS for a variety of protein analysis and immunocytochemical approaches. Although the standard concentration is 10 mM Tris, 150 mM NaCl, different protocols may utilize different concentrations of Tris in their version of TBS, so be sure to prepare the standard concentration for the protocol utilized in your own laboratory (Table 1.13). Tris-based buffers use concentrated HCl to achieve the desired pH.

Table 1.13. Tris-buffered saline (TBS) (pH 7.15) from stock solutions (10 mM Tris, 150 mM NaCl)

	Final (1x)	*Mix for 5x*
Tris-HCl pH 7.15	10 mM	50 ml of 1 M
NaCl Distilled water to 1 liter	150 mM	150 ml of 5 M

Add Na azide to 1% to prevent microbial contamination. Dilute one volume of stock and four volumes of distilled H_2O.

1x TBS (50 mM Tris-HCl [pH 8.1]) from Powder. Dissolve 48.4 g of Tris base and 70.1 g of NaCl in sterile distilled H_2O. Adjust the pH to 8.1 with concentrated HCl. Adjust the volume to 8 litres with distilled H_2O (Table 1.14). Store at room temperature.

Table 1.14. Tris-buffered saline (TBS) (10 mM Tris, 100 mM NaCl)

	Volume	*Final Concentration*
1 M Tris-HCl (pH 7.9)	10 ml	10 mM
5 M NaCl	20 ml	100 mM

Adjust the volume to 1 liter with H_2O.

Tris-EDTA (TE) (pH 7.4-8.0)

This standard buffer is used to resuspended and store DNA. It can be prepared by using stock solutions of 1 M Tris-HCl at pH values ranging from 7.4 to 8.0. (Table 1.15) Store at room temperature. Check your desired protocol for the optimum pH range.

Table 1.15. Preparation of Tris-EDTA

Component and Final Concentration	*Amount of stock to Add Per 100 ml*
10 mM Tris-HCl	1 ml of 1 M (pH 7.4-8.0 at 25°C)
1 mM EDTA	200 μl of 0.5 M (pH 8.0)
H_2O	98.8 ml

Table 1.16. 10x TE, pH 8 (100 mM Tris-HCl, 10 mM EDTA)

	Final (1x)	*Mix for 10x*
Tris-HCl (pH 8.0)	10 mM	100 ml of 1M
EDTA	1 mM	20 ml of 500mM (pH 8.0)

Distilled water of 1 liter. Sterilize by autoclaving. Store at room temperature. The following buffers are all generated from stock solutions.

TEN

(10 mM Tris-HCl, 1 mM EDTA, 500 mM NaCl).

10 ml of 1 M Tris-HCl (pH 8.0), 20 ml of 500 mM EDTA (pH 8.0).

100 ml of 5 M of NaCl.

Distilled water to 1 liter. Sterilize by autoclaving.

STE

(10 mM Tris-HCl, 1 mM EDTA, 100 mM NaCl).

10 ml of 1 M Tris-HCl (pH 8.0), 20 ml of 500 mM EDTA (pH 8.0).

20 ml of 5 M of NaCl.

Distilled water to 1 liter. Sterilize by autoclaving.

STET

(10 mM Tris-HCl, 1 mM EDTA, 100 mM NaCl, 5% Triton X-100)

10 ml of 1 M Tris-HCl (pH 8.0), 20 ml of 500 mM EDTA (pH 8.0).

20 ml of 5 M of NaCl.

Distilled water to 1 liter. Sterilize by autoclaving. Prior to making up to 1 liter, add Triton X-100 as needed to a smaller volume of the buffer.

Commonly Used Buffers

Buffer	*Synonym*	*pK*	*Molecular Weight*
Phosphate (pK_1)		2.12	98.0, free acid
Glycine-HCl		2.34	.53
Citrate (pK_1)		3.14	192.1, free acid
Formate		3.75	68.0, Na salt
Carbonate (pK_1)		3.76	106.0
Succinate (pK_1)		4.19	118.1, free acid
Acetate		4.75	8.20, Na salt
Citrate (pK_2)		4.76	214.1, Na salt
Succinate (pK_2)		5.57	162.1, *di*Na salt
MES[a]		6.15	195.2, hydrate
Carbanate (pK_2)	Bicarbonate	6.36	84.0, bicarbonate
Citrate (pK_2)		6.39	294.1, *Tri*Na salt, dihydrate
PIPES[b]		6.8	302.4
ACES		6.9	182.2

MPOS[c]		7.2	209.7
HEPES[d]		7.55	238.3
Phosphate (pK_2)		7.21	120.0, Na salt, monobasic
TES		7.7	229.2
Barbital	Barbitone veronal	7.78	128.1, barbituric acid
Triethanoamine		7.8	149.2
Tricine		8.15	179.2
Tris[c]		8.3	121.1
Bicine		8.35	163.2
Glycylglycine		8.4	132.1
Borate		9.24	201.2, Na tetraborate
CHES		9.5	207.3
Ethanolmine		9.5	61.1
Glycine-NaOH		9.6	97.1, Na salt hydrate
CAPS		10.4	221.3
Triethylamine		10.7	101.3
Phosphate (pK_2)		12.3	141.9, Na salt, dibasic

[a] *2-(N-morpholino)ethanesulfonic acid.*

[b] *Piperazine-N N′bis-(2-ethanesulfonic acid).*

[c] *3-(N-morpholino)propanesulfonic acid.*

[d] *N-2-hydroxyethylpiperazine-N′-2-ethanesulfonic acid.*

[e] *Tris(hydroxymethyl)aminomethane.*

NOTE

A buffer solution is one that resists pH change on the addition of acid or alkali. Such solutions are used in many biochemical experiments where the pH needs to be accurately controlled. From the Henderson – Hasselbalch equation, the pH of a buffer solution depends on two factors; one is the pK_a value and the other the ratio of salt to acid. This ratio is considered to be the same as the amount of salt and acid mixed together over the pH range 4-10, where the concentration of hydrogen and hydroxyl ions is very low and can be ignored. Let us take as an example acetate buffers consisting of a mixture of acetic acid and sodium acetate:

$$CH_3COOH \longrightarrow CH_3COO^- + H^+$$

$$CH_3COONa \longrightarrow CH_3COO^- + Na^+$$

Since acetic acid is only weakly dissociated, the concentration of acetic acid is almost the same as amount put in the mixture; likewise the concentration of acetate ion can be considered to be the same as the concentration of sodium acetate placed in the mixture since the salt is completely dissociated.

Unit **III**

DNA ISOLATION AND PURIFICATION

Part 1 PLASMID DNA

Contents

INTRODUCTION

Bacterial plasmids are small extra chromosomal, circular DNA, capable of autonomous replication. Most of the plasmids exist as double stranded circular DNA and can exist in any of the three forms-supercoiled (SC) DNA, covalently closed circular (CCC) DNA and open circular (OC) DNA. These three forms differ in their charge density and so can be separated by gel electrophoresis (Fig. 1.1).

Plasmids have a wide distribution throughout the prokaryotes. They are be divided into different categories depending on different criteria like conjugative or non-conjugative (depending on presence or absence of tra gene that promotes conjugation), relaxed or stringent (depending on number of copies per cell). Generally, conjugative plasmids are of relatively high molecular weight and are present as one to three copies per chromosome whereas non-conjugative plasmids are of low molecular weight and present as multiple copies per chromosome.

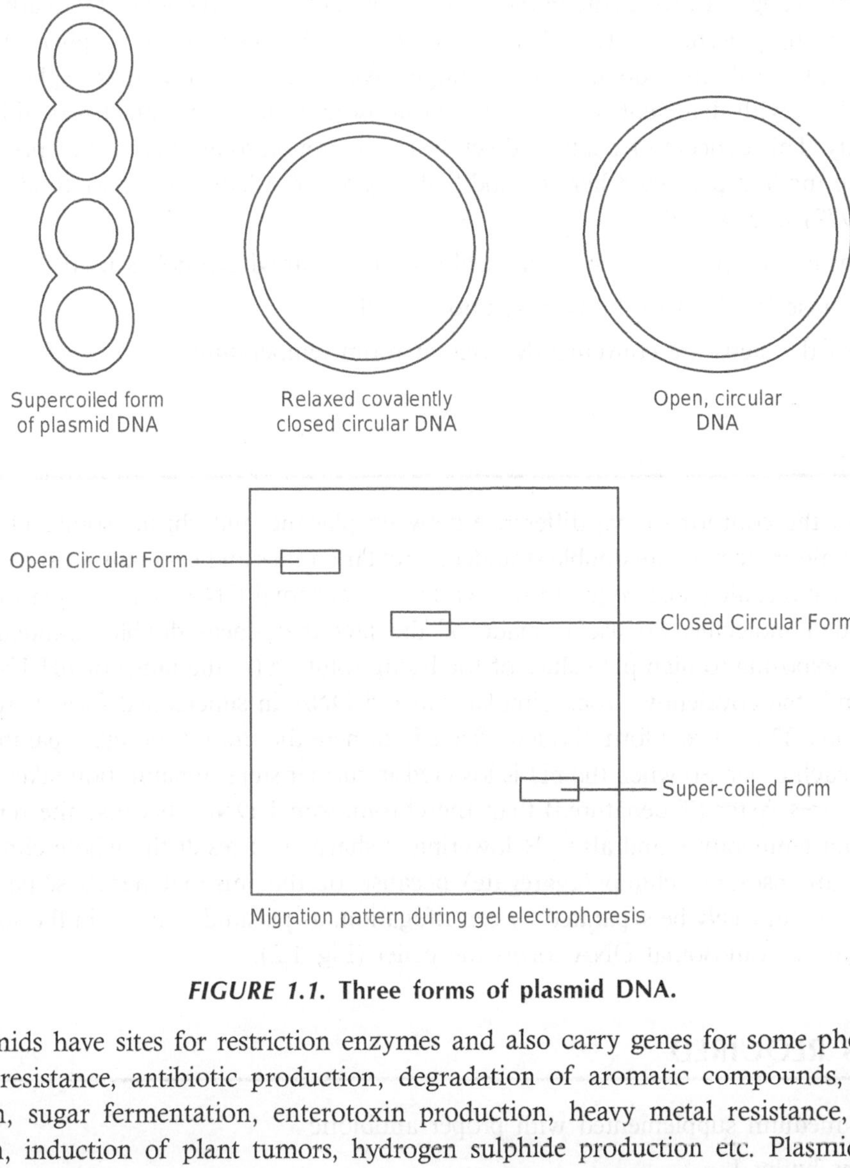

FIGURE 1.1. **Three forms of plasmid DNA.**

Plasmids have sites for restriction enzymes and also carry genes for some phenotype like antibiotic resistance, antibiotic production, degradation of aromatic compounds, haemolysin production, sugar fermentation, enterotoxin production, heavy metal resistance, bacteriocin production, induction of plant tumors, hydrogen sulphide production etc. Plasmids to which phenotypic traits have not yet been ascribed are called cryptic plasmids. With some modifications, bacterial plasmids serve as an ideal vector system for carrying out bacterial transformation because of the following properties—low molecular weight, ability to confer readily selectable phenotypic traits on host cells and unique restriction site. They have sites for restriction enzymes and also carry genes for antibiotic resistance which serve as screening markers.

Isolation of plasmid DNA is one of the first procedures to learn while going for recombinant DNA work. There are many methods for plasmid DNA isolation. All the methods have a common idea upon which they are suitably modified to suit different organisms. It includes controlled lysis of cells, removal of chromosomal DNA and debris and finally purification and concentration of plasmid DNA. Controlled lysis of cell is very important for getting intact plasmid as well as for separating it from very large fragments of chromosomal DNA. This is achieved by digesting with lysozyme in the presence of sucrose or glucose or by treating the cells with EDTA in the presence of Tris-Cl. It results in the cells, also called as spheroplast, which although partially wall free, do not burst straight away. Subsequently, these spheroplasts are subjected to lysis with detergent so that large chromosomal DNA and intact plasmid DNA are released gently. This concept of controlled cell lysis is common to all methods of plasmid DNA isolation. The method presented here is modified version of Alkaline Lysis Method (Birmboim and Doly, 1979) in a way like:

(*i*) Instead of phenol: chloroform, chloroform: isoamyl alcohol is used,

(*ii*) RNase is added to the lysis solution I and

(*iii*) All the steps are conveniently done at room temperature.

PRINCIPLE

It is based on the conformational difference between plasmid and chromosomal DNA of the bacteria. Plasmid molecules are double stranded, circular entities and generally exist as covalently closed circular molecule (supercoiled form), while chromosomal DNA molecule exist as a linear double stranded molecule. So use is made of the fact that linear double stranded DNA is denatured by exposing to high pH values of the lysing solution (in the range of pH 12-12.5). On the other hand, the covalently closed circular plasmid DNA in supercoiled form is resistant to these conditions. The relaxed form though affected but here the strands though separated slightly are closer to each other. So when the pH is lowered in further steps, renaturation takes place and plasmid renatures faster (if denatured) than the chromosomal DNA. Because the renaturation is done at cool temperature and also pH lowering is sharp, as a result the whole chromosomal DNA forms an insoluble clump (aggregate) because of the mis-matched base-pairing. The aggregate DNA can easily be separated by centrifugation, as plasmid remains in the supernatant while aggregate chromosomal DNA forms the pellet (Fig. 1.2).

REAGENTS REQUIRED

1. LB medium supplemented with proper antibiotic
2. Lysis buffer I

   25 mM Tris-Cl (pH 8.0)

   50 mM Glucose

   10 mM EDTA (pH 8.0)

   0.2 mg/ml RNaseA

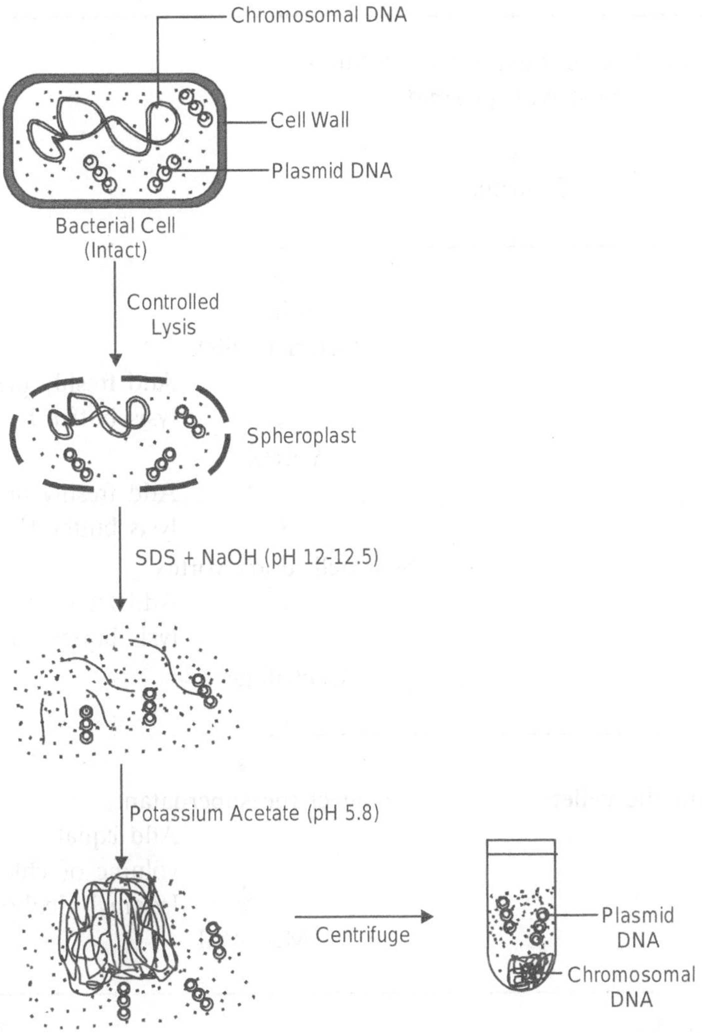

***FIGURE 1.2.* Diagramatic representation of plasmid DNA isolation from E. coli cells.**

3. Lysis buffer II

   0.2 N NaOH

   1% (w/v) SDS

4. Lysis buffer III

   3 M potassium acetate (pH 5.5)

5. Chloroform: isoamyl alcohol (24 : 1)
6. Isopropanol
7. Ethanol (70%)
8. TE (pH 8.0)

EXPERIMENTAL LAY-OUT

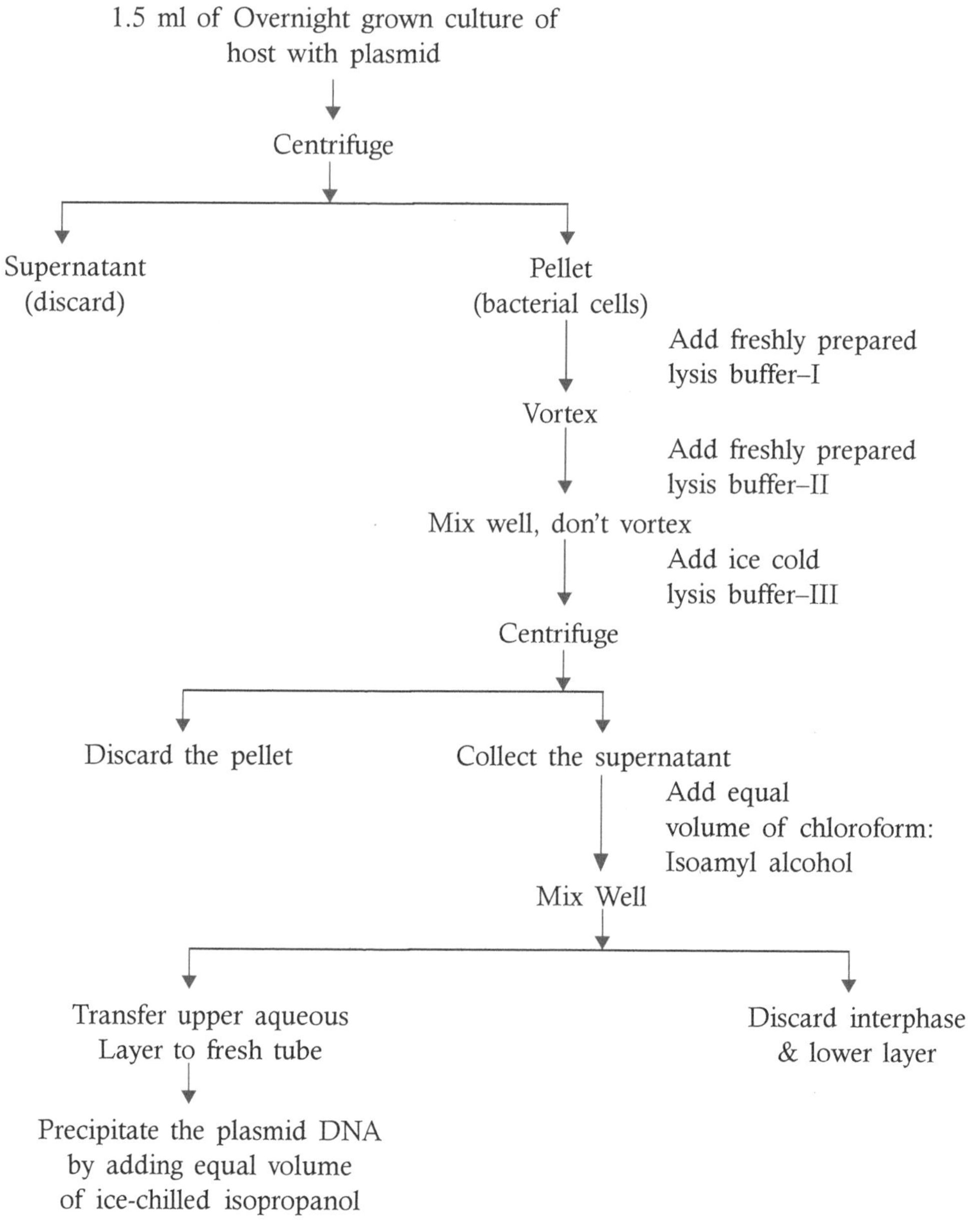

METHOD

1. Prepare a small overnight culture of the host with plasmid in 2 ml of a rich medium (LB) containing the appropriate antibiotic at 37°C with vigorous shaking.
2. Pour 1.5 ml of the above culture in a microfuge tube. Centrifuge in a microfuge for 30 seconds at maximum speed.

3. Discard the supernatant and dry the bacterial pellet. To this add 100 μl of ice cold lysis buffer I. Vortex vigorously.
4. Add 200 μl of freshly prepared lysis solution II to the above suspension. Mix the contents well by inverting the tubes rapidly. Do not vortex.
5. Add 150 μl of ice cold lysis solution III. Mix the contents by inverting the tubes several times.
6. Centrifuge the above lysate at maximum speed for 5 minutes.
7. Discard the pellet (chromosomal DNA) and collect the supernatant in fresh tube.
8. To the supernatant add an equal volume of chloroform: isoamyl alcohol. Mix and centrifuge at maximum speed for 2-3 minutes.Transfer the upper aqueous layer to a fresh tube.
9. Precipitate the plasmid DNA from the supernatant by adding equal volume of ice chilled isopropanol.
10. Centrifuge at maximum speed for 5 minutes.
11. Remove the supernatant and wash the pellet with 70% of ethanol several times. Take care of the pellet as it may not adhere tightly to the tube.
12. Keep the microfuge tube open so that ethanol is evaporated and pellet is dry.
13. Dissolve the pellet in 50 μl of TE and store at –20°C.

REMEMBER

1. Use gloves and goggles while working with sodium dodecyl sulphate (SDS). It is toxic, an irritant, and poses a risk of severe damage to the eyes. Do not breathe the dust.
2. Always use freshly prepared lysis solution II.
3. Take all the safety measures discussed in Unit I.

❑❑❑

Part 2 BACTERIAL (E.COLI) DNA

Contents

INTRODUCTION

The chromosomal DNA of E.coli is a large circular molecule of approximately 3.2 Kb size. The DNA is attached to the plasma membrane at many points. Being large in size, DNA is prone to mechanical breakage, however, if extraction is performed carefully, large fragments of chromosomal DNA can be obtained with an average length of 1–2 Kb.

PRINCIPLE

Principle of total genomic DNA isolation can be summarized in following five stages (Fig. 2.1.).

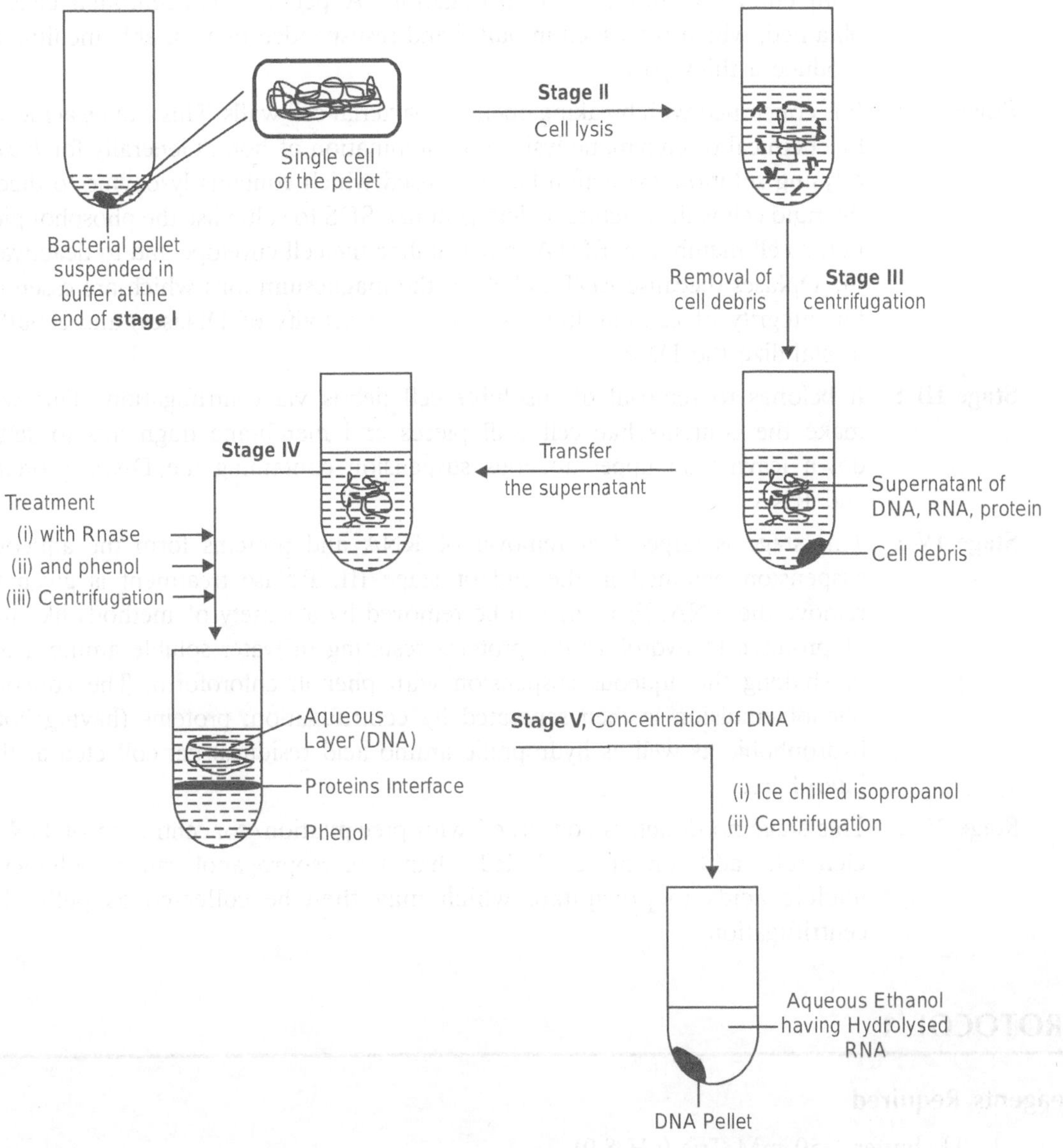

FIGURE 2.1. **Steps of total genomic DNA preparation.**

Stage I – Growth and harvesting of bacteria

Stage II – Breaking the cell wall

Stage III – Removal of insoluble cell debris

Stage IV – Removal of proteins and RNA

Stage V – Concentration of DNA.

Stage I : It belongs to growth and cultivation of bacteria in a suitable medium. It employs the standard techniques of bacterial cultures in a standard media and growing them overnight. After overnight period the bacterial cells are separated from culture medium via centrifugation. A pellet of concentrated cells is obtained, which is washed in buffer and resuspended in a suitable medium to produce a thick paste.

Stage II : It is concerned with breaking open the bacterial cell walls. This can be achieved by chemical or enzymatic lysis or a combination of both. Generally for *E.coil,* a lysing solution/extraction buffer is used which contains lysozyme to digest the rigid cell wall structure, a detergent like SDS to solubilise the phospholipids in the cell membrane, EDTA to destabilize the cell envelope and to deactivate the DNases (because EDTA chelates the magnesium ions which are essential for integrity of cell envelope as well as for activity of DNases) and a buffer to stabilize the DNA.

Stage III : It belongs to removal of insoluble cell debris via centrifugation. This will make the contents like cell wall pieces and membrane fragments to settle down leaving an upper aqueous suspension containing the DNA, proteins and RNA.

Stage IV : This stage is targeted at removal of RNA and proteins form the aqueous suspension obtained in the end of stage III. RNase treatment is given to remove the RNA. Proteins can be removed by a variety of methods like use of protease to hydrolyze the proteins resulting in water soluble amino acids or shaking the aqueous suspension with phenol: chloroform. The aqueous phenol emulsion is then separated by centrifugation; proteins (having both hydrophobic as well as hydrophilic amino acid residues) get collected at the interphase.

Stage V : This stage no doubt, is concerned with precipitation/concentration of DNA. Generally addition of ice-chilled ethanol or isopropanol causes polymeric nucleic acids to precipitate which may then be collected as pellet by centrifugation.

PROTOCOL 1

Reagents Required

1. TE buffer— 50 mM Tris (pH 8.0)
   50 mM EDTA
2. 250 mM Tris (pH 8.0)
3. Lysozyme—10 mg/ml
4. 0.5% SDS
5. 0.4 M EDTA
6. 1 mg/ml proteinase K

7. Phenol equilibrated with Tris (See Annexure VIII)
8. Sodium acetate
9. 95% ethanol
10. 200 μg/ml RNase
11. Chloroform

Experimental Lay-out

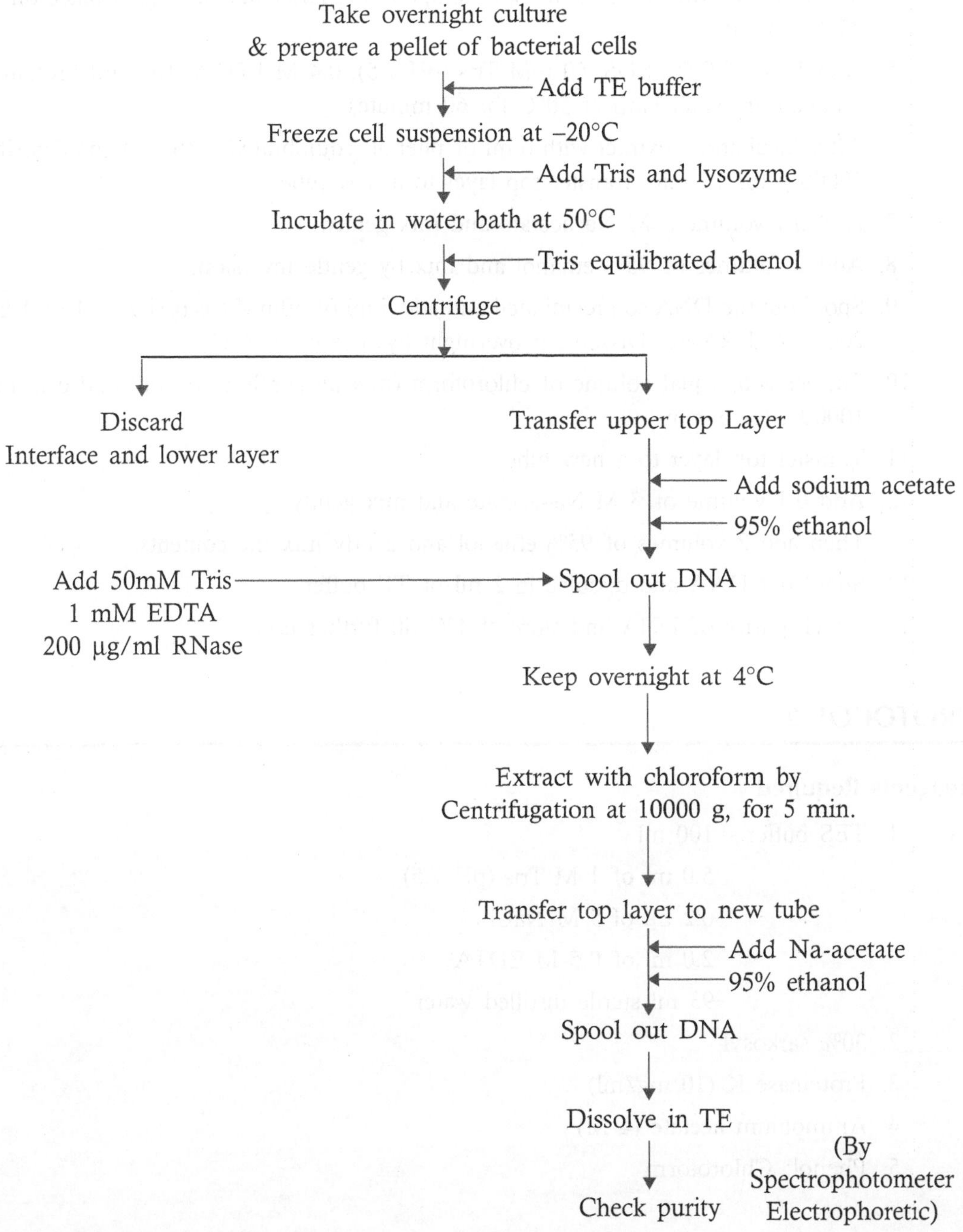

Method

1. Grow the bacterial cells in 500 ml of LB broth medium.
2. Subject the above overnight culture to centrifugation to obtain a pellet and dissolve the pellet in 5 ml of TE buffer (50 mM Tris (pH 8.0), 50 mM EDTA).
3. Freeze the above cell suspension at 20°C.
4. To the Frozen suspension add 0.5 ml of 250 mM Tris (pH 8.0) and lysozyme (10mg/ml) and thaw the contents at room temperature. After thawing again place on ice for 45–50 minutes.
5. Add 1 ml of 0.5% SDS, 50 mM Tris (pH 7.5), 0.4 M EDTA, 1mg/ml Proteinase K. Incubate in water bath at 50°C for 60 minutes.
6. After incubation, extract with 6 ml of phenol (equilibrated with tris) and centrifuge at 10000 g for 15 min, transfer top layer to a new tube.
7. Add 0.1 volume 3 M Na acetate and mix gently.
8. Add 2 volumes of 95% ethanol and mix by gentle inversion.
9. Spool out the DNA so precipitated and add 5 ml of 50mM tris (pH 7.5), 1 mM EDTA, 200 μg/ml RNase. Dissolve it overnight by rocking at 4°C
10. Extract with equal volume of chloroform (mix by gentle inversion) and centrifuge at 10000 g for 5 min.
11. Transfer top layer to a new tube.
12. Add 0.1 volume of 3 M Na–acetate and mix gently.
13. Then add 2 volumes of 95% ethanol and gently mix the contents.
14. Spool out DNA and dissolve in 2 ml of TE buffer.
15. Check purity of DNA and store at 4°C till further use.

PROTOCOL 2

Reagents Required

1. TES buffer – 100 ml
   - 5.0 ml of 1 M Tris (pH 7.5)
   - 0.2 ml of 5 M NaCl
   - 2.0 ml of 0.5 M EDTA
   - 93 ml sterile distilled water
2. 30% sarkosyl
3. Proteinase K (10 mg/ml)
4. Ammonium acetate (4 M)
5. Phenol: Chloroform

6. Ethanol
7. TE buffer (10 mM Tris, 0.1 mM EDTA)
8. RNase (10 mg/ml)

Experimental Lay-out

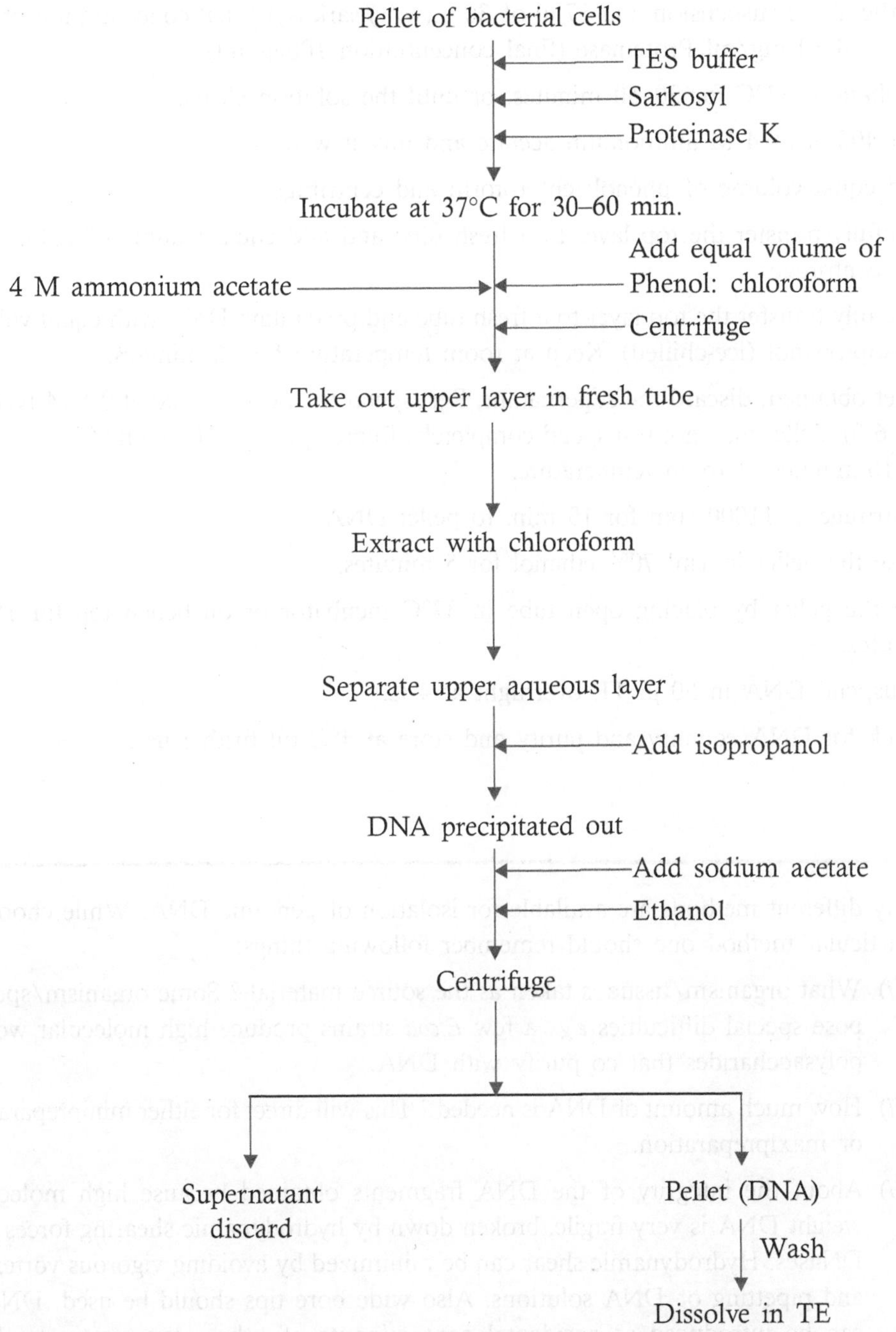

Method

1. Grow cells in 2–5 ml broth to late log phase.
2. Pellet 1–2 ml cells in microfuge tube.
3. Resuspend the cells in 400 μl TES buffer.
4. To the above suspension add 17 μl of 30% (w/*v*) sarkosyl (final concentration of 1%) and 5 μl 10 mg/ml Proteinase (final concentration 100μg/ml).
5. Incubate at 37°C for 30–60 minutes, or until the solution clears.
6. Mix 400 μl of 4 M ammonium acetate and mix it well.
7. Add equal volume of phenol: chloroform and centrifuge.
8. Carefully transfer the top layer to a fresh tube and add equal volume of chloroform and centrifuge.
9. Carefully transfer the top layer to a fresh tube and precipitate DNA with equal volume of isopropanol (ice-chilled), Keep at room temperature for 10 minutes.
10. Pellet obtained, discard the supernatant, Resuspend the pellet in 400 μl 0.1 M Na0AC (pH 6.0). Pellet may not resuspend completely, Reprecipitate DNA with 800 μl ethanol for 15 minutes at room temperature.
11. Centrifuge at 11000 rpm for 15 min. to pellet DNA.
12. Rinse the pellet in 1ml 70% ethanol for 5 minutes.
13. Dry the pellet by placing open tube in 37°C incubator or on bench top for 15–20 minutes.
14. Resuspend DNA in 50 μl TE overnight at 4°C.
15. Check for DNA equality and purity and store at 4°C till further use.

REMEMBER

1. Many different methods are available for isolation of genomic DNA. While choosing a particular method one should remember following things:
   (*i*) What organism/tissue is taken as the source material ? Some organism/species pose special difficulties *e.g.,* a few *E.coli* strains produce high molecular weight polysaccharides that co purify with DNA.
   (*ii*) How much amount of DNA is needed ? This will direct for either minipreparation or maxipreparation.
   (*iii*) About the integrity of the DNA fragments obtained because high molecular weight DNA is very fragile, broken down by hydrodynamic shearing forces and DNases. Hydrodynamic shear can be minimized by avoiding vigorous vortexing and pipetting of DNA solutions. Also wide bore tips should be used. DNases can be introduced as accidental contaminants of other reagents, particularly

RNase. To avoid this, molecular biology grade reagents (labeled DNase free) must be used or RNase solution can be made DNase free by boiling the RNase solution for 15 min.

(*iv*) How much time and expense are involved?

(*v*) Are facilities available for disposing off hazardous chemical like phenol, chloroform ? It not, that protocol should be used which does not include these chemicals.

2. Always remember to wear gloves, goggles while working with phenol/chloroform, SDS.
3. Take care of all the precautions mentioned in unit-I.

❑❑❑

Part 3 PLANT GENOMIC DNA

Contents

INTRODUCTION

Plant genomic DNA lies in the nucleus of the plant cell in the form of a DNA-protein complex known as chromosomes. A number of different methods exist to isolate the genomic DNA from plants. All these methods are based on the same concept including the five basic steps as that were discussed in prokaryotic genomic DNA isolation procedures (Part 2, Unit III), *i.e.*, (*i*) Harvesting of sample (*ii*) breaking open the cell (*iii*) removing insoluble cell debris (*iv*) removal of proteins (*v*) concentration or precipitation of DNA.

Plant cells are characterized by a cell wall surrounding them. To reach to the chromosomal DNA, a barrier of 3 envelopes has to be broken down—cell wall, cell membrane and nuclear membrane. So generally the harvested tissue is freezed in the liquid nitrogen followed by lyophilization and grinding. The low temperature prevents the nuclease activity (which degrade the DNA) as well as make the cell brittle so that grinding is easy.

Harvesting and lyophilization of leaf tissue:

1. Harvest young leaves from the plant. If older leaves have to be harvested, prefer without necrotic areas or lesions.
2. Cut the leaves into small pieces, put in a screen mesh bag, label it and keep it on the ice.
3. Put the bag in a Styrofoam container or other type of container able to hold liquid nitrogen. Quick freeze samples in liquid nitrogen. Samples should not be allowed to thaw until dried.
4. Transfer the sample to lyophilizer (should have temperature about –56°C and vaccum less than 15 microns). Sample should be dried in approximately 72 hours.
5. Leaf samples may be alternatively stored at –80°C until lyophilization.
6. The dried leaf sample after lyophilization may be stored in sealed plastic bags at room temperature for a short time (one month) or at –20°C for several years.

Grinding:

1. If the leaf sample is less than 4 g (fresh weight), grinding is done in pestle and mortar in the presence of a pinch of acid washed sand. But if leaf sample weigh more than 4–5 g, the sample should be grinded in a cyclone sample mill. Finer the grind, greater is the yield of DNA from a given amount of sample.

Alternate way of grinding:

1. Fresh leaf tissues can directly be grinded with liquid nitrogen in pestle and mortar.
2. Tightly capped ground sample can be stored at –20°C for several years.

Or

Can be immediately subjected to further treatments for DNA isolation.

After grinding there are a number of protocols to proceed. Each protocol has it's own applicability and choice of protocol depends on the nature of the sample plant species and many-a-times slight modifications are needed for different plant species. Here we will discuss the three commonly used protocols across the laboratories, the CTAB (cetyl trimethyl ammonium bromide) method, potassium acetate method and Urea-phenol method.

PROTOCOL 1: CTAB (CETYL TRIMETHYL AMMONIUM BROMIDE) METHOD

Principle

This is based on the method of Saghai-Maroof *et al,* 1984. CTAB (Cetyl trimethyl ammonium bromide) is a detergent (non-ionic in nature) having a strong affinity for carbohydrates, proteins, cell wall contents. So when CTAB is added to a solution of lysed cells, CTAB forms a complex with above listed macro molecules and further causes their precipitation on phenol extraction. Nucleic acids remain in the solution which can be separated out by centrifugation, followed by precipitation with ice-chilled ethanol or isopropanol. This method is versatile in its applicability like. To monocots, dicots, leaves, cotyledons, seeds, seedlings, embryos, tissue culture, callus etc.

Reagents

1. Extraction buffer

Stock	100 ml
dH_2O	73 ml
1 M Tris (pH 7.5)	10 ml
5 M NaCl	14 ml
0.5 M EDTA (pH 8.0)	2 ml
CTAB	1 g
14 M β-mercaptoethanol	1 ml

2. Wash I:

Stock	100 ml
100% Ethanol	76 ml
2.5 M sodium acetate	8 ml
dH_2O	16 ml

3. Wash II:

Stock	100 ml
100% Ethanol	76 ml
1 M Ammonium acetate	1 ml
dH_2O	23 ml

4. Chloroform: Isoamyl alcohol (24: 1)
5. Isopropanol
6. RNase A (10 mg/ml)
7. Phenol saturated with TE

Experimental Lay-out

Ground, lyophilized tissue

↓ ← Extraction buffer

Incubation at 65°C with gentle rocking

↓

Add chloroform isoamyl alcohol and centrifuge

↓

Take off upper aqueous layer

↓

Add chloroform: isoamyl alcohol to the aqueous layer and centrifuge

↓

To the aqueous layer add RNase A

↓ ← Mix

Add isopropanol/ethanol to precipitate the DNA

↓

Wash the isolated DNA with wash I and Wash II

↓

Dissolve in TE

Method

1. To approximately 400 mg ground, lyophilized tissue, add 9.0 ml CTAB extraction buffer (Warm) in a 15 ml polypropylene centrifuge tube. Mix well by inverting the tubes several times.
2. Incubate for 90 minutes with continuous shaking at 65°C in the water bath.
3. Remove the tubes from water bath and add 4.5 ml chloroform: isoamyl alcohol (24:1). Shake very gently for 5–10 min.
4. Spin in a centrifuge for 10 minutes at 2400–2500 rpm at room temperature.
5. Pour off top aqueous layer into new 15 ml tubes. Add 4.5 ml $CHCl_3$: isoamyl alcohol (24:1). Rock gently for 5–10 minutes.
6. Spin in centrifuge for 10 min at 2500 rpm at room temperature.
7. Pipette off the top aqueous layer into new 15 ml tubes containing 15–20 µl of 10 mg/ml RNase A (pre-boiled). Mix by gentle inversion and incubate for 30 minutes at room temperature.
8. Add 6 ml of isopropanol. Mix by gentle inversion.
9. Centrifuge for 5 min at 2500 rpm. Remove the aqueous solution as much as possible, air dry the tubes for 10 min., and add 1 ml TE (pH 8.0).
10. Rock gently at room temperature (or just leave in refrigerator) overnight.
11. Extract each sample with phenol (saturated with TE). Centrifuge sample for 10 min at 2500 rpm.

12. Transfer top layer to a new tube. Extract DNA with 1ml of chloroform: isoamyl alcohol (24:1). Centrifuge the sample for 10 min. at 2500 rpm. Transfer top layer into new tube.
13. Precipitate DNA by adding 50 μl of 5 mM NaCl and then 2.5 ml 100% Ethanol mix by gentle inversion.
14. Hook out the DNA into a new tube with 3–4 ml wash 1 solution. Leave for 20 min.
15. Rinse DNA on hook briefly in 1–2 ml of Wash II solution.
16. Transfer DNA to 1.5 ml eppendorf tube containing 0.5–1.0 ml TE. After DNA is completely dissolved, it can be checked for quality and concentration and then stored at – 20°C for future use.

Remember

1. This method will yield total cellular nucleic acids.
2. Always add β-mercaptoethanol just before you use the extraction buffer.
3. Don't forget to do the steps with phenol (saturated with TE) because it is done to remove the RNAase added in previous steps.

PROTOCOL 2: POTASSIUM-ACETATE METHOD

Principle

It is based on the method given by Tai & Tanksley in 1990. Principle is same as that of any genomic DNA isolation procedure *i.e.,* breaking opens the cell, precipitating out all the contaminating macromolecules like proteins, carbohydrates etc. and isolating the DNA from aqueous phase. Here in the aqueous phase, sodium dodecyl sulphate (SDS) is used.

Reagents

1. Extraction buffer (pH 8.0)

Components	**Amount (for 500 ml)**
100 mM Tris	6.05 g
50 mM EDTA	9.3 g
500 mM NaCl	14.6 g
1.25% SDS	6.25 g
Sterile Water	to 500 ml

(Dissolve all the reagents in water, except SDS, sterilize by autoclaving, add SDS slowly with stirring while solution is still hot, cool and adjust the pH to 8.0).

NOTE *Add 0.38 g of sodium bisulphate for every 100 ml of solution just before use.*

2. 5 M Potassium acetate
3. 3 M sodium acetate

4. TE buffer

Components	Amount (for 500 ml)
10 mM Tris	0.6 g
1 mM EDTA	0.185 g
Sterile Water	to 500 ml

Adjust the pH to 8.0 and autoclave.

5. RNAase A (10 mg/ml)
6. Isopropanol
7. Ethanol
8. Phenol: Chloroform: Isoamyl alcohol (PCI) (25:24:1)
9. Chloroform: Isoamyl alcohol (CI) (24:1)

Experimental Lay-out

Ground tissue

↓ ← Add extraction buffer (Warm)

Incubate in water bath at 65°C

↓

Add 5 M potassium acetate

↓

Mix vigorously and incubate on ice with shaking

↓

Centrifuge and filter out the supernatant

↓

Add ice-chilled isopropanol to the supernatant

↓

Incubate at Room temperature till DNA is precipitate out

↓

Centrifuge to collect the DNA pellet

↓

Dissolve in TE and add RNAase

↓

Give phenol: chloroform: isoamyl alcohol treatment

↓

Centrifuge

↓

To supernatant add ethanol/isopropanol to precipitate out DNA

↓

Dissolve in TE

Method

1. To the ground tissue, add 15 ml of extraction buffer (preheated to 65°C) with 0.38 gm sodium bisulphate added per 100 ml of buffer and mix well.
2. Incubate for 20–30 min at 65°C. Keep shaking the tubes every 5 min.
3. Add 5 ml of 5 M potassium acetate.
4. Mix vigorously and incubate on ice for 20 min. with shaking.
5. Centrifuge for 15 min at 3000 rpm.
6. Take out the supernatant into another 50 ml centrifuge tube.
7. To this add 2/3 volume of pre-chilled isopropanol and incubate at room temperature for 30 minutes or longer until DNA is precipitated.
8. Centrifuge for 10 min at 3000 rpm to collect DNA pellet.
9. Wash the pellet with 70% ethanol and air dry for 10 minutes.
10. Dissolve the pellet in 1 ml TE buffer and add 15 μl of RNase (10 mg/ml).
11. Incubate at 37°C for 30 min.
12. Make the volume to 2 ml by adding 1 ml of TE after RNase treatment.
13. Add equal volume (2 ml) of Phenol: Chloroform: Iso-amyl alcohol (25:24:1), mix well.
14. Centrifuge at 3000 rpm for 15 min and transfer the top aqueous layer to a new tube.
15. Extract once again with chloroform: Isoamyl alcohol (24:1).
16. Mix well and centrifuge at 3000 rpm for 15 minutes. Transfer the top aqueous layer to a new tube.
17. Add 1/10 volume of 3 M sodium acetate and 2 volume of pre-chilled absolute ethanol.
18. Mix gently and incubate at room temp for 30 min.
19. Hook out or centrifuge the DNA pellet for 3–5 min at 3000 rpm.
20. Add 10 ml of 70% ethanol and wash the DNA pellet and centrifuge for 1 min at 3000 rpm.
21. Discard the 70% alcohol and put the tubes upside down or paper towel to get rid of excess ethanol.
22. Resuspend the pellet in appropriate amount of TE buffer depending on the size of DNA pellet store at –20°C until use.

Remember

1. Take care of the precautions while handling SDS and phenol. (see unit I)
2. DNA obtained is good for RFLP and RAPD analysis.

PROTOCOL 3: UREA—PHENOL METHOD

Principle

This method uses 5% sarkosyl in the extraction buffer which is a good lytic agent. Principle is same as in previous methods, only detergents and their applicability sequence change.

Reagents

1. Extraction buffer (per 1000 ml)

Components	Amount
Urea	420 g
5 M NaCl	70 ml
1 M Tris-HCl (pH 8.0)	50 ml
0.5 M EDTA, (pH 8.0)	40 ml
5% Sarkosyl	100 ml

2. Phenol: Chloroform: Isoamyl alcohol (PCI): 25:24:1
3. Chloroform: Isoamyl alcohol (CI): 24:1
4. Ethanol
5. RNAase A (10 mg/ml)
6. 20% SDS
7. 5 M NaCl
8. TE buffer

Experimental Lay-out

Add extraction buffer, phenol and SDS to Feezer-dried leaf tissue

↓

Incubate and go for treatment with chloroform: isoamyl alcohol to denature the proteins

↓

Centrifuge and separate out upper aqueous layer

↓

Dissolve DNA in TE and treat with RNase A

↓

Go for extraction with phenol: chloroform: isoamyl alcohol

↓

Precipitate DNA from upper aqueous layer and dissolve in TE

Method

1. Add 5 ml of pre-warmed extraction buffer, 0.25 ml phenol and 0.2 ml 20% SDS to 3 g freezer-dried leaf tissue in 15 ml tube.
2. Incubate for 60 min at 60°C and mix by inverting the tubes.
3. Add chloroform: isoamyl alcohol and invert the tube gently for 10–15 minutes.
4. Centrifuge at 3000 rpm at room temperature for 15 min, transfer the top aqueous layer into a new 15 ml tube.
5. Add two volume of 100% ethanol (–20°C), mix well and hook out the DNA pellet, wash the DNA pellet with 70% ethanol.
6. Dry the pellet by disposing off the ethanol.
7. Add 2 ml TE buffer and dissolve.
8. Add 10 µl of RNase A (10 mg/ml), incubated at 37°C for 30 min, then add 2 ml TE buffer.
9. Add equal volume of phenol: chloroform: isoamyl alcohol (25:24:1), mix well by gentle inversion.
10. Centrifuge at 3000 rpm for 15 min and transfer the top DNA layer to a new tube.
11. Extract DNA with chloroform: isoamyl alcohol (24:1).
12. Add 5 M NaCl to a final concentration of 0.1 M.
13. Add two volume of 100% ethanol (–20°C) mix well and pellet the DNA. Wash the DNA pellet with 70% ethanol for two times.
14. Dry the pellet briefly and dissolve in 0.5 ml TE buffer and store at –20°C for future use.

❑❑❑

Part 4 ANIMAL GENOMIC DNA

Contents

INTRODUCTION

Isolation of genomic DNA animal cells is a fundamental requirement for animal genome characterization from genome sequencing, gene mapping and for the identification and isolation of genes for genetic engineering.

DNA is a chemically inert, stable and robust molecule that's why DNA outlasts most other intracellular components in locations like crime sites, mummies and other burial sites. Though chemically stable, it is a physically fragile molecule, vulnerable to hydrodynamic forces which may be generated from pipetting, shaking, or stirring. So genomic DNA can be obtained easily in fragmented form but to isolate as a complete molecule, it really becomes difficult. A number of cells can be used for the purpose of DNA isolation like muscle tissue, thymus tissue like monolayers of cells, fresh tissue or blood samples. There are a number of protocols available for isolating DNA from blood cells. The present protocol particularly uses proteinase K and phenol.

PRINCIPLE

Principle is the same as discussed in part-2 (*E.Coli* genomic DNA isolation) of this unit. For the present protocol, the blood cells are digested with Proteinase K in the presence of EDTA (to inhibit DNases). Membranes are solubalised as well as proteins are denatured with detergent like SDS (Sodium do-decyl sulphate). The nucleic acids are then purified by phase extractions with organic solvent (Phenols) followed by RNA removal (By RNase treatment). Shearing forces, as mentioned earlier break the DNA so that in final preparation size of fragments obtained rarely exceed 100–150 Kb in length.

REAGENTS/MATERIALS

1. EDTA (anti-coagulant).
2. Phosphate buffered saline (PBS) for frozen blood samples.
3. Ammonium acetate (100 mM).
1. Lysis buffer: 10 mM Tris-Cl (pH 8.0)
   0.1 M EDTA (pH 8.0)
   0.5% (W/V) SDS
   20 mg DNase-free RNase
2. Phenol equaliberated with 0.5 M Tris-Cl (pH 8.0)
3. T E (pH 8.0)
4. Tris-buffered saline (TBS)
5. Proteinase K
6. Blood samples (fresh of frozen)

EXPERIMENTAL LAY-OUT

Blood sample with anti-coagulant

↓

Collect white blood cells

↓

Incubate in lysis buffer

↓

Add Proteinase K

↓ ← Phenol equiliberated with 0.1 M Tris-Cl (pH 8.0)

Separate the two phases by centrifugation

↓

Total upper aqueous phase

↓ ← Ammonium Acetate and Ethanol

DNA pellet after Centrifugation

↓

Dissolve in TE

METHOD

1. Collect the blood samples freshly or take the frozen samples.
2. Collect the cells from the blood samples:

   From fresh samples:

   (*i*) Take approximately 20 ml of fresh blood in centrifuge tubes having 3.5 ml of EDTA.

   (*ii*) Remove the supernatant fluid carefully. Use a Pasteur pipette to transfer the buffy coat. It is population of white blood cells of heterogeneous density.

   (*iii*) Resuspend the buffy coat in 15 ml of lysis buffer. Incubate it for approximately 40–60 min at 37°C.

   From Frozen samples:

   (*i*) Take 20 ml of frozen blood and add EDTA to it.

   (*ii*) Thaw the blood in a water bath at room temperature and then transfer it to a centrifuge tube. Add equal volume of phosphate-buffered saline at room temperature.

   (*iii*) Centrifuge at 3500 g for 15 min at room temp.

   (*iv*) Remove the supernatant. Resuspend the pellet in 15 ml of lysis buffer and incubate for 1 hour at 37°C.
3. After lysis is complete, transfer the lysate to centrifuge tube (tubes should not be more than one third full).
4. Add Proteinase K (20 mg/ml) to a final concentration of 100 mg/ml. Use a glass rod to mix the enzyme solution gently.
5. Incubate in a water bath for 3 hours at 50°C with occasional mixing.
6. Cool the solution to room temperature and add an equal volume of phenol equilibrated with 0.1 M Tris-Cl (pH 8.0). Gently mix the two phases by slowly turning over the tube (approximately 10 minutes till the two phases are formed).
7. Separate the two phases by centrifugation at 500 g for 15 minutes at room temperature.
8. With a wide bore pipette, transfer the viscous aqueous phase to a fresh centrifugation tube.
9. Repeat the extraction with phenol two more times and collect the aqueous phase.
10. Transfer the aqueous phase to a fresh centrifuge and add 0.2 volumes of 0.1 M ammonium acetate, 2 volumes of ethanol and mix the contents thoroughly.
11. DNA is precipitated by above treatment. Collect the pellet of DNA by subjecting it to centrifugation at 5000 g for 5 minutes at room temperature.
12. Wash the pellet with 70% alcohol and let it dry (not completely).
13. Dissolve the pellet by adding 1–1.5 ml of TE (pH 8.0). Store the DNA solution at 40°C till further use.

REMEMBER

1. Remember that blood can be stored for several days at 0°C or indefinitely at –70°C before DNA is prepared.
2. Remember to take all the precautions while handling SDS and Phenol.
3. While transferring the aqueous phase into pipette, take care to do it very slowly to avoid disturbing the material at the interface and to minimize hydrodynamic shearing forces. If the DNA solution is very viscous then use a long pipette attached to an aspirator to remove the organic (lower phase).
4. As an alternative to EDTA, Acid Citrate dextrose solution B (ACD) can be used with following composition.

   0.48% W/V citric acid

   1.32% W/V sodium citrate

   1.47% glucose.

❑❑❑

Unit IV ELECTROPHORESIS

Part 1 A GENERAL ACCOUNT

Contents

- Introduction
- Principle
- Equipment Used
- Remember

INTRODUCTION

Electrophoresis means the movement of charged particles under the influence of electric field. Macromolecules of the cell, like DNA, amino acids, peptides, proteins, nucleotides can exist as electrically charged species at any pH and so show movement when electric field is applied. For separation of these charged molecules, a medium is needed through which, an electric current when passed, molecules travel at different rates depending on many factors (discussed later in this unit). The medium is in the form of a gel (colloid in a solid form) so the process is also known as gel electrophoresis. Now-a-days the most commonly used support media for electrophoresis are agorose and polyacrylamide.

Agarose Gels

Agarose is the primary stabilizing medium for electrophoresis of DNA. It is a natural polysaccharide isolated from agar which is obtained from various species of marine red algal (seaweed). Gelidium and Gracilaria species are the major source of agar. Agarose consist of 1, 3 linked β-D-galactopyranose and 1, 4 linked 3, 6 anhydro-α–L-galactopyranose. This basic agarobiose repeat unit (Fig. 1.1.) forms long chains with an average molecular mass of 120 Kda, representing about 400 agarobiose units. Charged groups like pyruvates and sulplates are also

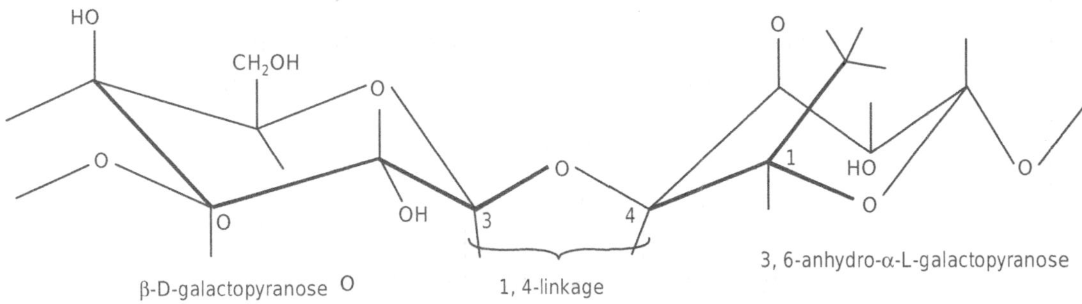

FIGURE 1.1. **Agrobiose.**

present in polysaccharide. Agarose is a preferred medium for using as gel-matrix because it allows rapid diffusion of high molecular weight (of 1000 Kda) macromolecules without significant restriction by the gel. Further agarose is non-toxic and unlike polyacrylamide, contains no potentially damaging byproducts of polymerization. Also, agarose gels are thermo reversible so low gelling-temperature and low-melting temperature agaroses permit easy recovery of samples, including sensitive heat labile materials. Following properties make agarose gels as an attractive support media:

(*i*) **Melting and gelling temperature:** Melting temperature increases as the gel concentration increases. Gelling temperature is also influenced by gel concentration, This is the reason that gelling or melting temperature are expressed at a given agarose concentration. This dependence is most pronounced at concentration less than 1%. This property is important because it is possible to vary gelling and melting parameters by using lower or higher concentration.

(*ii*) **Gel strength:** Gel strength is defined as the force, expressed in g/cm^3, that must be applied to fracture an agarose gel of a standard concentration. One of the main positive points of agarose is it's ability to show high gel strength at low concentration ($\leq 6\%$), The gel strength of a specific lot of agarose decreases over time because of spontaneous hydrolysis of agarose polysaccharide chain, this loss of strength can be particularly noticeable after five year from the manufacturing date.

(*iii*) **Methylation of agarose:** Uncharged methyl groups are also present on agarose polysaccharides. The extent of natural methylation is directly proportional to the gelling temperature. Unexpectedly, synthetically methylated agarose have lower, rather than higher gelling temperatures and the degree of synthetic methylation is inversely proportional to melting temperature.

Acrylamide/Polyacrylamide

The acrylamide monomers in the presence of smaller amounts of N, N', Methylene bisacrylamide (bisacrylamide) are polymerized to form cross-linked polyacrylamide gels (Fig. 1.2). The polymerization is initiated by addition of ammonium persulphate and the base N, N, N′, N′–Tetramethylene diamine (TEMED). One molecule of cross-linker is included for every 29 monomers of acrylamide. The polymerization is also initiated by oxygen so acrylamide solution needs to be shielded from exposure to air which is achieved by using vertical gel assembly invariably for setting of acrylamide gels.

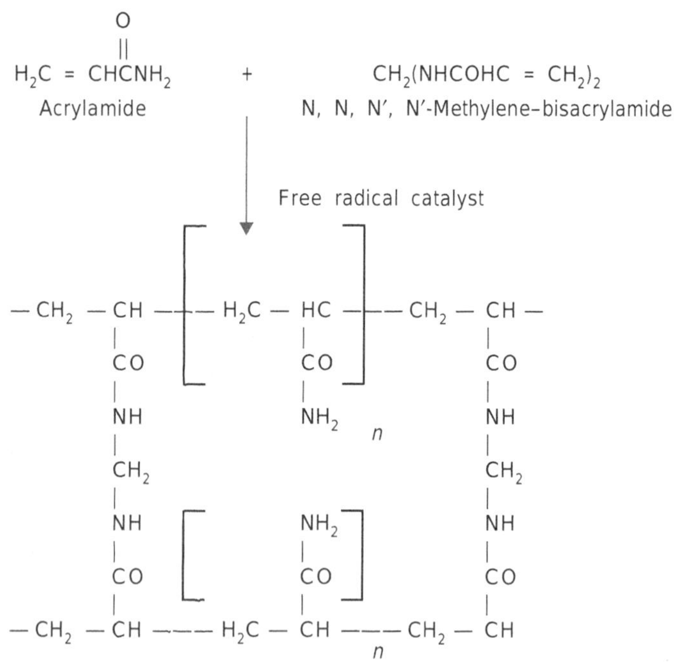

FIGURE 1.2. **Polymerization of acrylamide and bisacrylamide.**

Acrylamide gels of varying pore size can be prepared by changing the concentrations of both the acrylamide and bis-acrylamide. The strength generally varies from 3% to 30%. The low percent gels having large pore size are used for electrophoresis of proteins as well as DNA. Gels of between 10% and 20% acrylamide are used in technique such as SDS gel electrophoresis where the small pore size introduces the sieving effect for the separation of the proteins. Polyacrylamide gels are difficult to prepare and run as compared to agarose gels and needs expertise. However, they have some advantages over agarose gels like:

(*i*) These can accommodate much larger quantities of samples than agarose gels.
(*ii*) These can separate molecules whose lengths differ by as little as 0.2%
(*iii*) Molecules recovered from these gels is extremely pure.

PRINCIPLE

A potential gradient, E, is generated when a voltage (potential difference) is applied across the electrodes. This potential gradient can be defined by following equation:

$$E = V/d$$

V = Voltage applied
d = distance between the electrodes

Under this potential difference, the charged molecule move with velocity V which is given by following equation:

$$V = E.q/f$$

Where E = Potential gradient
q = Charge of molecule in coulombs
f = frictional coefficient

More commonly, the term electrophoretic mobility (μ) of an ion is used which is:

$$\mu = \nu \text{ (velocity of ion)/E (field strength)}$$

When a potential difference is applied molecules with different overall charges will separate according to their electrophoretic mobilities. Molecules with similar charges but with dissimilar sizes will also separate (because of different frictional forces).

EQUIPMENT USED

For electrophoresis to run, following two equipments are needed:

(*i*) Electrophoretic unit

(*ii*) Power Pack.

(*i*) **Electrophoretic Unit**. It is the equipment where the electrophoresis actually happens. Mainly two types of units are in routine use-vertical gel systems and horizontal gel systems. Depending on the specialized use, special electrophoretic units can also be designed.

Vertical gel system. This system is used for separating the DNA as well as proteins in polyacrylamide gels. The unit basically composed of a buffer reservoir having electrodes at two ends and glass plates (clamped together but held apart by spacers) in between which gel is casted. Dimensions of the glass plates as well as buffer reservoir vary with the nature of samples (*e.g.,* DNA, proteins) as well as with the manufacturer. A typical gel system used for separating the proteins is shown in Fig. 1.3. Gel dimensions are typically 12 cm × 14 cm, with a thickness of 0.5 to 1 mm. A plastic comb is placed in the gel solution and is removed after polymerization to provide loading wells for samples. When the apparatus is assembled, the lower

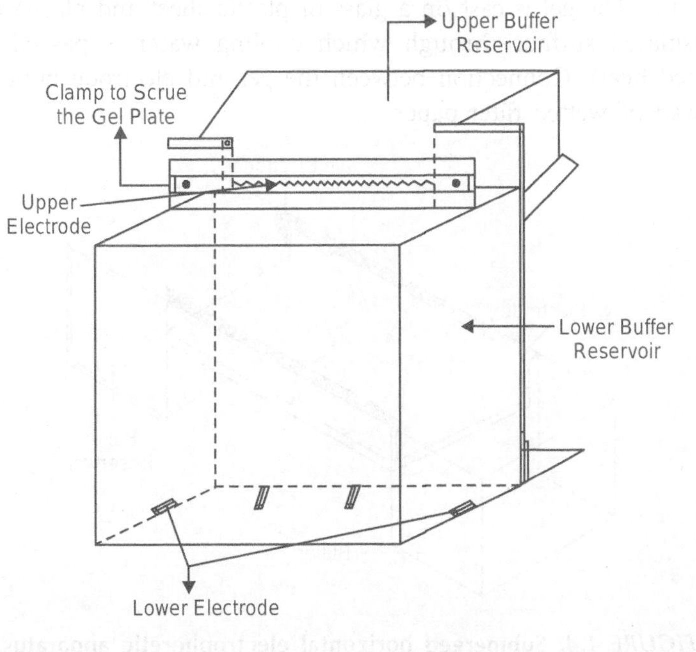

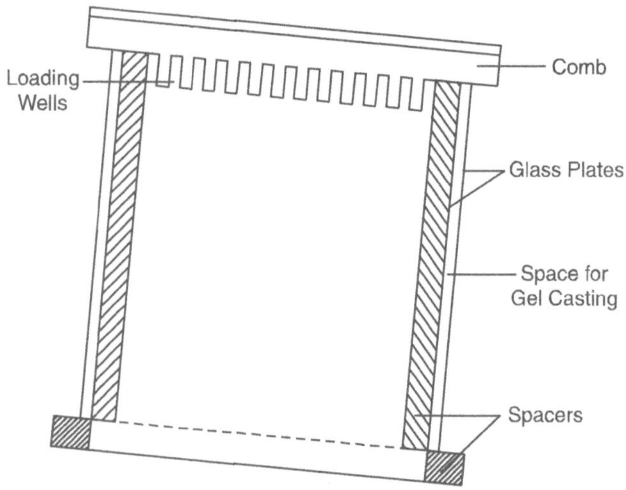

FIGURE 1.3. **Vertical electrophoretic units for proteins separation.**

electrophoretic tank buffer surrounds the gel plates and affects some cooling of the gel plates. The vertical gel used for separating the DNA, vary in dimensions from the above mentioned system. The dimensions of the gel can be 10 cm–100 cm like 60 cm × 80 cm etc. Spacers (usually Teflon, sometimes lucite) vary in thickness from 0.5 mm to 2.0 mm.

Horizontal gel system: This system is mainly used for immunoelectrophoresis, isoelectric focusing and electrophoresis of DNA and RNA in agarose gels. A typical horizontal gel system, shown in Fig. 1.4 is used for carrying out electrophoresis of DNA. This is also known as submerged horizontal gel electrophoresis because here gel is covered by a thin layer of buffer. Another type of horizontal system is shown in Fig. 1.5. The gel is cast on a glass or plastic sheet and placed on a cooling plate (an insulated surface through which cooling water is passed to conduct away generated heat). Connection between the gel and electrode buffer is made using a thick wad of wetted filter paper.

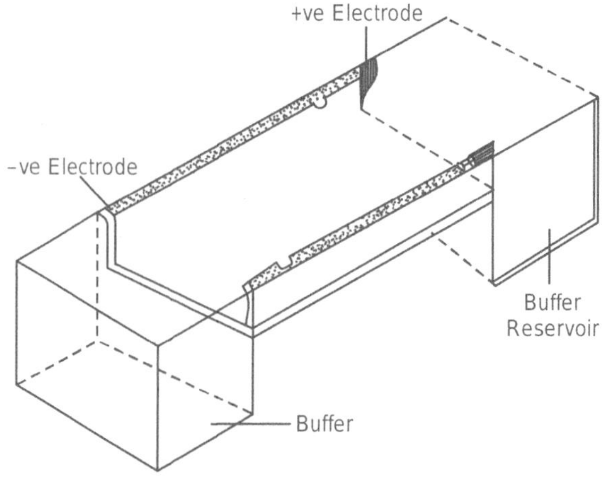

FIGURE 1.4. **Submerged horizontal electrophoretic apparatus.**

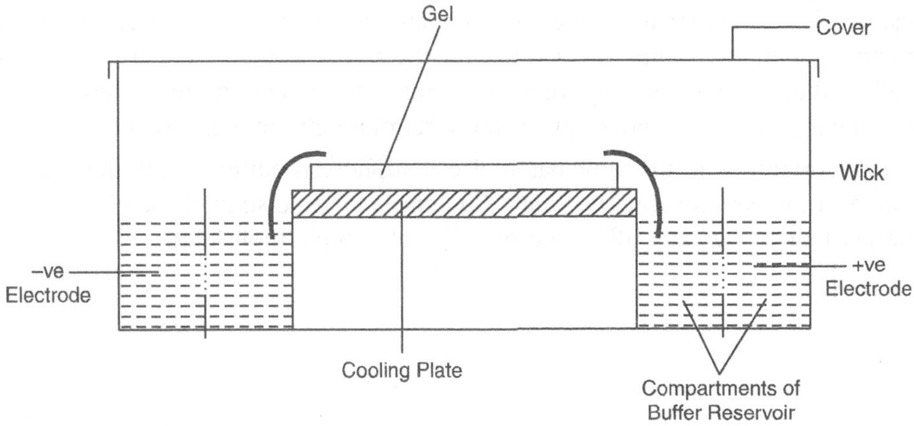

FIGURE 1.5. **Horizontal apparatus (non-submerged).**

(*ii*) **Power Pack**. The power pack supplies a direct current between the electrodes in the electrophoretic unit. All the electrophoresis is carried out in an appropriate buffer, which is essential to maintain a constant state of ionization of the molecules being separated. Any variation in pH would alter the overall charge and hence the motilities of the molecules being separated.

REMEMBER

1. One should be careful about the power/voltage at which the gel is run. It seems electrophoretic mobility can be increased by increasing the applied voltage (which in turn would increase the current flowing). But by doing so, one major problem for most forms of electrophoresis, *viz* generation of heat, arises. Heating of electrophoretic medium has the following effects:
   (*i*) Samples that are sensitive to heat may become unstable *e.g.,* proteins may get denatured, enzymes may lose the activity.
   (*ii*) Separated samples may mix because of formation of convection currents.
   (*iii*) A decrease in buffer viscosity and hence a reduction in the resistance of the medium.
   (*iv*) An increased rate of diffusion of sample and buffer ions leading to broadening of the separated samples.

   It might appear that if we run the gel at very low power (low current) heating problem can be overcome, but this can lead to poor separation as a result of the increased amount of diffusion resulting from long separation times. So compromise conditions have to be found with reasonable power settings, to give acceptable separation times, and an appropriate cooling system, to remove liberated heat.
2. Sometimes another phenomenon is also happening side by side of electrophoresis, viz electroendosmosis (also known as electrosmotic flow). It is due to the presence of

charged groups on the surface of the support medium *e.g.,* Commercially available agarose is not completely pure; it is contaminated with other polysaccharides, salts, proteins, sulphate groups (depending on the purity grade). Similarly surface of glass wall used in capillary electrophoresis contains silate (Si-OH) groups. These contaminants have charge on them and so affect the electrophoretic mobility of the desired sample.

3. It is important to use the same batch of electrophoresis buffer in both the electrophoretic tank and the gel preparation. Small differences in ionic strength or pH create fronts in the gel that can greatly affect the mobility of sample molecules.

❑❑❑

Part **2**

AGAROSE GEL ELECTROPHORESIS OF DNA

Contents

INTRODUCTION

For majority of DNA samples, electrophoretic separation is carried out in agarose gels. This is because most of DNA molecules/fragments which are routinely analyzed are larger than proteins and can not enter the polyacrylamide fine pores (except for special experiment like sequencing). Agarose gel is characterized by large pore size and is a promising medium for DNA fragments separation. The idea of using gel electrophoresis to analyze DNA was first of all given by Vin Thorne in 1966–67 to separate superhelical, nicked, and linear forms of polyomavirus DNA, radiolabeled with H^3 thymidine. Since then agarose gel electrophoresis of DNA has metamorphosized from very simple form to a very skilled and sophisticated technique.

The movement of DNA through Agarose gel is characterized by following points:

(*i*) At low voltage, the rate of electrophoretic migration of linear DNA fragments is proportional to the voltage applied. But at higher voltage, the effective range of separation of fragments decreases.

(*ii*) Smaller molecules move faster as compared to larger molecules because of the lesser frictional force

(*iii*) Concentration of agarose gels also play an important role in deciding the rate of migration of DNA samples, depicted by following equation:

$$\text{Log } \lambda = \log \lambda - K_r T$$

λ = Free electrophoretic mobility of DNA

K_r = Retardation coefficient

T = Concentration of gel

(*iv*) If the DNA is having some intercalating dyes, *e.g.,* ethidium bromide, the electrophoretic mobility of linear DNA is reduced by approximately 15–20%.

(*v*) The electrophoretic mobility of DNA is affected by the composition and ionic strength of the electrophoresis buffer. Several different buffers are available for electrophoresis of native double stranded DNA, *e.g.,* Tris-acetate (TAE), Tris-borate (TBE), or Tris phosphate (TPE) at a concentration of approximately 50 mM (pH 7.5–7.8). TPE and TBE are slightly more expensive than TAE, but they have significantly higher buffering capacity. Double stranded linear DNA fragments migrate approximately 10% faster through TAE than through TBE or TPE but the resolving power of these systems are almost identical.

MATERIALS/REAGENTS REQUIRED

1. Agarose solutions (1%)

   Take 1 gm of agarose in conical flask, add 100 ml of 1 X electrophoresis buffer and set for melting until a clear transparent solution is achieved.

2. 10 X TBE

Tris base	108 g
Boric acid	55 g
0.5 M EDTA (pH 8.0)	40 ml

   Add water to 1 Liter

3. 1 X TBE

   Take 10 ml of 10 × TBE and dilute to 100 ml with double distilled water.

4. Ethidium bromide (10 mg/ml in sterile double distilled water)

5. 10 X gel loading dye

Glycerol	5 ml
10 X TBE	1 ml
Bromophenol blue (Saturated)	1 ml
Xylene cyanol 10%	1 ml

   Mix well, autoclave and store in aliquots.

6. DNA samples
7. DNA size standards
8. Horizontal slab gel electrophoresis apparatus
9. Gel-sealing tape
10. Microwave oven
11. Power supply device capable of up to 500 V and 200 mA
12. Water bath preset to 55°C
13. UV Transilluminator

EXPERIMENTAL LAY-OUT

Set up the electrophoretic gel mould

↓ ← Pour the agarose solution to the gel mould

Keep for gel setting

↓

Add electrophoretic buffer to tank, covering the gel

↓

Load the samples

↓

Attach the tank with power supply

↓

Detection of DNA bands under UV transilluminator

METHOD

1. Prepare a 1% solution of agarose by adding 100 ml of 1 X TBE buffer to 1 gm of weighed agarose in a 250 ml conical flask.

2. Melt the agarose completely in a clear solution, using a microwave oven.
3. When melted, place the flask having melted agarose in water bath set at 55°C until it is cool enough to handle.
4. While the agarose is melting, tightly seal the end of the gel tray with plastic tape. After that keep the comb in position.
5. When the melted agarose is cool enough (up to 50°C), add 10 μl of ethidium bromide (10 mg/ml) and mix well.
6. Pour the cooled gel solution into the gel mould and allow the gel to set for 20 minutes.
7. When the gel hardens, remove the tape and comb. Place get tray into the box. Add buffer to the tank so that it covers the gel to a depth of about 2 mm.
8. Load the DNA samples carefully into the wells. In one well load a DNA size standard.
9. Close the lid of the gel tank and attach the electrical leads so that the DNA will migrate toward the positive end (red lead).

   Apply a voltage of 50 volts, bubbles will be generated at the electrodes within a few minutes and the bromophenol blue will migrate from the wells into the body of the gel.
10. Stop electrophoresis, when bromophenol blue crosses more than 2/3 of the length of the gel (in no case it should move out of the gels).
11. Visualize DNA bands on an UV trans-illuminator.
12. If require, this profile can be photographed through a red filter for documentation.

REMEMBER

1. 1% as used in above protocol, is the standard concentration, but if you are trying to resolve large fragments of DNA, you may use a lower concentration and for small fragments, a higher concentration would be appropriate.
2. Adding the ethidium bromide directly to the gel is an optional step. If you don't wish to add the dye here, you will have to stain the gel at the end. It is sometimes useful to have the dye in the gel while the gel is running because you can also interrupt the run, check the location of your DNA fragments, and then continue if you wish to run further. But ethidium bromide alters the conformation of the DNA, thereby altering the migration rate. So if you are to generate a restriction map and would like to measure fragment sizes accurately, it is always preferable to run the gel in the absence of the dye. In this case a brief destaining either with water or 1 mM magnesium sulphate solution is necessary.

   Caution: Ethidium bromide is a powerful carcinogen. Take all the precautions, while handling.

3. Use good quality sticking tape for sealing the edges and also bring the agarose solution temperature to 60°C before pouring.
4. Take care while inserting the comb that it remains 0.5–1.0 mm above the bottom of gel and let the gel polymerize completely before removing the comb.
5. Check the pH of buffer and use the same batch of electrophoresis buffer in both electrophoresis tank and the gel. Take care while pouring the gel that no air bubbles are trapped inside the gel or under and between the teeth of comb.
6. Poor staining of gel may occur even in the presence of ethidium bromide in gel because of the removal of ethidium bromide in extended electrophoresis as ethidium bromide migrate opposite to DNA. In this case again stain the gel by soaking for 30–45 minutes in a solution of ethidium bromide (0.5 µg/ml).

❑❑❑

Part 3

POLYACRYLAMIDE GEL ELECTROPHORESIS OF DNA

Contents

INTRODUCTION

Polyacrylamide gel electrophoresis (PAGE) of DNA is carried out when purification of double stranded DNA (with very small difference in size) is to be carried out or when small sized single stranded DNA molecules differing in size by very few nucleotides (might be 1 nucleotide only) are to be separated *e.g.,* in sequencing of DNA. In a sequencing experiment, whatever the sequencing method may be, the final analysis usually involves separating single stranded DNA molecules shorter than about 100 nucleotides and differing in size by only 1 nucleotide. To achieve this small pored gel is required so acrylamide gels are used instead of agarose gels. Further gels are polymerized in the presence of an agent (urea or formamide) that suppresses the base pairing in nucleic acids. So the electrophoresis is also known as denaturing polyacrylamide gel electrophoresis.

When electrophoresis of double stranded DNA particles is carried out, the urea is not added during polymerization and the process is known as non denaturing polyacrylamide gel electrophoresis. The present protocol is far non-denaturing PAGE.

MATERIALS/REAGENTS

1. Acrylamide: bisacrylamide (29 : 1)
2. Ammonium Persulfate (10% w/v)
3. Ethanol
4. KOH
5. Sigmacote
6. TEMED
7. 6 X gel loading buffer
8. Electrophoretic buffer
9. Fixing solutions

   Solution I : Mix 250 ml ethanol and 50 ml acetic acid and make up the volume to 500 ml with dH_2O.

   Solution II : Mix 125 ml ethanol and 50 ml acetic acid, make up the volume to 500 ml with dH_2O.

   Solution III : Mix 50 ml of ethamd and 2.5 ml of acetic acid, make up the volume to 500 ml with dH_2O.
10. Silver nitrate solution

    Dissolve – 1.9 g of silver nitrate in 1L & of dH_2O. Keep in a dark bottle
11. Solution IV

    Dissolve 7.5 g of sodium carbonate in 1 L of dH_2O
12. Developing solution

    Prepare a 0.75 M NaOH solution (30 g of NaOH in 1 L water), add 7 ml of formaldehyde and mix well
13. Bulldog paper clips
14. Electrophoretic apparatus

    (Glass plates, comb and spacers)
15. Gel sealing tape
16. Micropipette with drawn-out plastic tip
17. Petroleum jelly
18. Syringe
19. Big plastic trays
20. Shaker

EXPERIMENTAL LAY-OUT

Wash the plates

↓

Treat one plate with sigmacote

↓

Assemble the plates with spacers in between (3 sides) with tight clamps

↓ ← Pour the acrylamide solution

Insert the comb and keep for polymerization

↓

After it, take out the comb and rinse the well with 1XTBE buffer

↓

Set in an electrophoretic tank filled with buffer

↓

Load the samples and attach to power supply

↓

Run till the process is complete

↓

Remove the plates and go for staining of gel

Table 3.1. Volume of reagents

Polyacrylamide Gels (%)	*29% Acrylamide Plus 1% Methylene Bis Acrylamide (ml)*	*H_2O (ml)*	*5 x TBE (ml)*	*10% Ammonium Per Sulphate (ml)*
3.5%	11.6	67.7	20.0	0.7
5.0%	16.6	62.7	20.0	0.7
8.0%	26.6	52.7	20.0	0.7
12.0%	40.0	39.3	20.0	0.7

METHOD

1. Wear the gloves and wash the glass plates and spacers with warm detergent solution, rinse thoroughly with water and again rinse with KOH.
2. Dry the plates. Select one plate and treat its one surface with sigmacote (This will prevent the gel from sticking tightly to one plate and reduces the possibility of gel tearing).
3. Assemble the glass plates together with spacers on three sides and clamp them together with bulldog paper clips.
4. According to the size and thickness of the mold, calculate the volume of gel required and prepare the gel solution with the desired polyacrylamide percentage as given in Table 3.1.
5. Add 35 μl of TEMED for each 100 ml of acrylamide: bisacrylamide solution and mix the solution by gentle swirling.
6. Draw the solution into a 50-cc syringe. Invert the syringe and expel any air that has entered the barrel. Introduce the nozzle of the syringe into the space between the two glass plates. Expel the acrylamide gel solution from the syringe, filling the space almost to the top.
7. Place the glass plates against a stand at an angle of 10° to the bench top.
8. Immediately insert the appropriate comb into the gel (take care that air bubbles are not trapped) & clamp it with bulldog clips.
9. Allow it to polymerize for 40–60 minutes at room temperature.
10. After polymerization is complete, carefully pull the comb. With the help of syringe rinse out the wells with 1X TBE.
11. Attach the gel to the electrophoresis tank the notched plate should face inward toward the buffer reservoir.
12. Fill in the tank with electrophoretic buffer.
13. Use a syringe to flush out the wells and load the DNA samples with 6X-gel loading buffer into the wells.
14. Connect the electrodes and turn on the power to start the process.
15. When the dyes have migrated the lower half of the gel, turn off the electric power, disconnect the loads and discard the electrophoresis buffer from the reservoirs.
16. Detach the glass plates, lay the glass plates on bench (sigmacote plate uppermost) and with the help of plastic wedge lift a corner of the upper glass plate (the gel should remain attached to lower plate) Pull the upper plate smoothly away and remove the spacers.
17. Go for staining of the gels for detecting the position of bands in the polyacrylamide gels.
18. Ethidium bromide staining of acrylamide gel is less sensitive compared to agarose gels. On the other hand staining of acrylamide gel with silver nitrate is very sensitive.

Ethidium bromide staining:

19. Immerse the plate with the gel in a solution of ethidium bromide (10 mg/ml in running buffer) for about 30 minutes. Remove the plate and cover the gel with saran wrap. Invert the plate on a trans illuminator so that saran wrap is beneath the gel.

Silver nitrate staining of the gel:

20. Take a clean plastic tray, place the plate with gel (facing upwards) into tray and pour 500 ml of solution I, incubate at room temperature for 30 min.
21. Rinse the gel with water and incubate in solution II and III similarly as above; each treatment for 30 min followed by rinsing with dH_2O.
22. To the fixed gel in the tray add 500 ml of silver nitrate solution and stain for 2 hour with shaking.
23. Pour off the silver nitrate solution and rinse the gel 5 times with dH_2O.
24. To the stained gel in the tray, add 500 ml of developing solution, watch for development of bands. If the development is enough immediately pour off the developing solution and rinse with dH_2O.
25. Add 500 ml of solution IV to the gel and incubate for 30-40 min at room temperature. This will sharpen the bands. Photograph the gel. For photographing no UV light or special filter is needed. White light is sufficient.

REMEMBER

1. It should be remembered that acrylamide also gets polymerized by exposure to air, so the two glass plates should be tightly sealed with spacers in between clamped very tightly. There are many types of electrophoresis apparatuses available commercially, and the arrangement of the glass plates and spacers differ slightly from manufacturer to manufacturer. Whatever the design, the aim is to form a watertight seal between the plates and the spacers so that unpolymerized gel solution does not leak out.
2. Don't forget to wash out the wells thoroughly as soon as the comb is removed because small amount of acrylamide solution trapped by the comb will polymerize in the wells, producing irregularly shaped surfaces that will give rise to distorted bands of DNA.
3. Take care of the volume of the DNA samples while loading. Mixing of the samples should not take place.
4. Non-denaturing polyacrylamide gels are run at voltages between 1v/cm and 8v/cm. If the electrophoresis is carried out at a higher voltage, differential heating in the centre of the gel may cause lowing of the DNA bands or even melting of strands of small DNA fragments.
5. Always treat one plate with gel-repelling agent *e.g.,* sigmacote and try to use the same plate for this purpose in subsequent experiments.
6. Acrylamide, bis acrylamide are neurotoxic so take all the precautions while handling them. (As mentioned in Unit I).
7. Load gels carefully and use tight fitted combs for uniform wells. Also check that teeth of comb are uniform and not damaged especially when using shark toothcomb during DNA sequencing gel electrophoresis. Do not over load the gels.

❑❑❑

Part 4

POLYACRYLAMIDE GEL ELECTROPHORESIS OF PROTEINS

Contents

INTRODUCTION

Polyacrylamide gel electrophoresis is undoubtedly the most commonly used methods for characterizing proteins. The procedures are relatively rapid and sensitive (requiring only μg quantities of protein) and the detection methods are convenient and sensitive. Over the years, a number of modifications on the basic electrophoretic techniques have been evolved. As a result, a number of systems are available now (having difference in the principle of separation of protein molecules) like:

(*i*) **Separation on the basis of charge:** Isoelectric focussing gels.

(*ii*) **Separation on the basis of size:** SDS-Polyacrylamide electrophoresis or Denatured gel electrophoresis.

(*iii*) **Separation on the basis of shape:** Native Polyacrylamide electrophoresis or Non-SDS/Non-denatured gel electrophoresis.

(*iv*) **Separation on the basis of charge:** Two dimensional polyacrylamide gel electrophoresis (Charge as well as size).

Two more special types are as:

(*v*) **Cellulose acetate electrophoresis:** Cellulose acetate has the advantage over paper in that it is a much more homogeneous medium, with uniform pore size, and does not adsorb proteins in the way that paper does. There is therefore much less trailing of protein bands and resolution is better, although nothing like as good as that achieved with polyacrylamide gels.

(*vi*) **Continuous flow electrophoresis:** It is used for separation in free solution in large scale productions. Electrophoresis takes place continuously as the separating material is carried upwards by a flow of carrier buffer though annular space between two vertical concentric cylinders.

In this section, we will discuss the protocol of only one type of polyacrylamide gel and that is of the SDS – PAGE.

PRINCIPLE

SDS CH_3-$(CH_2)_{10}$-CH_2OSO_3-Na^+ is an anionic detergent and is used to denature the native proteins. On average; one SDS molecule binds for every two amino acids. The original native charge on the molecule is therefore completely swamped by the negatively charged SDS molecules while still maintaining the rod like structure. The samples to be separated are not in fact loaded directly into the main separating gel. When the main separating gel (normally about 10 cm long) has been poured between the glass plates and allowed to set, a shorter (approximately 1 cm) stacking gel is poured on top of the separating gel and it is into this gel that the wells are formed and the proteins loaded. The purpose of this stacking gel is to concentrate the protein sample into a sharp band before it enters the main separating gel. This is achieved by utilizing differences in ionic strength and pH between the electrophoresis buffer and the stacking gel, and involves a phenomenon known as isotachophoresis. The stacking gel has a very large pore size (4% acrylamide), which allows the proteins to move freely and concentrate, or stack, under the effect of the electric field. The band sharpening effect relies on the fact that negatively charged glycinate ions (in the buffer) have a lower electrophoretic mobility than do the protein – SDS complexes, which in turn have lower mobility than the chloride ions (Cl^-) of the loading buffer and the stacking gel. When the current is switched on, all the ionic species have to migrate at the same speed otherwise there would be a break in the electrical circuit. The glycinate ions can move at the same speed as Cl^- only if they are in a region of higher field strength. Field strength is inversely proportional to conductivity, which is proportional to concentration. The result is that the three species of interest adjust their concentrations. So that (Cl^-) > (protein-SDS) > (Glycinate). There is only a small quantity of protein-SDS complexes, so they concentrate in a very tight band between glycinate and Cl^- boundaries. Once the glycinate reaches the separating gel it becomes usefully ionised in the higher pH environment and its mobility increases (The pH of the stacking gel is 6.8 that of separating gel is 8.8). Thus the interface between glycinate and Cl^- leaves behind the protein-SDS complexes, which are left to electrophoresis at their own rates. The negatively charged protein-SDS complexes now continue to move towards the anode, and because they have the same charge per unit length, they travel into the separating gel under the applied electric field with the same mobility. However as they pass through the separating gel the proteins separate, owning to the molecular sieving properties of the gel. Small proteins move easily, large proteins move slowly because of frictional resistance.

REAGENTS

1. Acrylamide (30% stock): Dissolve 29.2 g acrylamide and 0.8 g bis acrylamide in distilled water and make up to 100 ml. Store under dark conditions in dark colored bottles at 4°C.
2. Resolving separating buffer (pH 8.8) - 1.5 M Tris-HCl.
3. Stacking gel buffer, (pH 6.8) - 1.0 M Tris–HCl.
4. Electrophoresis buffer (pH 8.3) - Dissolve 3 g Tris, 14.4 g glycine and 1 g SDS in 1000 ml of distilled water, store at 4°C.
5. Ammonium per sulphate (10%).
6. TEMED (N, N, N′, N′, Tetramethylene diamine).
7. Sample buffer (loading buffer).
8. Staining reduction – Dissolve 200 mg coomassie brilliant blue R 250 in 50 ml methanol/ethano, 7 ml acetic acid and 43 ml distilled water.
9. Destaining solution – Add 7 ml acetic acid to 30 ml methanol/ethanol and 63 ml distilled water.

EXPERIMENTAL LAY-OUT

Wash the plates and assemble tightly with Spacers in between

↓ ← Pour in the mould ← Prepare 10% resolving gel

Polymerization for 30–40 min

↓ ← Pour in the mould ← Prepare 5% stacking gel

Insert a comb immediately and set for polymerization

↓

Remove the comb after polymerization

↓ ← Load the samples

Run for electrophoresis

↓

Run the gel until bromophenol blue reaches the bottom

↓

Turn off power and go for staining the gel

Table 4.1. Solutions for preparing 10% of resolving gels

–δ Solution Component	*5 ml*	*10 ml*	*15 ml*	*20 ml*	*25 ml*
H_2O	1.90	4.00	5.90	7.90	9.90
30% acrylamide	1.70	3.30	5.00	6.70	8.30
0.5% Tris-HCl	1.30	2.50	3.80	5.00	6.30
10% SDS	0.05	0.10	0.15	0.20	0.25
10% APS	0.05	0.10	0.15	0.20	0.25
TEMED	0.002	0.004	0.006	0.008	0.01

Table 4.2. Solutions for preparing 5% stacking gel for SDS-PAGE

–δ Solution Components	*2 ml*	*4 ml*	*6 ml*	*8 ml*	*10 ml*
H_2O	1.40	2.70	4.10	5.50	6.80
30% acrylamide	0.33	0.67	1.00	1.30	1.70
10.5 M Tris-HCl (pH 6.8)	0.25	0.50	0.75	1.00	1.25
10% SDS	0.02	0.04	0.06	0.08	0.10
10% APS	0.02	0.04	0.06	0.08	0.10
TEMED	0.002	0.004	0.006	0.008	0.01

METHOD

1. Wash the plates and rinse with methanol, dry and assemble them according to the manufacturer's instruction (generally with 1.5 or 1.0 mm thick spacers).
2. Depending on the volume of the mould, prepare the acrylamide mixture for 10% resolving gel as in Table 4.1.
3. Mix the components and without delay pour the acrylamide mixture into the glass mould and overlay with distilled water and isopropyl alcohol.
4. Keep for polymerization for about 30–40 min.
5. After it, remove the overlay and wash the gel top with distilled water.
6. Prepare 5% stacking gel (Table 4.2).
7. Mix the components and pour over the resolving gel.
8. Immediately insert a comb and allow polymerizing the gel.
9. Prepare samples required (volume of sample (100 mg protein/lane) + (equal volume of sample buffer.)

10. Heat the samples in boiling water bath for 2 min to denature the proteins, keep them on ice to retain the denatured state.
11. Remove the comb from gel and rinse the wells with distilled water.
12. Mount the gel in the electrophoretic apparatus.
13. Add electrophoresis buffer to the top and bottom reservoirs of the electrophoretic apparatus.
14. Load the samples along with marker proteins into the lanes/wells.
15. Attach the apparatus to power supply unit and apply 8 V/cm for stacking gel (70 V) and 15 V/cm for resolving gel (150–200 V)
16. Run the gel until bromophenol blue dye reaches the bottom of resolving gel.
17. Turn the power supply off, remove the sandwich plates from the apparatus and place on a paper towel and remove the plates apart using a spatula.
18. Immerse the gel in 5 volumes of staining solution and stain for 4 hrs at room temperature.
19. Destain the gel.
20. Rinse in 7% acetic acid. Visualize the bands in an illuminator.

REMEMBER

1. The concentration used of the resolving gel depends on the molecular mass of proteins. *e.g.,* gels of 15% polyacrylamide are sufficient for proteins in the range 100000 to 10000. For proteins of 150000 molecular mass or more, gel of 10% or even 7.5% is required. So always select a concentration of resolving gel after knowing the molecular mass of the proteins.
2. A pure protein should give a single band on an SDS-Polyacrylamide gel. So if two or more bands are there the molecule is made up of 2 unequal subunits and so should be purified.
3. The staining with coomassie brilliant blue can detect a band containing as little as 0.1 μg of protein, such sensitivity is sufficient for many works. However, there are cases when greater sensitivity would be an advantage and this can be achieved by using a silver stain, such stain is at least 100 times more sensitive then the coomassie stain. The sensitivity is comfortable to autoradiography of radiolabeled proteins.
4. Use protein concentrations in range of 1.0 mg/ml for coomassie blue staining, use fresh and appropriate concentration of dye and use the optimum conditions for that particular enzyme during staining. Some isoenzymic bands appear only for short time and tend to disappear with time than take care to score them immediately on appearance.
5. Sometimes slow polymerization or no polymerization of gel may occur; it may be due to very low temperature, absence of catalyst or due to presence of oxygen.

   To overcome this, raise the temperature of room/place to 27–30°C where gel is being polymerized, check catalyst concentration and degas the solutions before pouring the gel.

6. Many a time smiling gels are obtained because of non-uniform temperature of gel or due to presence of air bubbles in gel. In this case degas the solution before pouring and take precaution not to trap air bubble in gel during pouring/polymerization.
7. Always use same lot of sample buffer for loading in all the wells because a small difference in density of samples in different wells results in unequal dye migration.
8. Sometimes dye mobility is low than expected. This can be due to improper pH of the different buffers used during electrophoresis or air bubbles may have trapped in gel while preparing the gel. The another reason for this may be that gel may be thicker than normally run.
9. Always use very high quality grade reagents for electrophoresis, use protease inhibitors during sample preparation and partially purify the sample by dialysis or by passing through a small sephdex G100 column.
10. Always check the resolving gel, buffer systems and pH of the solutions before running electrophoresis. Never run the gels too fast or too slow.
11. Take care while overlaying the water on resolving gel surface or use saturated n-butanol for overlaying the gel surface. Also keep the volume and density of samples equal in all the wells.
12. Provide optimum current and low ionic strength buffer where possible. If high ionic strength buffer is to be used then provide outside cooling *i.e.,* run gels at 4°C in a refrigerator/cold room.

❑❑❑

Unit **V**

CHECKING QUALITY AND QUANTITY OF DNA

Part 1 SPECTROPHOTOMETRIC METHOD

Contents

INTRODUCTION

Reliable measurement of DNA concentration is important for many applications in molecular biology. One of the important methods for DNA quantitation is through the Spectrophotometric measurement of the absorption at 260 nm. Corrections are usually required, necessary to account for interfering substances like RNA (absorbing at the same wavelength) and proteins (absorbing at 280 nm). Such corrections commonly require the measurement of the absorbance, by the interfering substance, at a wavelength remote from that for the compound under test, plus a knowledge of the absorbance at the test wavelength *e.g.,* $A_{280/260}$ ratio for proteins in the presence of nucleic acids.

PRINCIPLE

Purines and pyrimidines in nucleic acids absorb UV light. As described by the Beer–Lambert law, the amount of energy absorbed at a particular wavelength is a function of the concentration of the absorbing material.

$$I_t = I_i\ 10^{Edc}$$

Where I_t = intensity of transmitted light

I_i = intensity of incident light

E = molar absorption coefficient (also called as molar extinction coefficient)

d = optical path length (in cm)

c = concentration of absorbing material (mole/liter)

E is numerically equal to the absorbance of a 1M solution in a1cm light path and is so expressed in $M^{-1}cm^{-1}$. Absorbance data are collected using a UV spectrophotometer and are generally expressed as the absorbance A(log I_t/I_i). When d = 1cm, A is called the optical density or OD at a particular wavelength, X.

$$OD_x = Ec$$

E = absorption coefficient

c = concentration of absorbing material

For double stranded DNA, one A_{260} or OD_{260} unit corresponds to a concentration of 50 mg/ml, with path length of 1 cm. For nucleic acid the E decreases as more and more purines and pyrimidines are stacked in a polynucleotide chain. The value of E decreases in the following series – free base > oligonucleotides > single stranded nucleic acids > double stranded nucleic acids. Molar absorption coefficient of double stranded DNA at 260 nm is 6.6, whereas the molar extinction coefficient of single stranded DNA and RNA is 1.4. However, the absorption coefficient of both DNA and RNA are affected by the ionic strength and the pH of the solution. Accurate measurements of concentration can be made only when the pH is carefully controlled and the ionic strength of the solution is low.

The absorption/extinction coefficients of nucleic acids are the sum of the extinction coefficient of each of their constituent nucleotides. For large molecules of DNA, isolated in laboratory during a routine experiment, it is practically impossible to sum up the coefficient of all the nucleotides.

So an alternative formula is there to know the concentration of DNA sample.

$$\frac{\text{DNA concentration of solution A}(\mu g/ml)}{\text{DNA concentration of standard }(\mu g/ml)} = \frac{\text{OD at 260 nm of the diluted sample}}{\text{OD at 260 nm of standard}} \times \text{Dilution times}$$

Where,

- Dilution times is the value by which the original DNA sample is diluted.
- Standard value of DNA concentration is 50 µg/ml.
- OD_{260} is standard (50 µg/ml) in a path length of 1 cm is equal to 1.

Putting these values in the above formula :

$$\frac{\text{DNA concentration of solution A}(\mu g/ml)}{50\ \mu g/ml} = \frac{OD_{260}\ \text{of diluted sample}}{1} \times \text{Dilution times}$$

So, the final formula that comes out is :

DNA concentration of solution A ($\mu g/ml$) = OD_{260} of diluted sample × Dilution time × 50 $\mu g/ml$

MATERIALS/REGENTS

1. UV spectrophotometer
2. Quartz cuvettes
3. DNA sample (let's say a pellet of DNA dissolved in 50µl of sterilized *dd*H_2O)
4. 70% Ethanol
5. Double distilled water.

EXPERIMENTAL LAY–OUT

Set the spectrophotometer for absorbance by nucleic acids at 260 nm

↓ ← Make dilution of the original sample

Note down the OD of this diluted sample at 260 mm

↓

Put the values in the formula and calculate the concentration of the DNA sample.

METHOD

1. Turn the machine on using the on switch, wait for the machine to go through its start up routine.
2. Set up the programme for measuring absorbance of nucleic acids at 260 and 280 nm, following all the instructions given in the manual of the spectrophotometer.

3. Make a dilution of the DNA sample. Let's say you have dissolved the pellet of DNA (obtained at end of DNA isolation experiment) in 50 μl of ddH_2O. Now you have a 50 μl of DNA solution (say solution A) with unknown concentration which you want to know.

   For this prepare a diluted sample *e.g.,* take 5 μl of DNA solution A and to this add 1ml of ddH_2O, which means it is diluted 200 times.

   So the dilution times/factor = 200

4. Note down the OD_{260} of the diluted sample. Let us assume it is 0.112 obtained.

5. Put the values obtained at the end of point 3 and 4 in formula obtained at the end of principle.

$$\text{DNA concentration of solution A } (\mu g/ml) = OD_{260} \text{ of diluted sample} \times \text{dilution times} \times 50\ \mu g/ml$$

$$\text{DNA conc. } (\mu g/ml) = 0.112 \times 200 \times 50$$

$$= 1120\ \mu g/ml$$

$$= 1.12\ \mu g/\mu l$$

6. The above concentration is the concentration of DNA in the per μl of original sample.

For Quality Estimation

1. The ratio of the absorbance at 260 nm and 280 nm (OD_{260}: OD_{280}) is used as a measure of purity of isolated DNA.

2. For pure double stranded DNA solution the ratio should be between 1.7 and 1.9.

3. Higher ratios are often due to RNA contamination and lower ratio is due to protein or phenol contamination.

4. Though 260/280 ratio is the most widely used criterion. But 230/280 ratio is generally more sensitive to contamination with proteins. Absorbance ratio of 230/280 nm should be around 1. Higher ratios indicate contamination with protein. Routine procedure is to measure the absorbance from 190 to 320.

   Some common contaminants along with absorbance peak are:

	Substances	Peak absorbance (nm)
(*i*)	EDTA, Polysaccharide, ethanol	230
(*ii*)	DNA, RNA	260
(*iii*)	Phenol	270
(*iv*)	Proteins	280
(*v*)	Cell debris	320

REMEMBER

1. The material of the cuvette depends on the wavelength at which absorbance is noted. In the ultraviolet region, cuvette of quartz material are preferred. Never use glass cuvette for this purpose.
2. The specific absorption coefficients of both DNA and RNA are affected by the ionic strength and the pH of the solution. Accurate measurements of concentration can be made only when the pH is carefully controlled and the ionic strength of the solution is low.
3. It is difficult to measure the absorbance of small volumes of solution and the method is reliable only over a fairly narrow range of concentrations (5 μg/ml to 90 μg/ml).

❑❑❑

Part 2 ELECTROPHORETIC METHOD

Contents

- Introduction
- Principle
- Materials/Reagents
- Experimental lay-out
- Method
- Remember

INTRODUCTION

Among the various electrophoretic techniques available, agarose gel electrophoresis is used for estimation of the size, quantity as well as quality of DNA samples. The process is also known as gel quantitation and qualitation.

PRINCIPLE

The technique is based on the comparison of a known quantity of DNA to an unknown quantity of DNA on agarose gel. The gel is stained with ethidium bromide and is visualized under UV transilluminator. The brightness of a particular fragment, as well as the distance traveled, can be used to estimate the quantity and size of a particular fragment as compared to a known sample such as lambda cut with Hind III (the standard marker ladder). The known sample (10 µl of 0.1 µg/µl) is loaded into lane one. Since each fragment length is known, it is possible to estimate what percentage of total lambda DNA is present in that fragment. The unknown sample is loaded into next lane. The brightness of the band of unknown concentration sample is compared with the brightness of bands of standard marker ladder. This can lead to estimation of the quantity of DNA in the unknown concentration band.

REAGENTS/MATERIALS

1. Standard size ladder (lambda DNA/Hind III digest)
2. Unknown size/quantity sample of DNA (let's say DNA A)
3. Electrophoresis apparatus
4. Power supply
5. Gel tray and comb
6. Micropipette
7. Loading dye (6X)
8. Agarose
9. 1X TBE buffer
10. Ethidium bromide
11. Transilluminator
12. 1.5ml eppendorf tubes.

EXPERIMENTAL LAY-OUT

Prepare 0.8% agarose gel

Take unknown concentration (DNA A) → load in Lane 2 → ← load in Lane 1 ← Lambda DNA/HindIII Sample

↓

Run at 70 volts for 2 hours

↓

Stain the gel in Ethidium bromide for 6min.

↓

Destain in water for 2 minutes

↓

Observe bands in Transilluminator

↓

Take the photographs and analyze the Results by comparing fluorescence.

METHOD

For Checking Quantity

1. Prepare 0.8% of agarose gel by adding 0.8 gm of agarose to 99.2 ml 1X TBE buffer and microwave on medium low for 2–3 minutes or until clear (Don't seal the container, take care that the container is big enough not to let the contents spill out side). Pour the contents into gel cast tray with comb. Let the gel set for some time. When ready remove the gel, sealing tapes and comb. Put the gel into electrophoretic chamber.
2. Load 10 µl of sample A with 1µl of 6X loading dye in lane 2. In lane 1 add 10 µl of the Lambda/Hind III digest (of 0.1 µg/µl concentration) plus one µl of loading dye (11 µl).
3. Run at 70 volts for 2 hours for best results or 100 volts for one hour if time is an issue.
4. Stain the gels for approximately 6 minutes in ethidium bromide and de–stain in water for 2 minutes.
5. Observe in transilluminator and photograph the gel.
6. Observe the lambda/Hind III lane. Following fragments are obtained with slowest band of 23,130 bp and fastest band (not visible) of 125 bp:

   23130
   9416
   6557
   4361
   2322
   2027
   564
   125

8. Since a known quantity of the lambda Hind III was loaded into the lane it is possible to estimate the quantity in just one fragment. By knowing the total length of the lambda a percentage can be determined.

   23130/48502 = 48%
   9416/48502 = 19%
   6557/48502 = 13%
   4361/48502 = 9%
   2322/48502 = 5%
   2027/48502 = 4%
   564/48502 = 1%
   125/48502 = 0.2%

9. Since 1 µg of DNA was loaded into the lane 1 to begin with (as 10 µl of 0.1 µg/µl concentration was used which means 10 µl of sample contains 1 µg of DNA), it can now be used to determine the quantity of the lambda/Hind III digestion products.

e.g., 23130 is the slowest and the brightest band. Since 48% of the total DNA is in band 231,30 and there was 1 μg of total DNA, so 48% of 1 μg = 0.48 μg of DNA in band 23130. Similarly quantities of other fragments can be calculated.

10. Then compare the brightness of the known bands to the bands of sample A. If the brightness of a known band is close to the brightness of the unknown, then an estimation of the total number of micrograms can be made.

For Checking Quality

1. To check for equality, load about 5 μl of undigested DNA on the gel along with 0.5 mg of molecular weight standard such as lambda DNA cut with Hind III. When checking several DNA samples, at least two gel lanes of the standard (on both the sides of the samples) will do.
2. Compare the migration of the samples on the gel with that of the standard. The largest fragment in a lambda—Hind III measures 23.1 Kb.
3. The bands for the sample must appear as a distinct solid fluorescence. A diffused band moving toward lower molecular weight indicates degradation of DNA fragments.
4. If some fluorescence is observed in the wells also, may be that DNA is not dissolved completely or may be some proteins are still in the sample. If a band is appearing quite lower in the gel, it is indicating towards RNA contamination.

REMEMBER

1. Estimation of DNA concentration through gel electrophoresis may utilize undigested unknown samples for comparison with undigested concentration standards. However, undigested samples tend to display equally intense fluorescent bands so that subtle difference in their concentrations is difficult to appear on the gel.
2. Avoid adding ethidium bromide directly to the gel when going for checking equality and quantity of DNA.

❑❑❑

Unit VI RESTRICTION ENDONUCLEASES

Part 1 A GENERAL ACCOUNT

Contents

INTRODUCTION

Restriction endonucleases are enzymes, produced by many bacterial species that cut the DNA into discrete fragments by acting only at specific DNA sequences. The discovery of these enzymes started with the observations of phage infection pattern in some bacterial strains in late 60s. It was noticed that a phage DNA survival in a particular bacterial host depends on the last host where it has propagated *e.g.,* if phage is infecting a bacterial strain (say *Haemophilus influenzae* strain Rd), the phage DNA will be digested by bacterial defense enzymes. However, in this process of digestion, if some phage lambda are able to escape from the host and if these phages are re–plated on the same host *i.e., Hamophilus influenzae* strain, the phages will no longer be digested/restricted. Such phage infection cycles were studied for many generations and in diverse bacterial strains using radiolabeled phage DNA. All these studies could conclude together that some or most of the bacterial strains are blessed with a unique **restriction–modification** system. This system has restriction as well as modification enzymes. Modification involves methylation of certain bases at a very limited number of sequences within DNA which constitute the recognition sequences for the restriction endonucleases. Such pattern of methylation enables the restriction enzymes to discriminate between the host's own DNA and foreign DNA. This could explain that why those phages which somehow escaped from restriction by host restriction enzymes in the first infection cycle, were not restricted by the enzymes if re–infected in the same host; their DNA being replicated in the presence of modifying enzymes and so like the host DNA, phage DNA is also having a characteristics methylatio pattern and so protected from restriction system. Isolation and purification of restriction endonucleases has enabled the scientists to cut the DNA in a required manner. Discovery of these enzymes, which led to Noble prizes for W Arber, H Smith and D Nathans in 1978, was one of the key breakthroughs in the development of "**Recombinant DNA Technology**". There are three groups of restriction endonucleases, each distinguished by a slightly different mode of action:

Type I: They contain three different types of sub unit and require Mg^{2+}, ATP and S–adenosylmethioine (SAM) for DNA cleavage. Their recognition sites are complex and DNA cleavage occurs at non specific sites, 400–7000 bp from the recognition sites.

Type II: They contain only one type of subunit, and Mg^{2+} alone is required as cofactor. DNA cleavage occurs at specific sites within or adjacent to the enzyme's recognition site.

Type III: They contain two subunits, requires Mg^{2+} and ATP as cofactors, DNA cleavage occurs distal to the recognition site, by 25–27 base pairs.

Type II endonucleases perform site specific cleavage of DNA and are used in recombinant DNA technology.

The site at which the enzyme cuts is called the recognition site and it is usually palindromic in nature *i.e.*, the sequence is same on both strands of double stranded DNA reading in opposite direction. The recognition sequences are generally 4, 5 or 6 base pairs in length. The restriction sequences for some of the most frequently used restriction endonucleases are listed below:

	Enzyme	*Organism*	*Recognition Sequence*	*Blunt for Sticky End*
1	Bgl II	*Bacillus globigit*	AGATCT	Sticky
2	Bam HI	*Bacillus amyloliquefaciens*	GGATCC	Sticky
3	Pvu I	*Proteus vulgaries*	CGATCG	Sticky
4	Pvu II	*Proteus vulgaries*	CAGCTG	Sticky
5	Hinf 1	*Haemophilus influenzae Rf*	GANTC	Sticky
6	Hind III	*Haemophilus influenzae Rd*	AAGCTT	Sticky

7	Taq 1	*Thermus aquaticus*	TCGA	Sticky
8	Sau 3 A	*Staphylococcus aureus*	GATC	Sticky
9	Alu 1	*Arthrobacter lutens*	AGCT	Blunt
10	Eco R I	*Escherichia coli*	GAATTC	Sticky
11	Hae III	*Haemophilus acgyptius*	GGCC	Blunt
12	Not I	*Nocardia otitidis caviarum*	GCGGCCGC	Sticky

NATURE OF ENDS PRODUCED

The nature of ends produced by a restriction endonuclease is of considerable importance in the design of a gene cloning experiment. Blunt ends are produced when a single cut is made in the middle of the sequence *e.g.,* PvuII and AluI

–NNAGCTNN– —AluI→ –NNAG– –CTNN–
–NNTCGANN– –NNTC– –GANN–

Blunt ends

On the other hand most of the restriction enzymes cut in a staggered manner where the two DNA strands are not cut at the same position. As a result the resulting DNA fragments have short single stranded overhangs at each end. Such ends are termed as sticky ends or cohesive ends as base pairing between them can stick the DNA molecule back together again *e.g.,* Eco RI

–NNGAATTCNN– Eco RI–NNG→ AATTCNN–
–NNCTTAAGNN– –NNCTTAA GNN–

Sticky ends

CHANCES OF OCCURRENCE OF RECOGNITION SEQUENCE IN A DNA MOLECULE

Calculation of such frequency is based on the assumption that the nucleotides are ordered in a random fashion and the four different nucleotides are present in equal proportions (*i.e.,* GC content = AT content).

Formula is 4^n where n = no. of bases in recognition sequence.

Examples: A tetranucleotide molecule should occur once every $4^4 = 256$ nucleotides, A hexanucleotide sequence should occur once every $4^6 = 4096$ nucleotides.

Practically speaking the assumptions made above are not entirely valid. Mathematics may give an idea of how many restriction sites to be expected in a given DNA molecule, only experimental analysis can provide the true picture.

SOME DEFINITIONS

Isochizomers: This term is given to those enzymes which are isolated from different sources but recognize the same target. Some isochizomers cut at the same point (*e.g.,* Mbo I and Sau 3A I) while some pairs cut at different places (*e.g.,* Sma I and Xma).

Isocaudomers: These are the restriction enzymes isolated from different sources, recognize different sequence but the products obtained after restriction have similar sequence from opposite ends.

❑❑❑

Part 2

RESTRICTION ENZYME DIGESTION AND SIZE DETERMINATION OF DNA FRAGMENTS

Contents

INTRODUCTION

This experiment demonstrates the activity of restriction endonucleases and analysis of the result using agarose gel electrophoresis. The DNA to be restricted is lambda DNA, which has approximately N = 48, 500 base pairs. In this exercise lambda DNA is digested with the restriction enzyme Eco RI having the recognition sequence GAATTC. Subsequent gel electrophoresis of the restriction products will illustrate the different fragmentation patterns produced by the enzyme.

Size determination of DNA fragments. It is frequently necessary to determine the size of a particular piece of DNA. This is done by running a molecular weight standard on a separate lane of the same gel. Here the size standard ladder is lambda digested with Hind III enzyme. The standard's fragments are all of known size and can be used to construct a standard curve relating some function of size versus migration distance. From this curve, molecular size of unknown fragments can be obtained by simply extraploting the migration distance. For this purpose agarose gels are used with adjustments in agarose percentage, in order to accommodate the possible range in size of DNA pieces that might be encountered. Following table shows some typical agarose gel percentage appropriate for various DNA sizes.

Agarose (%)	*Effective Range of Resolution of Linear DNA Fragments (Kb)*
0.5	30–1
0.7	12–0.8
1.0	10–0.5
1.2	7–0.4
1.5	3–0.2

PRINCIPLE

Eco RI is a Type II endonuclease recognizing specific sequence GAATTC and cutting DNA within it. Fragments with protruding ends are produced and because of different size, get separated by agarose gel electrophoresis and are visualized after staining with ethidium bromide, a dye which intercalates between bases of DNA and gives orange coloured fluorescence under UV light.

REAGENT/MATERIALS

1. Lambda DNA (0.5 mg/ml in TE buffer)
2. Eco RI
3. Size standard ladder (lambda/Hind III)
4. Core buffer (supplied by enzyme supplier)
5. Microfuge tubes
6. Pipette tips
7. Tracking dye
8. Ethidium bromide stock (10mg/ml)
9. Agarose
10. 1 X TBE Buffer
    10.7 gm Tris base (0.089 M)
    5.5 gm Boric acid (0.089 M)
    4 ml 0.5 M EDTA pH 8.0 (0.008 M)
11. Goggles
12. Gloves (latex)

13. Ruler
14. Semi–log graph paper

EXPERIMENTAL LAY–OUT

Setting up of restriction digestion mix

↓

Incubate at 37°C for 45–60 minutes

↓

Denature enzyme by heating

↓

Run on 0.75% agarose gel
along with DNA size standard

↓

Proceed with agarose gel Electrophoresis

↓

Go for size determination of DNA based on the electrophoretic migration pattern of the digest.

METHOD

1. Set up a restriction digestion in a 1.5 ml microfuge tube as follows:

Lambda DNA	3 μl
10X core buffer	3 μl
ddH_2O	17.5 μl
Eco RI	2 μl

Centrifuge briefly in a microfuge to mix.

2. Incubate the tubes for 45–60 minutes at 37°C.
3. Denature proteins by heating in a 70°C water bath for 5 min followed by immediate cooling on ice.
4. Prepare a 0.75% agarose gel in TBE (1X).
5. Add 2 μl of tracking dye to each digest and load gel lane 1 with 15 μl of lambda with Eco RI and another lane 2 with 15 μl of the standard size ladder.
6. Proceed with agarose gel electrophoresis as described in Unit III, Part 2.
7. The sizes of the DNA fragments of size standard ladder in lane 2 are known.

   Hind III digest of DNA are: 23130, 9416, 4316, 2322, 2027, 546, 125

   Take care that 125 bp fragment is not run off the gel.
8. Plot a graph of migration distance us molecular size of the bands of ladder DNA. With the help of a ruler, note down the distance traveled by a particular band starting from the well. Plot this distance migrated against the respective molecular size. Do this for all the bands of the ladder. A graph of the type as shown in Fig. 2.1 is obtained.
9. Now with the help of ruler, note down the distance traveled by the unknown size fragment in lane 1. By extrapolating this distance on the standard curve, molecular size of the unknown fragments can be obtained.

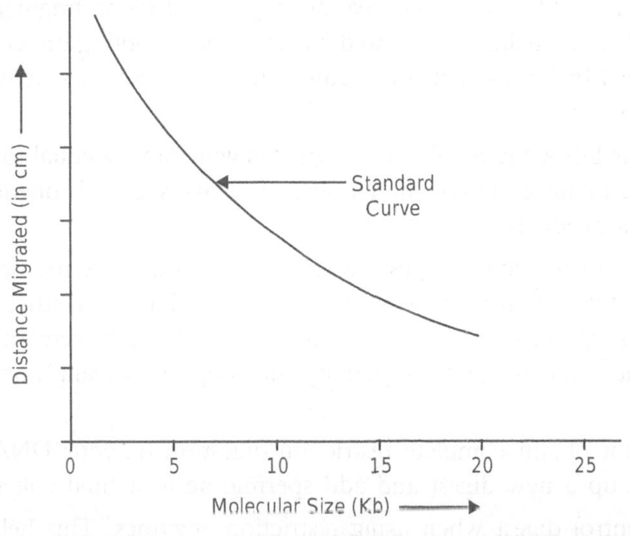

FIGURE 2.1 **A standard curve obtained by plotting distance travelled by a particular band versus molecular size of that band.**

REMEMBER

1. Minimum amount of DNA detectable by UV transilluminator is about 10mg. Therefore with restriction endonucleases producing 6 fragments per molecule, about 2–3 mg/ml of DNA is required in order to visualize the bands. For endonucleases with 75 sites, a higher concentration of DNA should be used in order to assure that detectable amounts of each fragment are produced.

2. DNA used for restriction analysis must be free of agents such as EDTA, SDS, phenol and ethidium bromide, which may interfere with or inhibit the activity of the enzyme.
3. Ethanol, glycerol and dimethyl sulfoxide (DMSO) could change the dielectric constant of restriction buffer and their addition/presence should be kept to a minimum.
4. Optimum reaction condition for each enzyme is different. Each enzyme has its own preferred salt, pH and Mg^{2+} optimum. Usually the manufacturer of the enzyme provides the optimal reaction conditions.
5. Because the restriction enzymes can be denatured by adverse conditions, they should be handled carefully. Detergents, high temperature, repeated freezing and thawing, acid or alkaline pH, oxygen, vortexing or violent shaking would denature and inactivate the enzymes.
6. Restriction enzymes are stored in –20°C deep freezer. They should be removed from the freezer on an ice just before use and the unused portion be immediately returned to the freezer. While using, the enzymes should be always kept on ice.
7. The enzyme should be pipetted by placing the end of the tip at the edge of the enzyme stock solution. Dipping pipet tips down into the enzyme solutions leads to inaccurate pipetting because the enzyme solution coats the outside of the pipette tips.
8. Stop solutions are used to stop the enzyme reaction. It usually contains urea and EDTA. Urea is added to prevent reannealing of restriction fragments. Heating to 65°C also prevents reannealing if followed by immediate cooling on ice. EDTA is added to complex the Mg^{2+} ions that are required for endonucleases activity, thus preventing further digestion.
9. With circular DNA the number of fragments generated is equal to number of cleavage sites. In case of linear DNA, the number of cleavage sites is one less than the number of fragments produced.
10. Never make up restriction digests with the restriction enzyme composing more than 1/10 of the final volume. This is due to the fact that the restriction enzyme proteins are stored in glycerol. At concentrations above 10%, glycerol not only inhibits the digestion but also can cause star activity—leading to aberrant, non–specific cuts of the DNA.
11. If you cannot obtain complete restriction digestion of your DNA after adding extra enzyme, set up a new digest and add spermidine to a final concentration of 2 mM.
12. Set up a control digest when using restriction enzymes. This helps you know if the enzyme is working properly and you set the reaction up well. Use a plasmid DNA which generates known fragmentation patterns.

❑❑❑

Part **3**

RESTRICTION MAPPING OF LAMBDA DNA WITH THREE ENZYMES VIZ ECO RI, BAM HI AND HIND III

Contents

- Introduction
- Principle
- Reagents/Materials
- Experimental lay–out
- Method
- Remember

INTRODUCTION

Restriction mapping is a technique that can yield information about the cloing site as well as about the position of restriction sites for particular restriction enzymes on a given DNA molecule. To construct a restriction map, a series of restriction digestions must be performed. Firstly, the number and sizes of the fragments produced by each restriction endonuclease must be determined by gel electrophoresis, followed by comparison with size markers. This information must then be supplemented by a series of double digestions, in which the DNA is cut by two restriction endonucleases at once. It may be possible to perform a double digestion in one step, if both enzymes have similar requirements for pH, Mg^{2+} concentrations, etc. Alternatively, if the requirements are different, the two digestions may have to be carried over one after the other, adjusting the reaction mixture after the first digestion to provide a different set of conditions for the second enzyme. Ambiguities can usually be resolved by partial digestion, carried out under conditions that result in cleavage of only a limited number of the restriction sites on any DNA molecule. Partial digestion is achieved normally either by a short incubation period, so the enzymes does not have time to cut all the restriction sites or by incubation at a low temperature (*e.g.,* 4°C rather than 37°C), which limits the activity of the enzyme. The result of a partial digestion is a complex pattern of bands in the gel. Along with the standard fragments, produced by total digestion, additional sizes are seen. These are molecules that comprise two adjacent restriction fragments, separated by a site that has not been cleaved. Their sizes indicate which restriction fragments in the complete digest are next to one another in the uncut molecule.

In the present experiment restriction mapping of lambda DNA with three enzymes viz Hind III, Eco RI, and Bam HI will be done. The recognition sequences for Hind III, Eco RI and Bam HI are respectively as—AAGCTT, GGATTC and GAATTC.

PRINCIPLE

It is based on the principle of restriction endonucleases activity and agarose gel electrophoresis. The selected three enzymes will restrict the DNA individually as well as in combination with each other (double digest). The various numbers of fragments are obtained which are separated or agarose gel (see unit III, part 2 for principle of agarose gel electrophoresis). The fragments obtained by single digest, double digest as well as partial digest, lead to calculation of relative distance of restriction sites of different enzymes.

REAGENTS/MATERIALS

1. Lambda DNA (0.5 mg/ml in TE buffer)
2. Eco RI
3. Hind III
4. Bam HI
5. Core buffers (supplied along with enzymes)
6. Microfuge tubes
7. Pipet Tips
8. Tracking dye
9. Ethidium bromide stock (10 mg/ml)
10. Methylene blue
11. Agarose
12. 1 X TBE buffer
13. DNA Size standard ladder
14. Goggles

15. Gloves
16. Pipettors
17. Electrophoresis Chamber
18. Power Supply
19. Water bath
20. Transilluminator
21. Microfuge

EXPERIMENTAL LAY-OUT

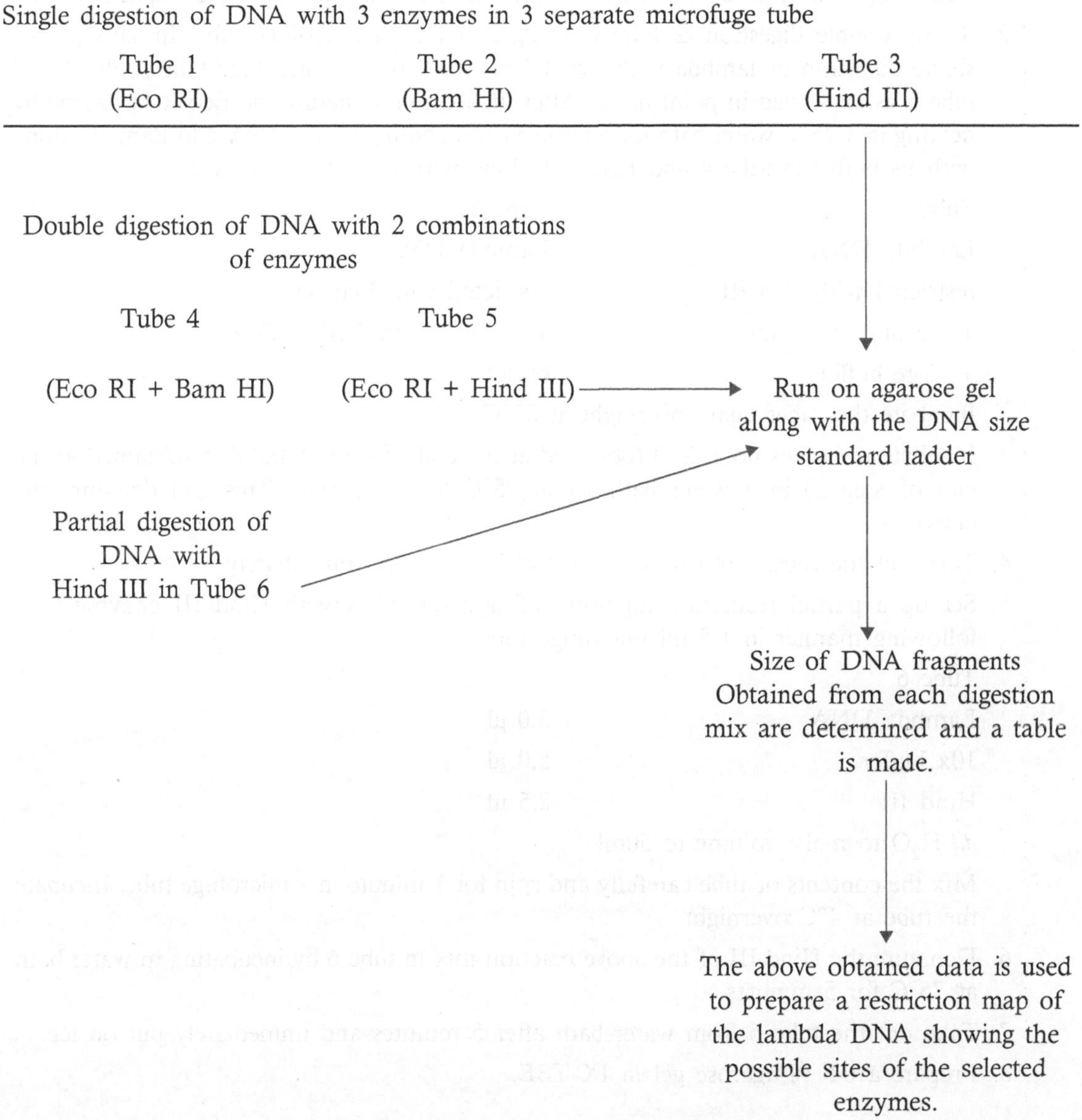

METHOD

1. Set up single restriction digestion as follows using three 1.5 ml microfuge tube:

	Tube 1	Tube 2	Tube 3
Lambda DNA	5 μl	5 μl	5 μl
10 X Core buffer	5 μl	5 μl	5 μl
Eco R–I	2.5–5 μl	0	0
Bam H I	0	2.5–5 μl	0
Hind III	0	0	2.5–5 μl
dd H_2O to make volume to	50 μl	50 μl	50 μl

   Centrifuge briefly in a microfuge to mix and incubate the tubes over night at 37°C.

2. Set up double digestion as follows using two 1.5 ml microfuge tube. Initially go for single digestion of lambda with Eco RI in two separate microfuge tubes, tube 4 and tube 5 as explained in point no. 1. After incubation, denature the Eco R I enzyme by heating in a 75°C water bath for 5 minutes and cooling on ice. Now add Bam HI along with its buffer in tube 4 and Hind III along with its buffer in tube 5.

Tube 4	Tube 5
Lambda DNA	Lambda DNA
restricted with Eco RI	restricted with Eco RI
+ 2.5 μl of Bam HI	+ 2.5 μl of Hind III + Core
+ Core buffer	buffer

   Incubate the tubes again overnight at 37°C.

3. Incubate the tubes no 1, 2, 3 (obtained at the end of step 1) and 4, 5 (obtained at the end of step 2) in a water bath set at 75°C for 5 minutes. This will denature the enzymes.
4. Take out the tubes from water bath and keep on ice immediately.
5. Set up a partial restriction digestion of lambda DNA with Hind III enzyme in a following manner in 1.5 ml microfuge tube.

Tube 6	
Lambda DNA	5.0 μl
10x buffer	5.0 μl
Hind III	2.5 μl

   dd H_2O to make volume to 50ml

   Mix the contents of tube carefully and spin for 1 minute in a microfuge tube. Incubate the tube at 4°C overnight.

6. Denature the Hind III of the above reaction mix in tube 6 by incubating in water bath at 75°C for 5 minutes.
7. Take out the tube 6 from water bath after 5 minutes and immediately put on ice.
8. Prepare a 0.8 % agarose gel in IX TBE.

9. In the gel, load the samples in following manner.

Lane 1	Lane 2	Lane 3	Lane 4	Lane 5	Lane 6	Lane 7
DNA Size standard ladder	tube 1	tube 2	tube 3	tube 4	tube 5	tube 6

10. Note down the size of fragments in each lane and prepare a table.

How to make the map out of this data:

For making the method more clear let us assume that above method has been done carefully and results are as follows:

Enzyme	*Number of Fragments*	*Sizes (Kb)*
Eco RI (tube 1)	2	24.0, 24.5
Bam HI (tube 2)	2	15.0, 33.5
Hind III (tube 3)	3	1.5, 17.0, 30.0
Eco RI + Bam HI (tube 4)	3	9.0, 15.0, 24.5
Eco RI + Hind III (tube 5)	4	1.5, 6.0, 17.0, 24.0
Partial digestion with Hind III (tube 6)	6	30.0, 31.0, 48.5

How to Proceed

1. The intact linear DNA molecule is of length 48.5 Kb.
2. The number of restriction sites for two enzymes *i.e.,* Eco RI and Bam HI is one because two fragments are obtained in single digestion.

   Eco RI fragments 24.0, 24.5.

   Bam HI fragments 15.0, 33.5.

   For knowing the relative position, see the double digestion products of Eco RI + Bam HI.

   Eco RI + Bam HI products 9.0, 15.0, 24.5

   Just by looking at the above products it becomes clear that 24.5 Kb fragment of single digestion of Eco RI is still intact in double digestion. So it is clear that Bam HI site lies in the 24.0 Kb fragment of Eco RI. The only possible arrangement is below:

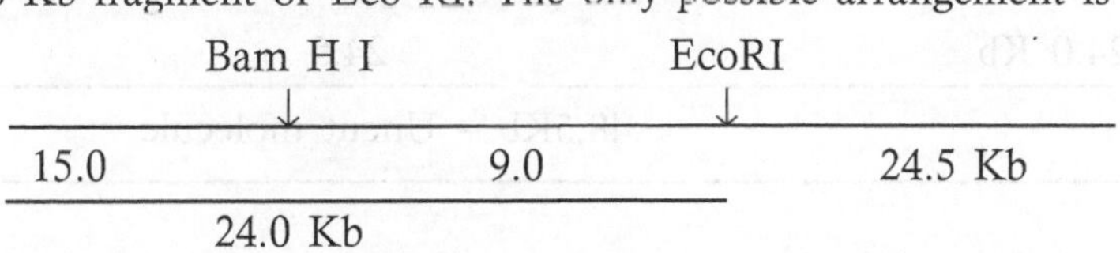

3. There are two restriction sites for Hind III as three fragment are obtained

   Hind III products 1.5, 17.0, 30.0

   The double digestion of DNA by Eco RI + Hind III are: 1.5, 6.0, 17.0, and 24.0.

   It is clear that 24.0 Kb fragment of Eco RI–single digestion is still intact here, it means the two sites of Hind III are lying in the 24.5 Kb fragment (1.5 + 6.0 + 17.0 = 24.5 Kb).

4. Ambiguities are still there about the relative distance between two Hind III sties and one Eco RI site. For this we have to go for partial digestion products of Hind III which are as:

   1.5, 17.0, 18.5, 30.0, 31.0, 48.5

   These products can be explained as:

   48.5 Kb fragment – Uncut lambda DNA

   1.5, 17.0, 30.0 – Products of complete digestion of lambda with Hind III.

   18.5 and 31.5 – Products where one restriction site is skipped. This is explained in following manner:

   - 18.5 kb fragment can be obtained if 1.5 and 17.0 kb fragments of complete digestion lie side by side and restriction site separating them is skipped somehow in partial digestion.
   - 31.5 kb fragment can be obtained if 1.5 and 30.0 fragment (of complete digestion) lie side by side and restriction site separating them is skipped somehow in partial digestion.

   Above two points lead to following arrangement:

	Hind III ↓		Hind III ↓	
30		1.5		13.0

5. Now come back to double digestion of Eco RI and Hind III to know the relative distance between their sites. The products are – 1.5, 17.0, 6.0, and 24.0.

   So by comparing the Hind III relative sites obtained at the end of point 4 with this, it can be concluded that the 30 Kb fragment of Hind III is having one Eco RI site resulting in two fragments 24.0 and 6 Kb. So the relative position of Hind III and Eco RI can be mapped as:

	Eco RI ↓		Hind III ↓		Hind III ↓	
24.0		6.0		1.5		17.0

6. Joining together the maps obtained at the end of point 2 and point 5, the complete map can be obtained which is:

	Bam HI ↓		Eco RI ↓		Hind III ↓		Hind III ↓	
15		9		6		1.5		17

   24.0 Kb — 24.5

   48.5Kb – Uncut molecule

REMEMBER

1. While going for double digestion, remembers to take care of the fact that their buffer requirement may be different. So in that case go initially for single digestion and add the second enzyme in it after first digestion is complete as done above.
2. Always take care of all the points mentioned under 'Remember' heading of Part 2, Unit VI.

❑❑❑

Unit VII LIGATION

Part 1 A General Account

Part 2 Ligation of a DNA Fragment, Carrying Gene of Interest, to a Plasmid for Constructing a Recombinant DNA Molecule

Part 1 A GENERAL ACCOUNT

Contents

INTRODUCTION

Ligation means joining DNA molecules together and this process is carried out by enzyme DNA ligase. Ligases catalyze the formation of phosphodiester bonds between the directly adjacent 3'-hydroxyl and 5' – phosphoryl termini of nucleic acid molecules. The substrates may be DNA or RNA, and the cofactors that generate high energy intermediates in the reaction may be ATP or NA^{D+}, depending on the type of ligase (Fig. 1.1). Under in vivo Conditions, DNA ligase is required for enzymatic completion of lagging strand synthesis during replication of DNA, and also involved in genetic recombination and DNA repair. In vitro, DNA ligases are used chiefly to create novel combinations of nucleic acid molecules and to attach them to vectors in molecular cloning.

TYPES OF DNA LIGASE

DNA ligases used in molecular cloning are either of bacterial origin or bacteriophage encoded. All eubacteria, whether thermophilic or mesophilic, contain a single ligase gene that encodes an NAD^+ dependent enzyme. During the first step of a ligation reaction, the diphosphate linkage

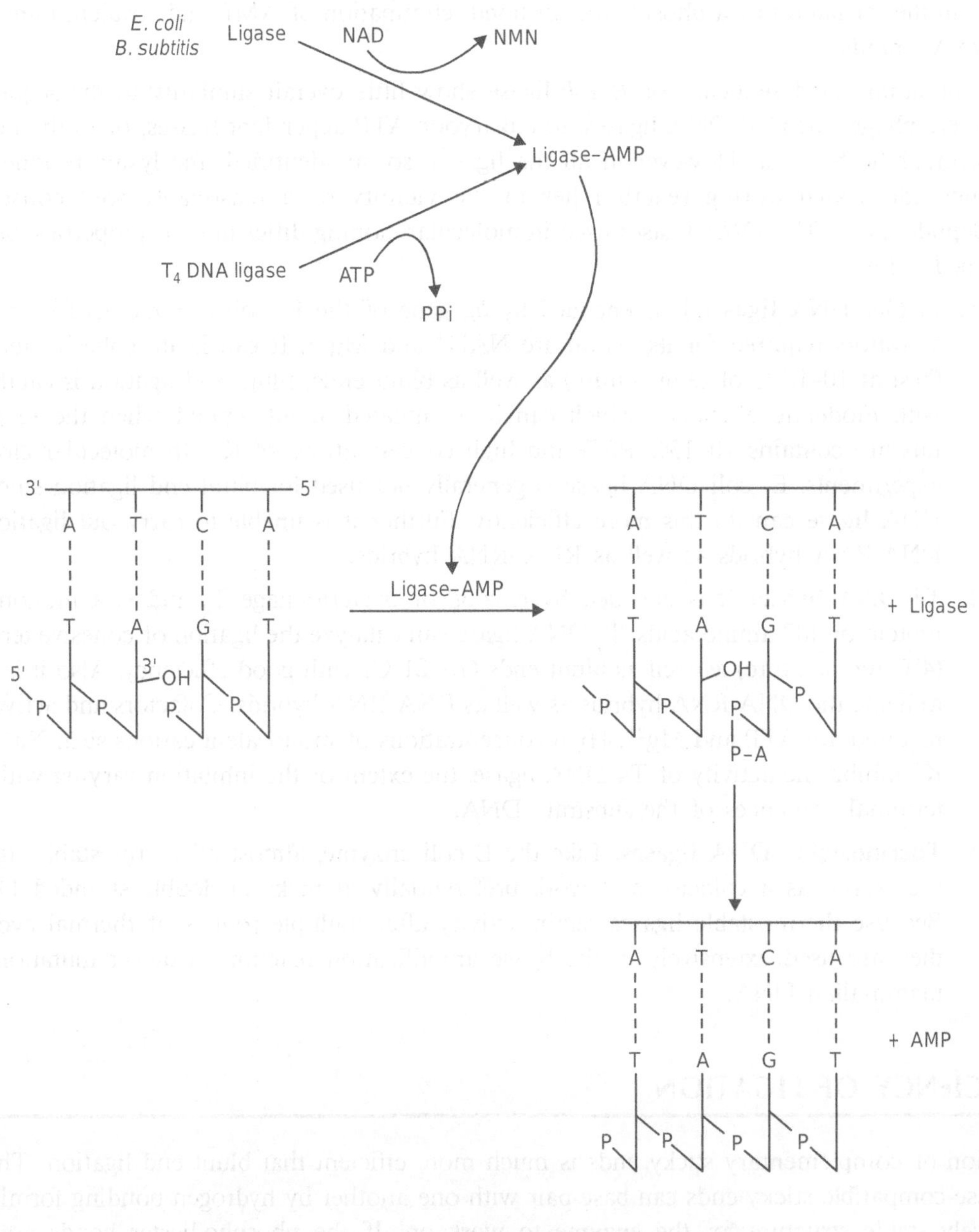

FIGURE 1.1 **Mode of action of DNA ligase enzyme.**

of NAD$^+$ is used as a phosphoanhydride and the adenyl group is transferred to the lysine residue of the enzyme. In case of ATP dependent ligases of eukaryotes and bacteriophages, the high energy intermediate is formed more conventionally, by hydrolysis of the α/β pyrophosphate in ATP and transfer of adenyl group to a lysine residue of the enzyme. From then on, the mechanism of the reaction catalyzed by the two types of enzymes is similar. The adenyl residue is transferred to the 5'-phosphate at the terminus of one DNA molecule, which is then open to nucleophilic attack by a hydroxyl group at the 3' terminus of an immediately adjacent DNA molecule. This results in the formation of a phosphodiester bond, elimination of AMP and covalent joining of the DNA strands.

The amino acid sequence of *E. coli* ligase show little overall similarity to the sequences of bacteriophage–encoded DNA ligases, to eukaryotic ATP-dependent ligases, or to the ligases of thermophilic bacteria. However in all the ligases, so far identified, the lysine residue that becomes adenylated during reaction lies in the vicinity of a reasonably well–conserved hexapeptide motif. The DNA ligases used in molecular cloning differ in their properties, briefly described below:

1. ***E. Coli* DNA ligase:** It is encoded by *lig* gene of the E. coli genome. Cofactors and activators required for its action are NAD^+ and Mg^{+2}. It can ligate cohesive termini (best at 10–15°C of temperature) as well as blunt ends. Blunt end ligation is catalyzed with moderate efficiency which can be stimulated about tenfold when the reaction mixture contains 10–15% PEG and high concentrations of K^+. In molecular cloning experiments, E. coli DNA ligase is generally not used for blunt end ligation since T_4 DNA ligase can do this more efficiently. Further it is unable to carry out ligation of DNA-RNA hybrids as well as RNA–RNA hybrids.
2. **T4 DNA ligase:** It is encoded by *gene 30* of bacteriophage T_4 and is a monomeric protein of 487 amino acids. T_4 DNA ligase can catalyze the ligation of cohesive termini (4°C temperature) as well as blunt ends (15–21°C) with good efficiency. Also it is able to ligate the DNA-RNA hybrids as well as RNA-RNA hybrids. Cofactors and activators required are ATP and Mg^{2+}. High concentrations of monovalent cations such Na^+ and K^+ inhibit the activity of T4 DNA ligase, the extent of the inhibition varying with the terminal sequences of the substrate DNA.
3. **Thermostable DNA ligases:** Like the E coli enzyme, almost all thermostable ligases use NAD^+ as a cofactor and work preferentially at nicks in double stranded DNA. Because thermostable ligases retain activity after multiple rounds of thermal cycling, they are used extensively in the ligase amplification reaction to detect mutations in mammalian DNA.

EFFICIENCY OF LIGATION

Ligation of complementary sticky ends is much more efficient that blunt end ligation. This is because compatible sticky ends can base-pair with one another by hydrogen bonding forming a relatively stable structure for the enzyme to work on. If the phosphodiester bonds are not synthesized fairly quickly, then the sticky ends will fall apart again. These transient, base-paired structures do, however, increase the efficiency of ligation by increasing the length of time the ends are in contact with one another. On the other hand, in blunt end ligation the ligase is unable to catch hold of the molecule to be ligated, and has to wait for chance associations to bring the ends together. If possible, blunt end ligation should be performed at high DNA concentrations. Further efficiency can be increased if blunt ends are somehow converted into sticky ends. This can be done by many of the methods explained below:

(*i*) **Using restriction endonucleases:** Sticky ends can be produced by digesting both the vector and the DNA to be cloned with the same restriction endonuclease, or with different enzymes that produce the same sticky ends. However, it is not always possible to do this. A common situation is where the vector molecule has sticky ends, but the DNA fragments to be cloned are blunt ended. Under these circumstance the coming three strategies/methods can be used.

(*ii*) **Linkers:** Linkers are short pieces of double stranded DNA of known sequence synthesized in the test tube. It is blunt ended but contains a single restriction site of a particular restriction endonuclease (Fig. 1.2) Firstly the blunt ended linkers are attached to blunt ended DNA molecules by blunt end ligation catalyzed by T_4 DNA ligase. The efficiency of this process is increased by adding the linkers into the ligation mixture at a high concentration. After that blunt ended DNA molecule attached with blunt ended linkers is subjected to restriction endonuclease activity (of which linker is having a single site). This results in sticky ends of the linker DNA molecule which could be used further in cloning experiments.

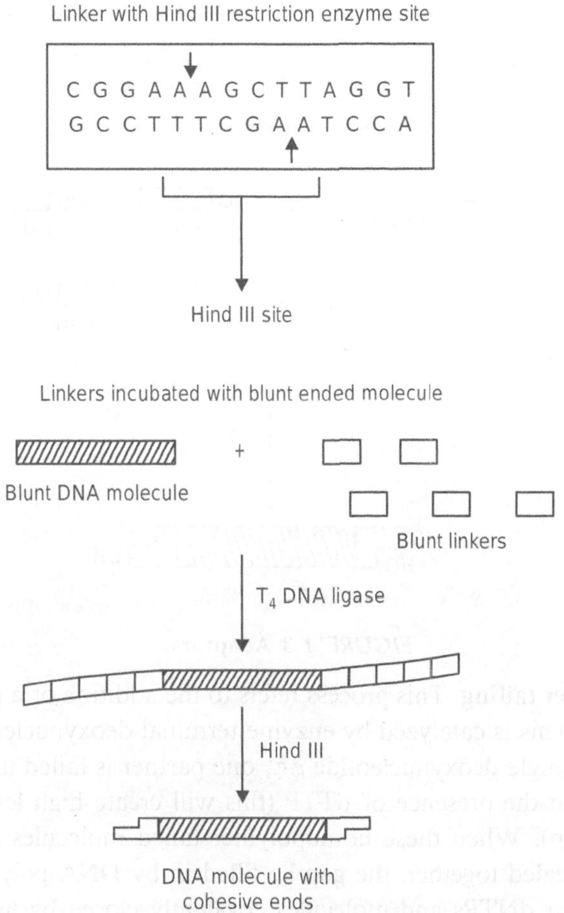

FIGURE 1.2 **Linkers.**

(*iii*) **Adaptors:** The major drawback of linkers is that what if the target DNA is also having the restriction site of the same enzyme having site on the linker. If this was the case then the restriction step needed to cleave the linkers for producing the sticky ends would also cleave blunt ended molecule (gene of interest). Adaptors give a solution to this problem. Adaptors are short synthetic oligonucleotide having one end blunt and one of and sticky. Adaptor molecules are synthesized in such a way that the blunt end is the same as natural DNA, but the sticky end is different. The 3' OH terminus of the sticky end is same as usual but the 5'-P terminal is modified (Fig. 1.3). It lacks the phosphate group and is in fact a 5'-OH terminus. DNA ligase is unable to form a phosphodiester bridge between 5'-OH and 3'-OH ends. This avoids the formation of a adaptor dimer molecule and adaptors can therefore be ligated to a DNA molecule but not to themselves. After the adaptors have been attached the abnormal 5'-OH terminus is converted to the natural 5'-P form by treatment with enzyme polynucleotide kinase producing a sticky end fragment that can be inserted into an appropriate vector.

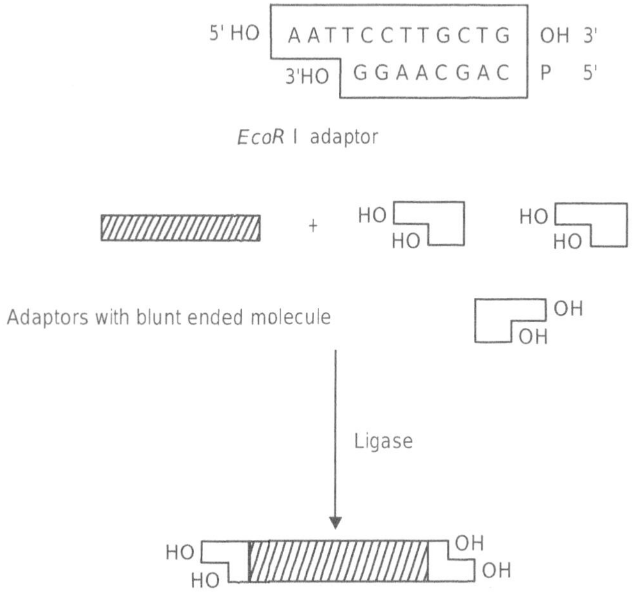

FIGURE 1.3 **Adaptors.**

(*iv*) **Homopolymer tailing**. This process refers to the addition of a tail of same nucleotides to the ends. This is catalyzed by enzyme terminal deoxynucleotidyl transferase in the presence of single deoxynucleotide *e.g.,* one partner is tailed in the presence of dATP while other in the presence of dTTP (this will create high level of complementarity between them). When these homopolymer tailed molecules are incubated together, they get annealed together, the gap be filled in by DNA polymerase in the presence of all the four dNTPs and molecules covalently closed by ligase (Fig. 1.4)

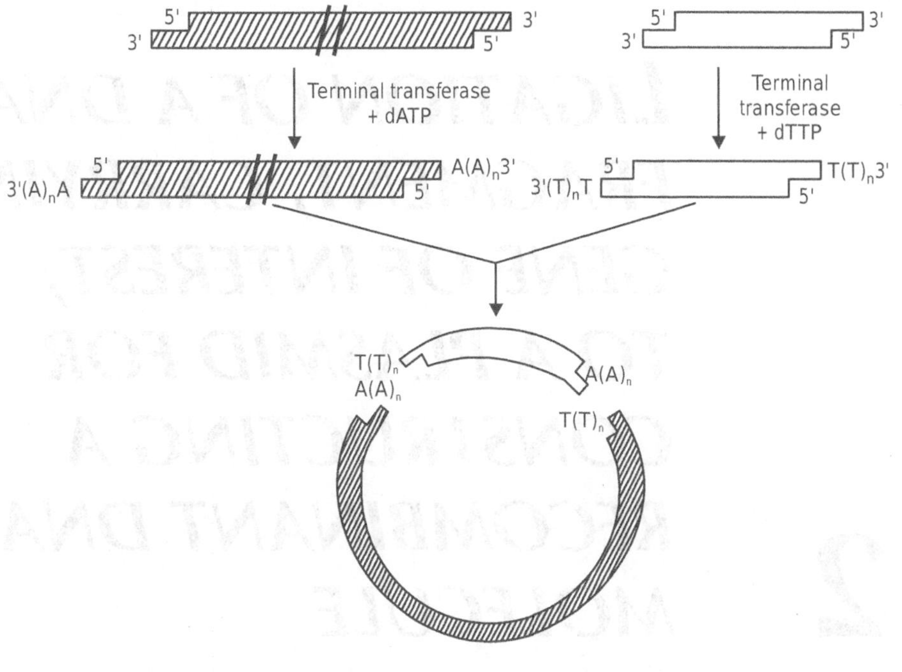

FIGURE 1.4 **Homopolymer tailing.**

❑❑❑

Part 2 LIGATION OF A DNA FRAGMENT, CARRYING GENE OF INTEREST, TO A PLASMID FOR CONSTRUCTING A RECOMBINANT DNA MOLECULE

Contents

INTRODUCTION

Most plasmid vectors in common use contain multiple cloning sites (MCS) that have recognition sequences for many different restriction enzymes. Given the large variety of multiple cloning sites currently available, it is almost always possible to find a vector carrying restriction sites that are compatible with the terminal of a particular fragment of foreign DNA.

The method given below explains in simple terms how to carry out ligation of such vector and foreign DNA having compatible ends.

PRINCIPLE

Both insert and vector DNA are restricted with appropriate restriction enzyme to generate compatible ends for cloning. The vector DNA is treated with calf intestinal alkaline phosphatase (CIAP) to remove 5' phosphate groups to prevent recircularization of the vector during ligation. The insert DNA can be purified after restriction by electrophoresis on low melting agarose gel and then eluted from the gel. The vector and insert DNA are joined with the help of DNA ligase to form recombinant vector.

REAGENTS/MATERIALS

1. 0.5 M EDTA
2. 1% agarose in 1 X TBE
3. TE saturated phenol-chloroform
4. Chloroform: isoamyl alcohol
5. 7.5 M ammonium acetate
6. Phenol: Chloroform: isoamyl alcohol (25:24:1)
7. Sodium acetate 3M, pH 5.2
8. Ethanol 100% and 70%
9. T_4 DNA ligase 10 X buffer

Tris-Cl, pH 7.8	300 mM
$MgCl_2$	100 mM
DTT	100 mM
ATP	100 mM

10. Calf intestinal alkaline phosphatase (CIAP) 10 X buffer:

Tris-Cl, pH 9.0	500 mM
$MgCl_2$	10 mM
$ZnCl_2$	1 mM

EXPERIMENTAL LAY-OUT

Restriction of plasmid DNA (vector) with appropriate enzyme

↓

Incubation with CIAP in buffer initially at 37° and then at 55°C

↓

Stop the above reaction

↓

Extract the vector DNA with phenol:
Chloroform: isoamyl alcohol

↓

Precipitate the DNA

↓

Pellet the Dephosphorylated vector DNA

↓

Insert DNA → Ligation (Incubating the two with T_4 DNA ligase)

↓

Recombinant vector

METHOD

(*a*) Restriction of Vector DNA for Cloning

1. Take 5 mg of plasmid DNA. Restrict it with 5–15 units of appropriate restriction enzyme in 5 ml of 10 X restriction buffer (for that particular restriction enzyme), adding deionized water to final volume of 50 ml.
2. Iucubate at appropriate temperature for 1–3 h. Run an aliquot on a 1% agarose gel to check completion of restriction.

(*b*) Dephosphorylation of Vector DNA

The vector cut as above with a single restriction enzyme, should be treated with calf intestinal phosphatase (CIAP) as described below.

1. To the above restriction mix add the following directly:

CIAP 10 X buffer	–	10 μl
CIAP	–	0.01 unit/pmol ends
Nuclease free water	–	to final volume of 100 μl

   CIAP may be diluted in 10 X CIAP buffer immediately before use and discard unused diluted enzyme.

   e.g., 1 μg of 1 kb DNA fragment = 1.52 pmol DNA, or 3 pmol ends.
2. Iucubate at 37°C for 30 min, then at 56°C for 15 min. Add another 0.014 unit CIAP/ pmol ends and repeat incubations at both the temperatures.
3. Stop the reaction by adding 2.0 μl of 0.5 M EDTA and heating at 65°C for 20 min.
4. Extract the DNA by adding an equal volume of phenol: Chloroform: isoamyl alcohol (25:24:1), vortex for 1 min and centrifuge at 12000 X g for 5 min.
5. Carefully remove the top aqueous phase to a fresh tube.
6. Add an equal volume of chloroform: isoamyl alcohol (24:1), vortex and centrifuge at 12000 X g for 2 minutes. Remove aqueous phase to a fresh tube.
7. Precipitate the DNA by adding 0.5 volume of 7.5 M ammonium acetate and 2 volumes of 100% ethanol. Place the sample on ice or at 20°C for 30 min.
8. Pellet the DNA by centrifugation at 12000 x g for 5 min, wash with 70% ethanol, air dry and dissolve in appropriate amount of TE buffer.

(*c*) Ligation of Plasmid and Insert DNA

1. Set up the following reaction. Generally 50–200 ng vector DNA is used in a typical ligation mixture.

Vector DNA	–	100 ng
Insert DNA	–	50 ng
10 X Ligase buffer	–	1 μl
T_4 DNA	–	1 μl
Nuclease free water to final volume	–	10 μl

2. Iucubate the reaction mixture either at 22°C for 4 to 6 hr (may be overnight) for blunt ends or a at 22°C for 3 h or at 4°c for 16 h (overnight) for protruding ends.
3. Confirm the ligated DNA by running over the gel.

REMEMBER

1. Dephosphorylation of the vector is recommended only when:
   (*a*) The DNA insert to be cloned is only available in small amounts
   (*b*) When the transformants are to be screened by restriction enzyme digestion of minipreparations of plasmid DNA.
   (*c*) When cloning blunt ended fragments of DNA.
2. To obtain the maximum number of correct ligation products in cloning blunt end target fragments, the two components of DNA in the ligation reaction must be present at an appropriate ratio. If the molar ratio of plasmid vector to target DNA is high, then the ligation reaction may generate an undesirable number of circular empty plasmids, both monomeric and polymeric; if too low the ligation reaction may generate an excess of linear and circular homo and heteropolymers of varying sizes, orientations, and compositions.
3. For this reason, the orientation of the foreign DNA and the number of inserts in each recombinant clone must always be validated by restriction endonuclease mapping or some other means. As a general rule, acceptable yields of monomeric circular recombinants can be obtained from ligation reactions containing equimolar amounts of plasmid and target DNA, with the total DNA concentration < 100 μg/ml.

❑❑❑

Unit VIII TRANSFORMATION

Part 1

TRANSFORMATION—A GENERAL ACCOUNT

Contents

INTRODUCTION

The term transformation was first of all given to the process of uptake of exogenous DNA by the bacterial cells. This exogenous DNA in the surrounding media could be from any source (may be dead, lysed bacteria or may be eukaryotic source with a minimum length of about 500bp.). Many bacterial species are capable of such uptake naturally *e.g., Bacillus subtilis, Escherichia coli, Haemophilus influenzae, Pseudomonas aeroginosa* etc.

With the discovery of DNA manipulating tods and techniques (using restriction enzymes, vectors, ligase enzyme, means of gene isolation etc.) new challenges emerged in the form of transferring desired genetic information to desired organism (bacterial/plant/animal) directly under in vitro conditions. The process was termed transformation (encompasses transfer of genetic information to any organism (prokaryotic/eukaryotic) by artificial means) and currently this is one of the most popular technique of recombinant DNA technology used across the laboratories. This simple appearing idea does not seem to be that simple when practically performed in laboratory. A lot number of problems are faced like:

(1) The DNA may not be able to physically cross the cell membrane and cell wall (if present) of the host cell or it can be said that host cell may not be receptive for foreign DNA.

(2) Once inside the new host cell, again two basic problems arises:

(*a*) Detection of uptake which is dependent on gene expression, failure could be due to lack of accurate transcription or translation.

(*b*) The exogenous DNA may not be maintained in the transformed cells. It should get integrated in the host genome for propagation through generations.

TECHNICAL PLAN

Technical plan should be based on the approaches to overcome the above cited problems:

1. Regarding first problem, methods should be used which make the host cell competent. There is an enormous range of techniques available but only a limited number of methods will be suitable for any one organism. Some of the important methods are:

(*i*) Making the cells competent by treatments like electric shocks, calcium treatment etc.

(*ii*) Using protoplasts, instead of intact cells, of walled organisms like fungi, higher plants and some bacteria. Protoplasts prepared in this way are suitable for introducing exogenous DNA into the cytoplasm after certain simple treatments like use of polyethylene glycol and electroporation. There are two fundamental problems with using protoplasts like fragility of the protoplasts and problems encountered in the regeneration of whole cells after the exogenous DNA has been taken up.

(*iii*) Microprojectile bombardment of intact cells by metal particles coated with DNA. This method does not require complex cell cultures or pretreatments and avoids the problems of regenerating protoplasts.

(*iv*) Infection of cells with viruses resulting in transfer of genes from the virus to the host cell.

(*v*) Microinjection of DNA directly into a cell nucleus. This is a standard method for introducing genes into animal cell lines.

The above mentioned techniques are not all equally applicable. Some techniques are good for whole cells while others for protoplasts/naked cells. Some methods requires special equipment (*e.g.,* in electroporation, microprojectile bombardment and microinjection), while some require special pre-preparation. With some of the methods, we can use a large number of cells (for example when using naturally competent cells) whilst in others the technique is labour intensive and can only be carried out with a limited number of cells (*e.g.,* microinjection). To select one particular technique following points should be considered:

(*a*) Time required for preparation

(*b*) Any special requirement (equipment or reagent)

(*c*) Number of cells needed to be transformed

(*d*) After DNA transfer the percentage of recovery of viable cells

(*e*) The regeneration potential of viable cells.

The choice of technique is also guided by literature, some previous experience, the availability of equipment and also upon the informed guesswork.

Term **'transformation efficiency'** is commonly used during such comparison which is usually the number of transformants per mg of added DNA or the number of transformants per viable cells. One common problem is the regeneration to the complete organism, so in order to have a realistic idea of true transformation efficiency per viable cell, the regeneration frequency should also be considered.

2. The second problem in transformation is regarding the detection of the uptake of foreign DNA by host cell. Technical plan for transformation includes sufficient means which make this detection easy. There are various methods available.
   (*a*) Direct selection for the desired gene *e.g.,* If the gene transferred is providing resistance to some antibiotic, addition of that antibiotic in the medium will allow only the transformed cells to grow.
   (*b*) Reporter genes. If the gene transferred is not coding for any trait which may be utilized in direct selection method, alternatively reporter or marker genes may also be used along with the desired gene for transformation. A marker gene or reporter gene produces a phenotype, which permits either an easy selection or quick identification of the cells in which it is present. Thus marker genes are either selectable or scorable. A selectable marker governs a feature, which enables only such cells that possess it to survive under the selective conditions (*e.g.,* antibiotic resistance marker gene). On the other hand, scorable markers produce distinct phenotype, which allow an easy identification of the cells having them from those that do not contain them *e.g., gus* gene, *lux* gene, *mos* gene etc.
3. The last but very important target of a successful technical plan of transformation is to ensure the survival of DNA in new host cell.

   To achieve this DNA of interest has to be inserted into a suitable cloning vector, so that multiple copies of the DNA of interest are produced. Such multiple copies are termed as clones and the DNA of interest can be described as being amplified (cloned). A cloning vector is a DNA molecule that will accept an insert of another DNA molecule thus allowing:
   - Replication of the insert *i.e.,* by attachment to a suitable replica (a DNA molecule with its own origin of replication).
   - Amplification of the insert to generate multiple copies.
   - Selection of the transformed cells within a population of potential host cells.

   Now a day a variety of cloning vectors are available *e.g.,* plasmids, bacteriophages, cosmids, YAC, BAC, MAC and some specialized vectors. These cloning vectors are highly modified derivatives of natural vectors. They are produced by a variety of techniques including restriction enzyme technology and DNA ligation. These modifications make than more amenable for use in the laboratory. For example, easily selected marker genes have been inserted and many carry restriction enzyme sites placed in strategic positions within their DNA sequence. This latter feature enables us to cut these vectors in predictable positions and to insert the desired piece of DNA there. (For detailed study of vectors see 'Molecular cloning: A laboratory manual' by

Sambrook and Russell and Gene cloning by TA brown). When going for transformation experiment, a choice has to be made among the available vectors. Following factors play a role in such a choice:

(*i*) Nature of host cell

(*ii*) Detection/screening system applicable

(*iii*) The screening procedure envisaged to screen for inserts of interest

(*iv*) Amount of genetic material we wish to transfer or clone

(*v*) Required efficiency

(*vi*) Personal preference and expertise of the worker

(*vii*) Resources available.

Some commonly used vectors with important properties are discussed (in brief) below:

(*i*) **Plasmids:** Plasmids are double stranded covalently closed DNA molecules which exist as extrachromosomal element in many bacterial cells. Plasmids are characterized by the presence of an origin of replication and some plasmids even carry some selectable marker genes (*e.g.,* antibiotic resistant gene). They have a number of restriction sites which can be used for cloning purpose.

Naturally occurring plasmids need some modifications for being used as the desired cloning vector. They are modified, shortened, reconstructed and recombined both *in vitro* and *in vivo* to create plasmids of enhanced utility and also with specific functions *e.g.,* psc101, pBR322, PSVC series vectors etc. Plasmid derived vectors are advantageous in the sense that they are usually small and easy to handle with a large number of potential selectable markers for transformation. The major disadvantage is regarding the capacity for foreign DNA insertion which is rather small (in practice maximum 10-15 Kb in size). Transformation efficiency is also less than that of infection.

(*ii*) **Bacteriophages**: Bacteriophages are viruses of bacteria and follow a life cycle which is dependent on their host. They enter bacteria and undergo either lytic cycle or lysogenic cycle. Bacteriophage λ (size, 48.5Kb) have an origin of replication, genes for capsid structure, genes encoding enzymes for DNA replication (lysis/ lysogenic) and a very interesting feature *i.e.,* single stranded protruding cohesive sequences of 12 bases at both the ends (with sequence 5′GGGCGGCGACCT at one while other end is complementary to it). When outside a bacterial cell, DNA remains linear but as soon as it enters a host, it circularizes by annealing of two cohesive ends. The circular form is must for replication. The sealed cohesive ends are called cos sites, essential for cleavage and packaging of mature phage DNA into phage head.

The natural bacteriophage was modified by removing the non essential region (or stuffer region) and introducing polylinkers (a DNA segment having many unique restriction enzyme sites). They are advantageous in the sense that they can take larger inserts than plasmids (maximum 23Kb) and are easy to screen for specific genes using probes against plaques. The major disadvantage is that it is often hard to transform or transfect organisms other than the natural host. Further large inserts may become scrambled by recombination, or more than one insert may be cloned into a single site.

(*iii*) **Cosmids:** Cosmids are the synthetic vectors formed by joining selected sequences of plasmid with the selective sites of bacteriophage. They are produced by mutation, recombination and various recombinant DNA techniques. Cosmids generally have (*i*) cos sites from bacteriophage λ (*ii*) Sequences needed for finding of and cleavage by terminase so that under appropriate conditions they are packaged in vitro into empty phage particles (*iii*) origin of replication (mainly of plasmid origin) (*iv*) unique restriction sites (may be polylinkers) (*v*) a selectable marker from plasmid.

Cosmids are advantageous in the sense that very large insert (up to 40–45Kb) DNA can be cloned and so reduce the number of clones that need to be screened. Further, packaging imposes a size selection for recombination and provides a highly efficient means of infecting host bacteria. However, it is often technically difficult to isolate and restrict high molecular weight DNA to provide inserts of sufficient size. Further due to the large size, inserts may after become scrambled by recombination and so extensive further analysis has to be done to ensure that right DNA sequence has been cloned.

(*v*) **Phagemids:** It contains origin of replication from a phage, in addition to that of the plasmid. This vector is a multipurpose vector and can serve as either (*i*) a cloning vector (*ii*) an expression vector (*iii*) a riboprobe vector or (*iv*) a sequencing vector.

This type of vector are generally used for specialized purpose.

(*vi*) **Artificial chromosome vectors:** Artificial chromosomes are circular or linear vectors that are stably maintained in, usually, 1 to 2 copies per cell. There are several types of such vectors:

(*i*) Bacterial artificial chromosomes (BAC)

(*ii*) Pi-designed artificial chromosome (PAC)

(*iii*) Yeast artificial chromosome (YAC)

(*iv*) Mammalian artificial chromosomes (MAC)

(*v*) Human artificial chromosomes (HAC).

They are highly specialized vectors and used for cloning of very large fragments of DNA.

It should be clear that there is often a choice of more than one vector that may be used for a particular organism.

❑❑❑

Part 2 BACTERIAL PLASMID TRANSFORMATION

Contents

INTRODUCTION

The bacterial plasmid transformation experiment can be performed in three steps: (*i*) Preparation of competent cells (*ii*) Transformation of competent cells and (*iii*) Selection of transformants.

As with many breakthroughs in recombinant DNA technology, the key development as for as transformation was concerned occurred in the early 1970s, when it was observed that E coli cells that had been soaked in an ice-cold salt solution were more efficient at DNA uptake than unsoaked cells. A solution of 100mM calcium chloride is commonly used, although other salts like rubidium chloride are also effective. Some other cations like CO^{+++}, K^{+}, Mn^{++} are also suitable for this purpose.

PRINCIPLE

Actually such cations destabilize the cell envelope by disturbing the charge distribution on surface and so facilitate the DNA uptake. Possibly these cations causes the DNA to precipitate on to the outside of the cells, or perhaps the salt is responsible for some kind of change in the cell wall that improves DNA binding. The actual movement of DNA into competent cells is stimulated by briefly raising the temperature to 42°C.

REAGENTS/MATERIALS

1. LB broth

Bacto-tryptone	–	10 g
NaCl	–	10 g
Yeast Extract	–	5 g
double distilled water	–	to 1 liter

   Adjust pH to 7.5 using NaOH and autoclave
2. LB agar plates—add 15.0 g/l agar in LB broth, sterilize, cool, add proper antibiotic (let's say ampicillin) and pour in petriplates.
3. $MgCl_2$ 0.1 M
4. $CaCl_2$ 0.1 M
5. 50% glycerol/50m M $CaCl_2$ solution
6. E. coli DH5α
7. SOC medium

   2% tryptone

   0.5% yeast extract

   10 mM NaCl

   20 mM glucose (dextrose)

   1.5 mM KCl

   10 mM $MgCl_2$

   10 mM $MgSO_4$

   $MgCl_2$, $MgSo_4$ and glucose are prepared separately and added to the broth.

EXPERIMENTAL LAY-OUT

Log Phase E Coli cells (DH5α)

↓ Centrifuge

Pellet → (*i*) Resuspend in $MgCl_2$ at 4°C; (*ii*) Incubate on ice for 5–10 min.

Supernatant → Decant off

↓ Centrifuge

Pellet → (*i*) Wash with $CaCl_2$

Supernatant → Decant off

↓ Centrifuge

Pellet
(*i*) Suspend in $CaCl_2$
(*ii*) Incubate overnight
(*iii*) Add glycerol (ice cold)

Supernatant (Decant off)

↓

Competent Cells are ready

↓ Either

Subjected to Transformation | Stored at – 80°C

↓ (*i*) Plasmid DNA (*ii*) Heat Shock (*iii*) Incubate in SOC medium

Transformed E Coli Cells

↓

Plate in different aliquots of appropriate antibiotic containing plates.

↓

Selection of transformed cells

METHOD

(*i*) Competent Cell Preparation

1. Inoculate 10 ml LB broth with a loopful of an appropriate strain of E Coli cells (*e.g.,* DH5 α) from a single colony and grow at 37°C with shaking at 200 rpm overnight.
2. Subculture 5 ml of the overnight grown culture with 500 ml of LB broth in a 2 liter flask.
3. Grow with shaking at 37°C to an OD_{600} of 0.4/0.5 as measured by a spectrophotometer.
4. Chill the cells on ice, pour the culture into 250 ml plastic centrifuge bottles and spin at 2500 xg for 5 min, decant the supernatant.
5. Resuspend the pellets in 100 ml of cold 0.1 M $MgCl_2$ (1/5th of the original culture volume) and transfer the cell suspension to two 50 ml oakridge tubes. The cells must be kept at 4°C.
6. Incubate the cells on ice for 5 min.
7. Centrifuge the cells at 2500 xg for 5 min at 4°C. Decant the supernatant.
8. Wash the cell pellets with ice cold 0.1 M $CaCl_2$ (do not vortex). Centrifuge at 2500 x g for 5 minutes at 4°C and decant supernatant.
9. Suspend each pellet in 7 ml of ice cold 0.1 M $CaCl_2$.
10. Incubate the cells on ice overnight.
11. Add 3 ml of ice-cold 50% glycerol/50 mM. $CaCl_2$ to each tube and mix gently.
12. Aliquot 0.5 ml of cells into pre chilled tubes and quick freeze in liquid nitrogen and store the freezer cells at – 80°C.

(*ii*) Transformation

1. Add approximately 10 ng of DNA (in 10 to 25 μl volume) (ligation mix is diluted appropriately) into a 15 ml sterile round bottom test tube and place on ice. Rapidly thaw competent cells by warming between hands, and dispense 100 ml immediately into test tubes containing plasmid DNA, Swirl and place on ice for 10–30 min.
2. Heat shock cells by placing tubes for 2 min in 42°C water bath. Add 1 ml SOC medium to each tube Incubate for 1 h at 37°C at 250 rpm.

(*iii*) Selection of Recombinants

1. Plate transformed cells on LB plates containing appropriate antibiotics, 0.1 M IPTG and 40 mg/ml X-gal.
2. Incubate overnight at 37°C.
3. Check the colour of colonies. Recombinant will give white colonies while non-recombinants form blue colonics.

REMEMBER

1. Use protective gloves when handling hot agar and/or dry ice.
2. At the completion of the experiment, materials exposed to bacteria should be collected in metal trays or other containers suitable for autoclaving during the laboratory periods. All materials should be sterilized for 20 minutes before disposal; alternatively, the plates and all materials exposed to the cells, can be soaked in 10% bleach overnight and then discarded.

Part 3

AGROBACTERIUM—MEDIATED TRANSFORMATION IN PLANTS (e.g., NICOTIANA TABACUM)

Contents

INTRODUCTION

Agrobacterium tumefaciens is the naturally occurring soil bacteria which causes the crown gall disease to the plant. A closely related species Agrobacterium rhizogenes causes hairy root disease. In both the cases, the disease is caused due to the transfer of a DNA segment from the bacterium to the plant nuclear genome. The DNA segment (T-DNA) is the part of a large Ti plasmid found in virulent strains of Agrobacterium. Ti plasmid has following regions (Fig 3.1).

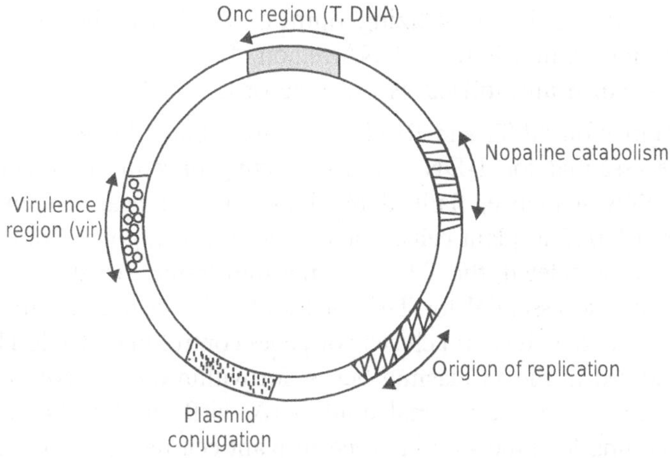

***FIGURE 3.1* A typical Ti plasmid**

(*i*) T-DNA region which is responsible for tumor induction so that mutations in this region lead to the production of tumors with altered morphology (shooty or rooty mutant galls).

(*ii*) A region responsible for replication.

(*iii*) A region for conjugation.

(*iv*) A region for virulence.

The T-DNA further consists of the following regions (Fig. 3.2):

(*i*) An 'onc' region consisting of three genes (two genes tms_1 and tms_2 representing shooty locus and one gene tmr representing rooty locus) responsible for the biosynthesis of two phytohormones, namely IAA (auxin) and isopentyladenosine-5' monophosphate (cytokinin).These genes encode enzymes responsible for the synthesis of these phytohormones, so that the incorporation of these genes in plant nuclear genome leads to the synthesis of these hormones in the host plant. The phytohormones in turn alter the developmental programme leading to the formation of crown gall.

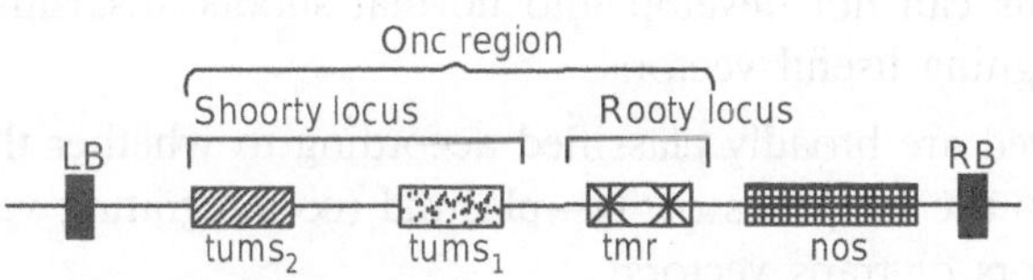

***FIGURE 3.2* The 'T-DNA' structure.**

(*ii*) An 'os' region responsible for synthesis of unusual amino acids or sugar derivatives which are collectively called opines. Opines are amino acid (arginine and pyruvate) derivatives which diffuse from the tumor into the surrounding soil where they may serve as a sole carbon source for Agrobacteria harboring a Ti plasmid. They also induce the tra operons which allows conjugal transfer of the Ti plasmid to other Agrobacterium. Two most common opines are octopine and nopaline. For the synthesis of octopine and nopaline, the corresponding enzymes octopine synthase and nopaline synthase are coded by the T-DNA. Many organisms including higher plants are incapable of utilizing opines, which can be effectively utilized by Agrobacterium. Out side the T-DNA region, Ti plasmid carries genes that catabolise the opines which are utilized as a source of C and N_2.

The T-DNA regions on all Ti and Ri plasmids are flanked by almost 25 bp direct repeat sequences which are essential for T-DNA transfer, acting only in cis orientation. It was also observed that any DNA sequence, flanked by these 25 bp repeat sequences in the correct orientation can be transferred to plant cells. The transfer is polar, going from right border to left border. If the left border is deleted, the T-DNA is not transferred. Besides 25 bp flanking border sequence, vir region, is also essential for T-DNA transfer. Vir region is capable of functioning even in trans orientation and is the main clusters of genes controlling infection by A. tumefaciens. The vir region is localized in the Ti plasmid but is not within the T-DNA. It consists of about 25 genes organized in seven transcriptional units: virA, virB, virC, virD, virE, virG and virF. These vir genes are coding for functions involved in plant cell recognition, bacterial attachment to plant cell, the excision, transfer and probably the integration of T-DNA into the target genome. While some vir gene regions are essential for tumorogenesis (virA, virB, virG, virD, others affect the efficiency of transfer and hence the host range (virC and vir E). The regulation of vir genes is complete. In the absence of any induction, virA and vir G are constitutively expressed at a low level. All vir genes are, however, inducible by a variety of plant phenolic compounds and by a group of plant monosaccharide that act synergistically. Vir gene induction also depends on low temperature (below 28°C) and acidic pH, and is further enhanced by a variety of other factors such as certain opines, phosphate starvation and flavanoid compounds. Regions on the bacterial chromosome chv A, chv B, att and psc A are also involved in both the attachment of the bacteria to the plant cell and chemotactic response of the bacteria towards the susceptible plant tissue.

Ti plasmid can not be used as such for transformation purpose because of following properties.

(*i*) Large size

(*ii*) Absence of unique restriction enzyme sites

(*iii*) Tumor induction property.

Since tumor cells can not develop into normal shoots, disarming of Ti plasmid was an essential step in designing useful vectors.

Vectors so derived are broadly classified according to whether the DNA to be introduced is physically linked to the vir genes of Ti—plasmid (co-integrating vectors or cis vectors) or is separate (binary vectors or trans vectors).

(*i*) **Co-integrating vectors**: Such vectors need the help by an intermediate vector for carrying out transformation. These vectors are based on a wild type Ti plasmid from which a portion of the T-DNA, coding for phytohormone autonomy of plant cells, is replaced by a novel sequence of DNA. Such disarmed Ti-plasmids are stable within Agrobacterium and they retain vir genes and all other elements necessary for transfer of sequences located between the border sequences. The foreign DNA to be inserted between the T-DNA borders is cloned on an intermediate vector that can be manipulated in E-coli and which has a region of homology with the sequences between the border repeats of the co-integrative vector. The intermediate vector carrying the DNA to be introduced is transferred from E-coli to Agrobacterium containing co-integrative vector by conjugation in the presence of helper plasmid. Once inside the Agrobacterium the plasmids lacking suitable origin of replication are lost. However, recombination occurs between the intermediate plasmid and the disarmed Ti plasmid so that the former can be rescued and retained in the bacteria. The Agrobacterium cells with recombined plasmid can be selected for by an appropriate genetic marker, such as antibiotic resistance.

(*ii*) **Binary vectors**: These are constructed based on the knowledge that T-DNA and vir genes could be separated on different plasmids within the same bacterial cell without the loss of the T-DNA transfer capabilities. The development of a binary vector strategy involves two plasmids—(*a*) a disarmed Ti plasmid containing vir genes (*b*) a separate small binary vector plasmid containing T-DNA and broad host range origin of replication so that it replicates in both E.coli and Agrobacterium. The desired foreign genes are inserted into the binary vector T-DNA between the left and right border sequences.

The Agrobacterium based method is the method of choice for many plant species because it offers several unique advantages in plant transformation:

(*i*) Simplicity of gene transfer.

(*ii*) A precise transfer and integration of DNA sequences with defined ends.

(*iii*) High transformation efficiency.

(*iv*) High probability of single copy integration with minimal rearrangement.

(*v*) High fertility of transgenic plants.

The present protocol describes the methodology for transformation of tobacco plant using Agrobacterium strain containing a binary vector. This binary vector contains a neomycin phosphotransferase (npt) gene for conferring resistant to kanamycin (selectable marker) as well as gus reporter gene.

PRINCIPLE

The principle is based on the natural mechanism of T-DNA transfer from Ti plasmid to the plant cell. In vivo, successful infection requires wounding of the plant tissue which induces a series of metabolic pathways and the subsequent accumulation of metabolites necessary for the wound healing process. Some of these metabolites represent effective Chemoattractants for agrobacteria. Amongst these metabolites, acetosyringone play on important role. At higher concentration it act as inducer of the vir gene complex by a signal transduction mechanism. This ultimately leads to excision of T-DNA and its integration into the plant nuclear genome.

Under in vitro conditions, the leaves are cut into small pieces (wounded) and incubated with agrobacterium in the presence of acetosyringone. It induces a signal transduction process in the vector DNA, which excise the gene of interest in T-DNA and transfers it to the plant cell.

REAGENTS/MATERIALS

(1) Agrobacterium strain LBA 4404 (pB1121)

(2) YEP medium

Yeast extract	10 g/l
Peptone	10 g/l
NaCl	5 g/l

Adjust pH to 7.0 with NaOH (IN)

For solid medium, add 8 g/l of agar. Dissolve all the constituents of YEP medium in water and adjust pH to 7.0. Autoclave the medium and add the antibiotic after proper cooling has taken place (kanamycin to be added as 0.5ml of stock solution). Prepare YEP solid medium in petri plates. Also maintain liquid broth without agar in flasks.

(3) Kanamycin

(4) Callusing medium (pH 5.8)

MS basal salt	4.5 g
BAP	1.0 mg
2, 4-D	0.5 mg
Agar	8.0 mg

Make up to 1000 ml with distilled water

(5) Shooting/selection medium (pH 5.8)

MS basal salts	4.5 g
Sucrose	30.0 g
B 5 vitamin	1 ml
BAP	1 mg
NAA	0.5 mg
Agar	8.0 mg
Kanamycin	50.0 mg

Make up to 1000 ml with distilled water.

(6) For B-glucuronidase (Gus) enzyme assay

Buffer I: 50 mM sodium phosphate buffer, pH 6.8 prepare

(*a*) 889.9 mg Na_2HPO_4 in 100 ml distilled water

(*b*) 780.0 mg $Na_2H_2PO_4$ in 100 ml distilled water

Mix 51 ml of b and 49 ml of a, and add 0.1 ml of triton X 100.

Buffer II (Gus solution): Dissolve 5.2 mg X – gluc (1 mM) in 0.1 ml of dimethyl formamide (DMFA) in a clean glass tube. Add 7.9 ml of buffer I (without Triton X – 100) and add 2 ml methanol.

EXPERIMENTAL LAY–OUT

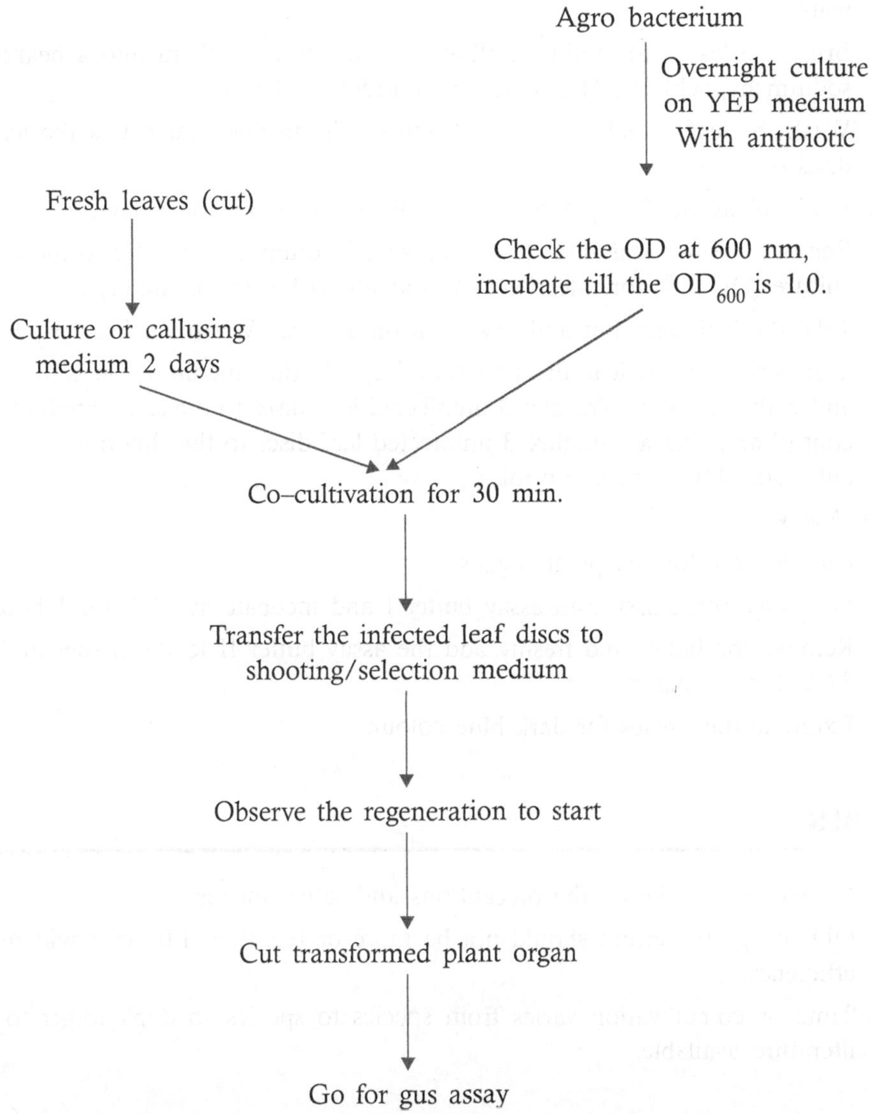

METHOD

1. Inoculate agrobacterium (LBA4404 with pB1121) from glycerol stock on petri dish containing YEP medium using a sterile inoculation needle by streaking method. Incubate at 28°C overnight.
2. Pick a single colony from the petri dish and inoculate into 20 ml of YEP liquid medium contained in 100-250 ml flask. Incubate on a rotary shaker with speed of 120 rpm at 28°C.

3. Check the OD_{600} and when OD reaches 1.0 remove the culture and pour in a petridish.
4. Collect young healthy green leaves from tobacco plants and wash with sterile distilled water.
5. Bring the leaves to laminar airflow unit and transfer them into a beaker containing sodium hypochlorite (1–2% active chloride) for 10 min.
6. Wash the leaf samples 3–4 times with sterile distilled water. Cut the leaf into small discs of 1 cm diameter.
7. Pre-incubate the leaf pieces on the callusing medium for 2 days.
8. Remove the 4–5 leaf discs from callusing medium and transfer to the agrobacterium culture (OD 1.0) for infection (co–cultivation) for 15–30 minutes.
9. Take the leaf discs out and dry them on a sterile Whatman filter paper.
10. Transfer the dried leaf discs to shooting/selection medium containing kanamycin^{50} and carbenicillin^{300}. Transfer 3 uninfected leaf discs to the same medium for negative control and transfer another 3 uninfected leaf discs to the shooting medium (without antibiotic) for positive control.

Gus Assay

1. Cut the transformed plant organs.
2. Submerge plant tissues in assay buffer I and incubate at 37°C for 1 hour.
3. Remove the buffer and freshly add the assay buffer II to the tissues and incubate at 37°C for overnight.
4. Examine the tissues for dark blue colour.

REMEMBER

1. Remember to take all the precautions and safety means.
2. OD of agrobacterium should not be more or less then 1.0, as it will result in poor efficiency.
3. Time of co-cultivation varies from species to species so don't forget to consult the literature available.

❑❑❑

Unit **IX**

POLYMERASE CHAIN REACTION

PART 1 **PCR—A GENERAL ACCOUNT**

PART 2 **PERFORMING AMPLIFICATION OF DNA USING PCR IN A ROUTINE LABORATORY EXPERIMENT**

Part 1

PCR—A GENERAL ACCOUNT

Contents

INTRODUCTION

The polymerase chain reaction is a technique that permits geometric and enzymatic amplification of specific sequences of DNA. Kary Mullis invented the PCR technique in 1985 while working as a chemist at the Cetus Corporation, a biotechnology firm in Emeryville, California. Mullis has often described how the concept of PCR came to him during a night drive along Highway 128 to his cabin in Anderson Valley, one April weekend in 1983. He was playing in his mind with a new way of analyzing mutations in DNA and suddenly realized that he had thought up instead a method of amplifying any DNA region by chance. To amplify a particular region of DNA, it needs to be repeatedly melted (separated into single strands by heating), annealed (with the primers) and extended (copied by DNA polymerase). The idea here was that you do this 20 times and get 2 copies, then 4, then 5, then 20, or about a million. The trouble with this was that the melting phase requires sufficiently intense heating that most enzymes are damaged (unfolded or denatured). Since PCR requires tens of cycles, enzyme denaturation was a problem. Mullis along with his co-workers at Cetus, developed the first generation cycling machine, Mr. cycle, in which fresh polymerase was added after every cycle but this was not an elegant solution.

With the isolation and purification of DNA polymerase from Thermus aquaticus (the thermophilic bacterium discovered by Thomas Brock in the 1960s in the hot springs of Yellow stone National Park), the development of second generation Thermocycler capable of heating and cooling reactions took place. The use of this thermostable polymerase in PCR was made by Saiki et al in 1988 for the first time and it greatly increased the efficiency of PCR and opened the door to automation of the method. By the end of 1980s, cloning was no longer the only way to isolate genes; PCR had become a fundamental cornerstone of genetic and molecular analysis. The simplicity and extraordinary power of PCR earned taq polymerase Science's molecule of the year in 1991 and the Nobel Prize in chemistry for Kary Mullis in 1993.

BASIC PRINCIPLES

PCR is based on geometric (not arithmetic) amplification of DNA, 30 cycles theoretically amplifies the target DNA 270 million fold. This allows amplification and therefore detection and analysis of a single DNA molecule. This amplification is dependent upon two factors:

1. The Three Component PCR Cycle Design (Fig 1.1)

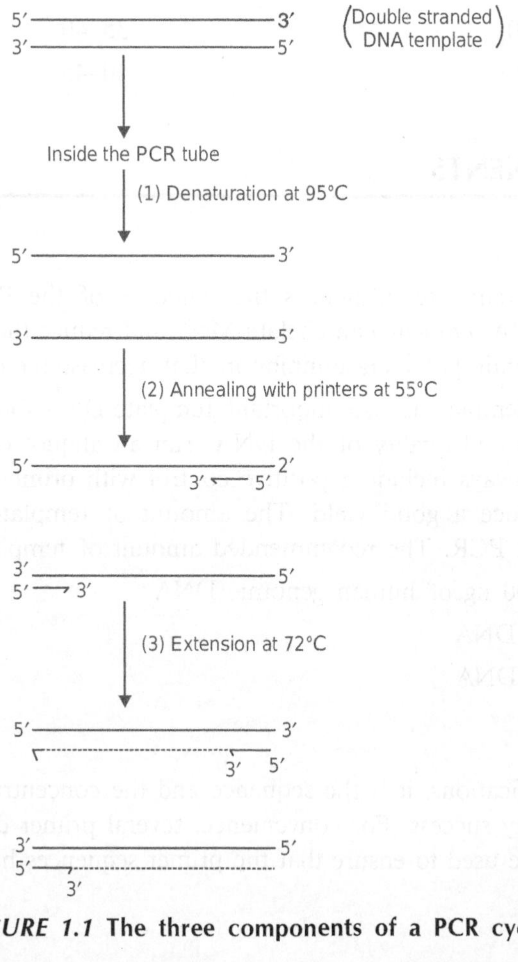

FIGURE 1.1 **The three components of a PCR cycle.**

(*a*) **DNA denaturation (typically 95°C)**

It separates strands of substrate DNA for each cycle. Complete denaturation is essential for efficient PCR. Time duration for this cycle is typically 30 sec to 2 min.

(*b*) **Primer annealing (typically 45–72°C)**

It allows primer binding to target DNA sequences. Temperature (Tm) varies depending on length and composition of primer as well as on complexity and concentration of substrate DNA. Time duration is typically 30 sec. to 2 min.

(*c*) **DNA extension (typically 72°C)**

It generates DNA product by extending hybridized primer. The optimal time is dependent on length of anticipated product, 1 min is generally adequate for product up to 2 Kbp.

2. Number of Cycles

It is dependent on amount of substrate DNA. If more of substrate DNA is available, less no of cycles is needed *e.g.,*

Molecules of substrate DNA	No. of cycles needed
15000	30–35
1000	35–40
50	40–45

REACTION COMPONENTS

1. Template

The quality of the template influences the outcome of the PCR. For instance, large amounts of RNA in a DNA template can chelate Mg^{2+} and reduce the yield of the PCR. Also, impure templates may contain polymerase inhibitors that decrease the efficiency of the reaction.

The integrity of the template is also important.Template DNA should be of high molecular weight. To check the size and quality of the DNA, run an aliquot on an agarose gel. When testing a new template, always include a positive control with primers that amplify a product of known size and produce a good yield. The amount of template in a reaction strongly influences performance in PCR. The recommended amount of template for standard PCR is:

- a maximum of 500 ng of human genomic DNA
- 1–10 ng bacterial DNA
- 0.1–1 ng plasmid DNA.

2. Primers

In most PCR amplifications, it is the sequence and the concentration of the primers that determine the overall assay success. For convenience, several primer design software programs are available. These can be used to ensure that the primer sequences have the following general characteristics:

- 18–24 bases long.
- no internal secondary structure.
- 40–60% G/C content.
- balanced distribution of G/C and A/T rich domains.
- Melting temperature (Tm) that allows annealing temperature of 55°–65°C (for maximum specificity use temperatures of 62°–65°C).

NOTE *Optimal annealing temperatures are often higher than the Tm of the primers (approx 5°–10°C) and have to be determined empirically. For both primers the Tm should be similar.*

Bases that do not hybridize to the template may be added at the 5' end of a primer (*e.g.*, for introducing restriction sites into the amplification product). Primer concentration between 0.1 and 0.6 μm are generally optimal. Higher primer concentration may promote mispriming and accumulation of non specific product. Lower primer concentrations may be exhausted before the reaction is completed, resulting in lower yields of desired product.

NOTE *For some systems, a higher primer concentration (up to 1mm) may improve results. When testing new primers, always include a positive control reaction with template that has been tested for function in PCR.*

3. DNA Polymerase

For most assays, the optimum amount of thermostable DNA polymerase (or a blend of polymerases) will be between 0.5 and 2.5 units/50 μl reaction volume. Increased enzyme concentration sometimes lead to decreased specificity.

4. $MgCl_2$ Concentration

The optimal Mg^{2+} concentration may vary from approximately 1 mM to 5 mM. The most commonly used Mg^{2+} concentration is 1.5 mM (with dNTPs at a concentration of 200 μm). Mg^{2+} influences enzyme activity and increases the Tm of double-stranded DNA; Excess Mg^{2+} in the reaction can increase non-specific primer binding and increase the non-specific background of the reaction.

5. dNTPs Concentration

Always use balanced solutions of all four dNTPs to minimize polymerase error rate. Imbalanced dNTPs mixtures will reduce Taq DNA polymerase fidelity.

If you increase the concentration of dNTPs, you must also increase Mg^{2+} concentration. Increase in dNTP concentration reduces free Mg^{2+}, thus interfering with polymerase activity and decreasing primer annealing. The final dNTPs concentration should be 50–500 μm (each dNTP). The most commonly used dNTPs concentration is 200 μm.

6. pH

Generally, the pH of the reaction buffer supplied with the corresponding thermostable DNA polymerase (pH 8.3–9.0) will give optimal results. However, for some systems, raising the pH may stabilize the template and enhance results.

7. Additives with Specific Use

In some cases, adding the following can enhance the efficiency or specificity of PCR:

(*a*) Betaine (0.5–2 M)

(*b*) Bovine serum albumin (BSA; 100 ng/50 μl)

(*c*) Detergents

(*d*) Dimethylsulfoxide (DMSO; 2–10%) (V/v)

(*e*) Gelatine

(*f*) Glycerol (1.5%) (V/v)

(*g*) Pyrophosphatase (0.001–0.1 units/reaction)

(*h*) Spermidine

(*i*) Formamide

TROUBLESHOOTING IN PCR

While carrying out PCR in the laboratory routinely, following problems may be encountered:

1. No product
2. Low fidelity
3. Non-specific bands
4. Smeared bands
5. Low yield.

Following points may help you to over come the problem and get your experiment working :

(*a*) ***Non-optimal Mg^{++} or No Mg^{++}***

No Mg^{+2}, no reaction as it is required by DNA polymerase to act. Role of Mg^{+2} in PCR is dual, promoting DNA/DNA interactions and forming complexes with dNTPs that are the actual substrates for Taq polymerase. Typically, the dependence of the PCR yield on Mg^{+2} is a bell curve with a broad maximum. When Mg^{+2} is too low, primers fail to anneal to the target DNA, when Mg^{++} is too high, the base pairing becomes too strong and the amplicon fails to denature completely when you heat to 94°C.

(*b*) ***Amount of Template Not Optimal***

The necessary amount of template varies from reaction to reaction; the amount of enzyme should be optimized for each template. For this:

- titrate the amount of template in the reaction.
- perform optimization experiment varying enzyme concentration by 0.50 increments in suggested range (0.5 to 5.0 units).

(*c*) ***Enzyme Inhibitors may be Present in the Reaction Like***

- 50 mM ammonium acetate
- EDTA
- >0.02% sarcosyl

- 0.5 M urea
- >5% DMF
- >10% formamide
- Heparin
- >20% PEG (polyethylene glycol)
- > 0.01% SDS (Sodium dodecyl sulphate)
- >10% DMSO (Dimethylsulfoxmide)
- >20% glycerol
- Residual phenol
- Pyrophosphate

(*d*) ***Primer Annealing Temperature (Ta) is too Low/too High***

At both sub and super optimal Ta values, non-specific products may be formed and the yield of products is reduced. In order to calculate Ta opt, an empirical formula has been given (Rychlik et al 1990):

$Ta^{opt} = 0.3Tm^{primer} + 0.7\ Tm^{product} - 14.9$

(*e*) ***Primers are Degraded or not Optimal***

Primers should be designed to avoid obvious errors like self-complementarity, complementarity between paired primers, or excessively long oligs (> 30bp). Of the available ones check the following:

- If Primers are short, and A-T rich, add 0.9–2.0% (V/v) DMSO (dimethyl sulfoxmide).
- If primers are GC rich, add 1–10% (V/V) formamide.
- Double check priming sequence, use primer design program if available.
- Check aliquots of primers on a gel to ensure they are not degraded.

(*f*) ***Incomplete Denaturation***

DNA reaction buffer with higher Mg^{2+} concentration (4–5 mM) may require a higher denaturation temperature to allow complete separation of the DNA template strands. It is recommended to use the supplied buffer without adding further magnesium. Following suggestions are for this:

- Increase initial denaturation temperature to 95–97°C.
- Denature DNA minus enzyme and buffer for 4–6 minute.
- Increase cycling denaturation time 15–30 seconds.
- Try hot start PCR.

(*g*) ***Machine Based Error***

Manufacturer may be contacted.

APPLICATIONS OF PCR

(A) Nucleic Acid Detection

(*a*) Mutations and polymorphism analysis of PCR products by sequences (RFLP, SSCP, restriction fragmentation, specific hybridization).

(*b*) Infectious agents (disease diagnosis).

(*c*) Chromosomal abnormalities.

(B) cDNA and Genomic Cloning

(*a*) Homology cloning.

(*b*) Cloning of flanking sequences-random primed PCR or inverse PCR.

(*c*) Differential display of mRNA:

- For state specific changes in gene expression (*e.g.,* differentiation, transformation, response to treatment, disease states)
- Utilizes multiple, specific upstream primers and 3 anchored, oligo dT downstream primers (T15C, T15A and T15G).

(C) Mutagenesis

Mutations introduced either by alteration of primer (with preservation of 3^1 end to allow annealing) or by random introduction. Mutated primers should include at least 15 bases of sequence that matches target DNA.

(D) Genome Sequencing

Using random primers (not a specific sequence), the entire genome of an organism can be amplified in pieces (within amplifiable regions of PCR). Once the pieces are amplified, they must be sequenced and then put back together to determine the genome sequence.

MODIFICATIONS OF PCR

PCR has been modified in a number of ways to suit specific situations applications, some important modifications of PCR are:

(*i*) Asymmetric PCR

(*ii*) Nested PCR

(*iii*) Inverse PCR

(*iv*) RT-PCR

(*v*) PCR for site directed mutagenesis

(*vi*) Hot start PCR

(*vii*) Touchdown PCR.

(*i*) ***Asymmetric PCR***

This variation uses one primer at a very low concentration as compared to the other (1:100). This results in generation of single strand copies of one strand. It is used for many purpose like SSCP (Single Stranded Conformation Polymorphism) marker identification, sequencing of DNA (by producing single stranded DNA) etc.

(*ii*) ***Nested (Drop-in-drop-out) PCR***

Here two amplification reactions are carried out for the target DNA. The target sequence is amplified using a pair of PCR primers, and the amplification product is reamplified using another pair of PCR primers complementary to the regions located immediately beyond the 3^1–ends of the first pair of primers. This is employed when the first pair of primer amplifies some sequences in addition to the target sequence.

(*iii*) ***Inverse PCR***

Here amplification of those DNA sequences occurs which are away from the primers and not those which are flanked by the primers. This is done by using primers (oligonucleotides) complementary to the 5^1-ends of the segment in question. This orients the free 3^1–OH of the primers outward of the sequence; as a result the newly synthesized chain grows away from the borders of the concerned segment.

(*iv*) ***Anchored PCR***

Sometimes sequences of only one of the two ends of the DNA, to be amplified, are known. In such cases, first only one primer is used for PCR so only one strand will be copied. After that a poly A tail is attached at the end of the newly synthesized strand. This newly synthesized strand with poly A tail at it's 3' end will then become template for the daughter strand synthesis utilizing an anchor primer with which a poly T sequence is linked to complement with poly A of the template. In the next cycle, both the original primer and anchored primer will be used for gene amplification.

(*v*) ***RT (Reverse Transcribed) PCR***

This is used to amplify mRNA/RNA sequences into DNA duplex. The mRNAs have always been a piece of great interest because they represent the expressed sequence only. The modification requires conversion of mRNA into cDNA first and then that cDNA is subjected to amplification by PCR. The mRNAs have poly A tail at the 3' end so the complementary cDNA is having poly T tail at 5' end. This poly T sequence makes the construction of one primer easy *i.e.,* primer have poly A sequence. The other end of cDNA is modified by terminal nucleotide transferase in the presence of dGTP so that G residues are added up on the 3' end. So the poly G sequence at the 3' end direct the synthesis of second primer *i.e.,* poly C. This approach can also be used to selectively amplify the desired cDNA molecule from among a mixture of cDNAs. In order to achieve this, the base sequence of the 3' end of the desired mRNA must be known *i.e.,* the sequence first preceding the 3' poly—A tail. This base sequence is utilized in the primer formation.

(*vi*) ***PCR for Site–Directed Mutagenesis***

PCR can also be helpful in causing mutation at the desired site in a DNA molecule. This is very conveniently done by using primers with the changed sequence. It seems that if this is the procedure, the mutations can be introduced only at the ends of the DNA where primers bind. But mutations can also be very well induced within internal sites of a gene using the overlap extension variation of PCR (Fig. 1.2). In this variation, the desired gene (Where mutation is to be introduced at internal site) is amplified in two reactions: (*i*) one with the 3' end primer and one primer (mutated) of internal site so that amplification result in one complete strand and other incomplete. (*ii*) In the other reaction, the second 5' end primer and other internal site primer

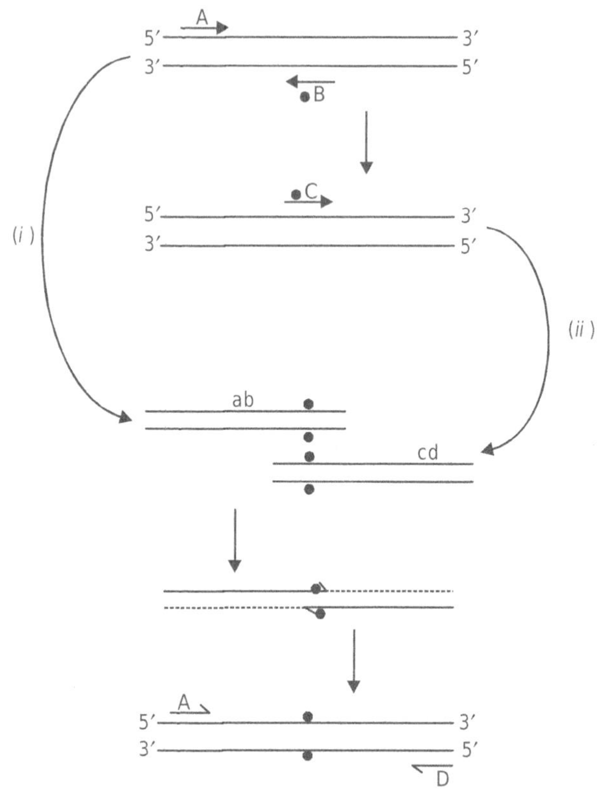

FIGURE 1.2 Mutation at the internal site by overlap extenrior method.

(mutated) is used for PCR. This also yields one complete and one incomplete strand of the gene. The incomplete strands from the two reactions are separated by electrophoresis and made into duplexes by mixing then together and annealed; the strands pair together in the complementary region represented by the internal primers, which contain the mutation. The synthesis step of PCR extends the 3' ends of the incomplete strands to yield complete DNA duplex copies of the gene containing the targeted mutation in internal site.

(*vii*) ***Hot Start PCR***

In this modification reaction mixture without Taq or Mg^{2+} is heated to the denaturation temperature (90–95°C) for 2–3 min. After this, the missing reagent is added to the PCR tubes by opening them. No doubt this increases the risk of contamination but this prevent the non specific amplifications that may occur due to presence of small amount of single stranded DNA in the PCR reaction mixture.

(*viii*) ***Touchdown PCR***

Here PCR is initiated at very high temperatures which allow only perfectly matched primer template DNA hybrids to form and support amplification. After that annealing temperature is dropped gradually with each cycle of PCR. Because the first few cycles will amplify only the target sequence due to the high annealing temperature, these specified products will become template for subsequent cycles, so annealing temperature becomes less critical during the later cycles. This variation increases the specificity of the PCR without lowering the efficiency.

❑❑❑

Part 2 PERFORMING AMPLIFICATION OF DNA USING PCR IN A ROUTINE LABORATORY EXPERIMENT

Contents

INTRODUCTION

The aim of this chapter is to help in performing a PCR experiment in practical class. Here lambda phage DNA is used as the substrate DNA material, as it is commercially available and relatively cheap.

PRINCIPLE

As explained in Part 1, Unit IX.

REAGENTS/MATERIALS

1. Electrophoresis experiment unit (power supply and gel chamber)
2. Eppendorf tubes
3. Micropipettes
4. Thermocycler
5. Lambda DNA (template DNA)
6. Primer pair
7. PCR ready mix (contains Taq polymerase, buffer, $MgCl_2$, nucleotides)
8. Sterile distilled water
9. DNA loading buffer
10. Running buffer (TBE or TAE)
11. Agarose
12. DNA marker ladder
13. Ice bucket
14. Mini-coolers
15. Reaction tube holder
16. Mixer.

TIME SCHEDULE FOR THE EXPERIMENT

	Activities	*Description*	*Time Consumption*	*Time for the Background*
1	Preparation of PCR reagents	Preparation of workbench	15 min.	
		Preparation of stock solution	30 min.	
		Defrosting of PCR reagents	5 min.	
		Setting PCR program in thermocycler	10min.	20 min.
		Pipetting steps	15 min.	
2.	PCR Cycling using thermocyder		60–90 min.	30 min.
3.	Electrophoresis	Preparation of agarose gel and samples	60 min.	
		Gel loading of samples	15 min.	
		Electrophoresis	30–45 min.	30 min.

EXPERIMENTAL LAY-OUT

Prepare the contents for PCR
In 0.5 ml eppendorf tube

↓

Set the required
Program in the
Thermocycler →

↓

Put the 0.5 ml eppendorf in PCR
Machine immediately after you have
Added the last reaction component

↓

After completion of PCR,
Store the products at – 20°C

↓

Prepare 1.5–2%
Agarose gel →

↓

Run the PCR products alongwith
DNA ladder on gel

↓

Stain the bands with ethidium
Bromide and visualize under UV
Transilluminator

METHOD

1. Set the PCR machine with the desired program

	Temp.	time
1. Denaturation	94°C	5min.
2. Denaturation	94°C	1min.
3. Annealing	35°C	2min.
4. Extension	72°C	3min.

Steps 2–4 repeated 35 times.

These conditions may vary from organism to organism.

2. Pipette the reagents into PCR reaction tubes (0.5 ml volume) as follows:

Comoponents	*Stock Concentration*	*Volume to be Pipetted*	*Volume to be Pipetted for Negative Control*	*Final Concentration Per 50 µl Reaction*
Readymix Taq-Polymerase Buffer $MgCl_2$ Nucleotides	2x	25 µl	25 µl	1.5 U Taq Pol 10 mM Tris- HCl 1.5 mM $MgCl_2$ 0.2 mM dNTP
Primer 1	25 pmol/µl	2 µl	2 µl	1 µM
Primer 2	25 pmol/µl	2 µl	2 µl	1 µM
Sterile Double distilled Water		19 µl	21 µl	
Target DNA Bacteriophage Lambda	100 ng/µl (1 U = 50 mg)	2 µl	–	200 ng/50 µl

3. Immediately after you have added the last component into the tubes, place the tubes in PCR and start the program.
4. Meanwhile prepare the 1.5% agarose gel.
5. After completion of PCR cycle, take out the tubes and store at –20°c.
6. Set the apparatus for agarose gel electrophoresis. Load the samples in following order :

Lane 1	Lane 2	Lane 3
Marker	PCR sample	Negativecontrol DNA
+ loading buffer	(15–18 µl) + loading buffer	(15–18 µl) + loading buffer

7. Run at 100 V for 40 min. (conditions may vary from sample.)
8. Stain the gel with ethidium bromide and visualize under UV transilluminator.
9. Analyze the results (you will have no bands visible in lane 3, negative control) of lane 2 by comparing with marker ladder.

REMEMBER

1. Take care of all the points mentioned under 'Trouble shooting in PCR' in part 1, unit IX.
2. The efficiency of amplification declines with our increase in the length of the target sequence, further it is also affected by primer length; it declines if the primer used for PCR is too long.

❑❑❑

Unit X MOLECULAR MARKERS

Part 1

MOLECULAR MARKERS—A GENERAL ACCOUNT

Contents

INTRODUCTION

Molecular markers (DNA markers) reveal neutral site of variation at the DNA sequence level. By neutral means, unlike morphological markers, these variations do not show themselves in the phenotype, and each might be nothing more than a single nucleotide difference in a gene or a piece of repetitive DNA. They have the big advantage that they are much more numerous than morphological markers, and they do not disturb the physiology of the organism.

Restriction enzymes, electrophoretic separation of DNA fragments, southern hybridization, the polymerase chain reaction (PCR), and labeled probes are the tools and techniques that allow us to access and use these markers. Recent advances in scientific instrumentation in particular the improvements in electrophoresis and analysis systems, have led to a new revolution in the development and use of molecular markers. Today, different laboratories have a long list of molecular markers that have been developed so far. Equally long is the list of applications of molecular markers. In this chapter some commonly used molecular markers are briefly discussed. Also in the end a brief list of applications of molecular markers is also given.

SOME IMPORTANT MOLECULAR MARKERS

1. Restriction Fragment Length Polymorphisms (RFLP)

Each restriction enzyme recognizes a specific and characteristic nucleotide sequence and cut the DNA at that restriction site. Because a single nucleotide alteration can create or destroy a restriction site, mutations cause variation in the number of sites. Thus, there is variation or polymorphism between individuals in the positions of cutting sites and the lengths of DNA between them, resulting in restriction fragments of different sizes (usually within 2–10 kb in range). The two different-sized fragments are alleles of one locus. The locus itself is identified by the probe used to detect it and polymorphism among individuals is revealed by hybridization pattern of the probes to the restriction digest of these individuals. RFLP is co-dominant since we can distinguish all the 3 morphs (*e.g.,* AA:Aa:aa). This makes the RFLP more informative than a morphological marker with full dominance, where we can identify only two phenotypes : (AA or Aa) and aa. RFLPs arise as mutations that alter restriction sites, but the events giving rise to then, over evolutionary time, are as stable as the mutations giving any other form of allelic variation *i.e.,* they are constant for practical purposes.

2. Random Amplified Polymorphic DNA (RAPD)

RAPD method produces characteristic amplification pattern from any nucleic acid template and use short oligonucleotides of arbitrary or semi arbitrary sequence to target a multiplicity of anonymous sites. In fact, arbitrariness lies with the sequence of the primer used, and the technique with this primer amplifies specific regions of a genome. Polymorphism is produced due to the complementary sequence for the primer used being present in one strain (say strain X) but not on the other (say Strain Y). Consequently an amplification product will be detectable as bands in strain X, while no band in strain Y (This being an ideal situation). The F_1 hybrid between the two strains (X x Y) will show the band, while in F2 a 3:1 ratio will be obtained. Thus RAPDs behave as dominant markers and the presence of band is regarded as the dominant allele.

RAPDS are the markers of choice when there is no or very little sequence information available for the species under consideration. RAPDs are preferable over RFLPs as it is cheaper and less labor intensive. Main drawback of RAPD analysis is that it is not possible to determine whether two separate bands represent two separate loci or two distinct alleles of a single locus. Further RAPD is not reproducible across various laboratories and stringent control of experimental conditions is needed for reliable and reproducible results.

3. Amplified Fragment Length Polymorphism (AFLP)

AFLP molecular markers are generated by a combination of restriction digestion and PCR amplification. In this technique, genomic DNA is first digested with one or more restriction endonucleases. Next an adapter of known sequence is ligated to the digested genomic DNA and amplification is carried out using primers with sequence specificity for the adapter. The resulting amplification products are typically observed by radio labeling one of the primers, followed by fragment separation on acrylamide gels. Observed AFLP patterns are moderately complex.

One of the strength of AFLPs is the sheer number of markers produced per assay. Multiplication bands are generated in each amplification reaction that contain DNA markers of random origin. AFLPs are quantitative in the sense that heterozygous and homozygous genotypes can be differentiated by the band profile. Also AFLPs have the capacity for simultaneous screening of many different regions distributed throughout the genome. The ability of this technology to generate many markers with minimum primer testing and system's high resolution are other features that make AFLPs attractive as genetic markers (primarily dominant). The major advantages of AFLP technology over the other arbitrarily primed PCR protocol is that up to 100 markers can be detected per PCR and it is more robust than others.

The major disadvantage of AFLP is its expensive nature, complex procedure and unknown gene function. Further, reproducibility is also a major concern.

4. Microsatellites (or Simple Sequence Repeats, SSRs)

Microsatellites are tandem repeats of DNA sequences of only a few base pairs (1–6 bp) in length, the most abundant being the dinucleotide repeats. Microsatellites are also known as short tandem repeats (STRs) or simple sequence repeats (SSRs). The DNA sequences flanking SSRs are known to be conserved and this have been exploited to design suitable primers for amplification of the SSR loci using PCR. Differences in the length of the PCR amplified products from different genotypes reveal the SSR polymorphism. Microsatellites are an abundant source of variation in many organisms. The high information content, co-dominant nature, definitive allele assignment and rapid analysis by PCR make microsatellites well poised for exploitation in varietal identification as well as in the various breeding applications including linkage analysis, agronomic trait selection, germplam assessment and also in various population genetics experiment.

SSR amplification reaction is generally driven by annealing of two specific primers to the complementary unique flanking sites of SSR region. Differences in the length of amplified products are due to change in the times core unit is tendemly repeated. Primers are generally 17–22 nucleotides long with GC content of approximately 50% and a Tm about 60°C. The ideal size for analysis of SSR amplification products is approximately 100–250bp. Agarose, polyacrylamide gel electrophoresis, denaturing PAGE and capillary electrophoresis, have been used to determine the size of the amplified SSR products.

SSR assay are more robust than other marker systems like RAPDS, AFLPs and also have a very high reproducibility factor. Unfortunately, this is not a universal technique and the availability of species specific SSR loci for only a small number of plant taxa is a major constraint to their ubiquitous adoption.

5. Single Nucleotide Polymorphism (SNP)

A SNP is a single base substitution of one nucleotide with another. Both substitutions have to be observed in the general population at a frequency greater than 1%. An example of a SNP is individual X having a sequence TATCGT while individual Y has sequence TAACGT, the polymorphism is T/A. If a SNP is found within a small, unique segment of DNA, it serves both as a physical landmark and as a genetic marker whose transmission can be followed from parent to progeny. Though relatively little work has been done on SNPs in plants, but it is sufficient enough to conclude that SNPs are abundant in plants and have the potential to provide the basis for a superior and hightly informative genotyping assay.

There are a number of methods for identifying SNPs with in a genetic locus like direct sequencing, single strand conformation polymorphism (SSCP), chemical cleavage of mismatches (CCM) and enzyme mismatch cleavage (EMC).

SNP are rapidly becoming the markers of choice because of their high frequency in the genome and low mutation rates compared with microsatellites. SNP research will provide fundamental understanding of many polygenic diseases, thus providing new therapeutic targets. Researchers may begin to reveal relevant genes associated with a disease, by studying SNP profile or haplotypes associated with a disease trait. Within agriculture also, SNP study is extremely important. SNP are highly amenable to automation and can potentially be used to create a high density map required for association mapping and genotyping. The relatively high frequency of SNP in the genome brings about the possibility of cloning genes via linkage disequilibrium. The major weakness of using SNPs is the requirement of huge resources for development as well as utilization of SNPs.

6. Transposable Elements based Markers

Transposable elements (TE) are jumping genes that are dispersed throughout the genome. In eukaryotes, TEs have been divided into two classes based on their transposition methods-class I transpose via an RNA intermediate (retrotrasposon) while class II transpose by DNA intermediates. Several retrotransposon populations are highly plastic, showing great variation in copy number and genomic localization between closely related species. Sequence specific amplification polymorphism approach (SSAP) known as transposos display has been the most popular transposon based marker method. Basically all types of transposon based marker method are anchored AFLP approach that amplifies the region between a retrotransposon specific primer and a nearby cleaved restriction site to which an oligonucleotide adaptor has been added.

The data analysis of SSAP closely resembles that of the AFLP. One advantage of SSAP over AFLP is that it generates relatively higher levels of polymorphism so that lesser number of experiments are needed to generate the desired number of markers. One major disadvantage of retrotransposon based markers is the requirement of sequence information from the terminal regions of the mobile elements. This is only available for a few of these mobile elements in a small number of plant species.

7. Inter-Simple Sequence Repeat (ISSR) Approach

A marker system called inter-simple sequence Repeats (ISSRs) has only recently been developed as an anonymous, RAPDs—like approach that accesses variation in the numerous microsatellite regions dispersed throughout the various genomes (Particularly the nuclear genome) and circumvents the challenge of characterizing individual loci that other molecular approaches require. Microsatellites are very short (usually 10-20 bp) stretches of DNA that are hypervariable, expressed as different variants within populations and among different species. They are characterized by mono-, di or tri nucleotide repeats, *e.g.,* AA— or AG—, or CAG—, that have 4–10 repeat units side-by-side. In ISSRs, specifically the di and trinucleotide repeat types of microsatellites are used, because these are characteristic of the nuclear genome (mononucleotide types are found in the chloroplast genome).

ISSR markers have been preferably employed for genotyping, assessment of genetic diversity, determining the genetic relationships and molecular/linkage mapping studies as they amplify large number of DNA fragments per reaction, representing multiple loci across the genome.

Potential applications of molecular markers: Molecular markers have become an essential tool for the modern biology. They have numerous advantages, of which some are listed below :

1. Preparation of linkage maps.
2. Characterization of genetic resources.
3. Phylogenetic studies.
4. Identification and mapping of quantitative trait loci (QTLs).
5. Marker-assisted selection in plant breeding.
6. Identification of breeding lines and varieties.
7. Gene tagging.
8. Identification of true hybrids.
9. Conservation of genetic resources.
10. Preparation of cytogenesis map.
11. Sequencing of genomes.
12. DNA Finger printing.
13. Forensic application.
14. Detection of adulteration of food-products.
15. Isolation of gene.

....and the list can go on.

❑❑❑

Part 2

RFLP (RESTRICTION FRAGMENT LENGTH POLYMORPHISM)

Contents

INTRODUCTION

See Part 1, Unit X.

PRINCIPLE

RFLP is based on the principles of restriction digestion of DNA as well as that of southern blotting. Every individual has a particular number as well as particular location of restriction enzyme sites on the chromosome (unique to that individual only). So when DNA of two or more individuals is digested with one or more restriction enzymes they will differ in the number as well as lengths of various fragments obtained (Figure 2.1 shows how the presence of an additional site will produce three fragments and its absence will produce only two fragment). The fragments obtained are separated by gel electrophoresis. These fragments are known as

RFLPs

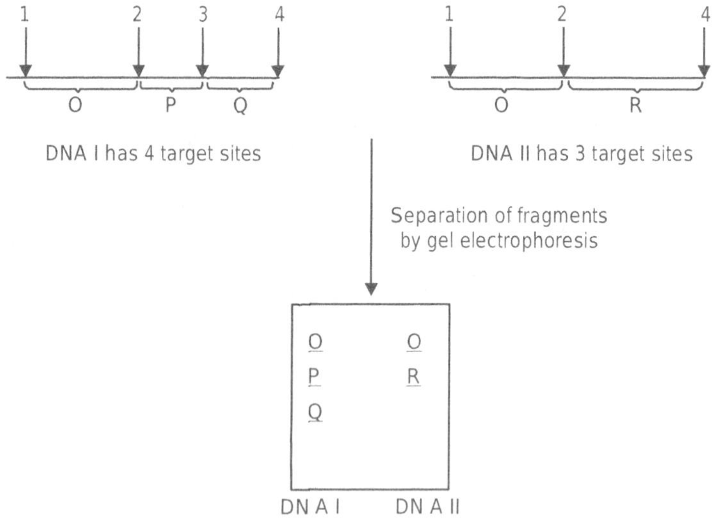

FIGURE 2.1. **Difference in the number of fragments obtained due to change in restriction site by mutation.**

so an RFLP can be defined as a sequence of DNA that has a restriction site on each end. The fragments are detected by subsequent southern blot hybridization to a radiolabeled DNA probe. Probe is a single stranded DNA molecule of known sequence and known function (preferably) which is radiolabeled (to show its presence on the X-ray film, sensitive to radioactivity of the probe). The locus specific probes, consisting of a sequence of unknown identity or part of the sequence of a cloned gene, are obtained by molecular cloning and isolation of suitable DNA fragments. Alternatively, probes developed for closely related species may be used.

RFLPs are generated due to point mutation, deletion, substitution, addition, interchanges and translocations in the restriction enzyme sites.

REAGENTS/MATERIALS

1. **For DNA isolation**

   See Unit III

2. **For restriction enzyme digestion**

   See Unit VI, Part 2

3. **For gel electrophoresis**

   See Unit IV, Part 2

4. **For southern blotting (based on Amersham protocol version 2.0).**

   (*i*) 5M HCl

   (*ii*) 5N NaOH

   (*iii*) 20 X SSC (pH 7.0)

   Dissolve 88.2 gm of sodium citrate (dihydrate) and 175.3 gm of sodium chloride in sterile water to one liter. Adjust the pH to 7.0 using HCl.

(*iv*) Dilution to working concentration

(*a*) Depurination buffer: 0.25 M HCl

Stock	500 ml
5 M HCl	25
Water	475 ml

(*b*) Alkali transfer buffer: 0.4 M NaOH

Stock	500 ml
5 M NaOH	40 ml
Sterile water	460 ml

(*c*) 2 X SSC

Stock	500 ml
20x55C	50 ml
Sterile water	450 ml

5. For probe labeling (based on method of Feinberg and Vogelstein, 1984, Anal Biochem, 137:266-267).

(*a*) OLB (oligolabeling buffer) TE (3mM Tris-HCL, 0.2 mM EDTA, pH 7.0)

Add 300 μl of 1 M Tris HCL and 40 μl of 0.5 M EDTA, 80–90 ml of ddH_2O. Bring pH to 7.0 with HCL and volume to 100ml.

(*b*) 100 mM dATP

(*c*) 100 mM dGTP

(*d*) 100 mM dTTP

(*e*) Tris-$MgCl_2$ solution (1.25 M Tris-Cl, 0.125 $MgCl_2$)

Dissolve 19.7 g Tris-HCl and 2.54 g $MgCl_2.6H_2O$ in 90 ml dH_2O. Bring pH to 8.0 and bring volume to 100 ml.

(*f*) Solution A

To 1.0 ml Tris-$MgCl_2$ solution, add 15 μl of 2-mercaptoethanol, 5 μl of dATP, 5 μl dGTP and 5 μl dTTP. Store at -20°C. Mark the tube with red and blue tops.

(*g*) Solution B

Dissolve 23.8 g of Heps in 35 ml ddH_2O. Bring final volume to 50 ml. Store 1 ml aliquots at -20°C.

(*h*) Solution C

Dissolve 50 units of hexamers pd (N_6) in 555 μl of OLB TE.

(*i*) OLB

Add 1110 μl of OLB TE in the bottle with 50 units of pd (N_6), then add 740 μl of Solution A and 1850 μl of solution C. Mix and store 200 μl of aliquot at -20°C.

(*j*) Bovine serum albumin, BSA, (1 mg/ml)

(*k*) ^{32}P-dCTP (3000 Ci/mM)

(*l*) Klenow (5.3U/μl).

6. For southern hybridization (based on Amersham protocol version 2.0).

(*i*) Denhardt solution (100 X):

10 g Ficoll

10 g Polyvinyl pyrollidone

10 g Bovine Serum Albumin

Make up the volume to 500 ml with sterile water. Filter through cloth. Store at –20°C in 10 ml aliquots.

(*ii*) SDS 10% (pH 7.2)

Dissolve 100 gm of SDS in 900 ml of sterile water. Heat at 68°C to assist dissolution. Adjust the pH to 7.2 by adding few drops of concentrated HCl. Make up the volume to 1 liter with sterile water.

(*iii*) Salmon sperm (SS) DNA, 10mg/ml

Dissolve 5 gm of deoxyribonucleic acid sodium salt type III from salmon testes in 500ml sterile water. Heat to 65°C with vigorous shaking until the consistency of raw egg albumin is reached. Autoclave twice to shear the SS DNA.

EXPERIMENTAL LAY-OUT

DNA isolation

↓

Digestion of the DNA with a restriction enzyme

↓

Separation of the restricted fragments by agarose gel electrophoresis

↓

Southern blotting

↓

Probe labeling → Southern hybridization

↓

Scoring of RFLPs by direct observation of autoradiogram.

METHOD

I. Restriction Digestion and Checking the Digestion

1. Digest the samples with appropriate enzyme in the same way as discussed in unit VI.
2. Before loading the digested samples to southern blotting, ensure that all the samples are well digested.
3. For checking the digestion of samples for blotting purpose, prepare a 1% gel in a gel tray.
4. Take 3–5 ml of digested samples and mix with 2 ml of 1x loading dye and load in the gel.
5. Run for 1.5–2 hours with 70 volts and check for the digestion under the UV transilluminator.
6. If some samples are undigested or partially digested, add 1 ml more of restriction enzyme and reincubate at 37°C for complete digestion.

II. Southern Blotting using Hybond N+ Membrane (based on Amersham protocol version 2.0)

1. After loading the samples along with marker, run the gel (0.8% agarose in 1 X TAE) overnight at 25–30 volt. It will take approximately 16–18 hr.
2. After taking the picture, rinse the gel briefly in distilled water.
3. Soak the gel in 0.25 M HCl with gentle shaking for 15 min to depurinate the DNA.
4. Rinse gel in distilled water and add 0.4 N NaOH to denature the DNA and keep under gentle shaking for about 15 minutes, when the dye regains its original colour.
5. Cut the paper towels to size (normally 20 cm x 10 cm).
6. Cut the Hybond N+ membrane to size and label with marker pen (date, initial, enzyme or blot no etc.)
7. Cut 3 MM Whatman paper (size 20.5 x 25.5 cm).
8. Soak sponges in about 1.5 liters of 0.4 N NaOH. Take care to squeeze and release the solution constantly so that inside sponge is completely and uniformly saturated.
9. Soak two sheets of Whatman paper in 0.4 N NaOH and place on the sponges. Remove air pockets if any by rolling a glass rod or plastic pipet over it.
10. Flip the gel and slide it on to the whatman paper and remove the air bubbles under the gel.
11. Prewet the labeled Hybond N+ membrane in transfer buffer and place on the gel. Remove the air bubbles by rolling the glass rod on the membrane.
12. Frame the gel with plastic strips to prevent short-circuits of the blot.
13. Soak two more sheets of whatman 3 mm paper in 0.4 N NaOH and place over the membrane. Remove the air bubbles by rolling the glass rod.
14. Place half a stack of absorbent paper towels on top of the 3 MM paper on each half of the blot to allow transfer of DNA by capillary action.

15. Place are glass plate on top of the paper towel and place a 0.75 to 1 Kg weight over the glass plate.
16. Once during the transfer, replace the wet paper towels and also add transfer buffer to the tray.
17. Allow the transfer to proceed for at least 6 hrs. Preferably overnight.
18. After blotting is over, remove the membrane using forceps and rinse the membrane briefly (about 1 minute) in 2 x SSC to remove any adhering agarose.
19. Proceed with hybridization step.

III. Probe Labeling (Based on the Method of Feinberg and Vogelstein, 1984, Anal Biochem, 137: 266–267)

1. For 50 ml reaction volume, take probe DNA (50–100 ng for insert and 150–200 ng for plasmids in a total volume of 30 ml) in 1.5 ml eppendorf tubes.
2. Denature the probe DNA by heating to 95–100°C for 5 minutes in a boiling water bath, chill on ice for 5 min and then centrifuge briefly.
3. Meanwhile prepare the cocktail in 1.5 ml eppendorf tube containing OLB, BSA and Klenow. Aliquot 16 ml of the mix into each of 1.5ml eppendorf tube containing denatured DNA. Be sure to keep all the tubes on ice-water before adding ^{32}P-dCTP.
4. Take tubes to radiation room, add 4 ml of ^{32}P-dCTP to each tube, mix and incubate at 37°C for minimum 1.5 hours. Cover with plexiglass.

   Sign for the amount of ^{32}P-dCTP used on sign-in sheet in radiation room.

IV. Southern Hybridization

(Based on Amersham protocol version 2.0)

A. *Prehybridization*

1. Set an incubator at 65°C with blower on and slow shaking.
2. Go for preparation of hybridization buffer as follows:

Contents	Volume	Final Conc.
20x SSC (pH 7.0)	250ml	5X
100x Denhardt Solution	50ml	5X
10% SDS	50ml	0.5%

   Make up the volume to 1 liter with sterile water.

3. Heat the hybridization buffer to 65°C in the hybridization chamber until the buffer becomes clear. (It may take 20–25 minutes).
4. Take the sheared salmon sperm DNA (ss DNA), boil for 10–15 minutes and immediately put on ice until you use it.
5. Get the plastic boxes of appropriate size to fit the filters (*e.g.*, small boxes for small filters so that buffer could be saved). Number of boxes should be equal to the number of probes to be used.

6. Get your filters ready and arrange the filters in each of the boxes. Write down which filter(s) correspond to which box (es).
7. Pour the required amount (*e.g.,* 40 ml for 20x10 cm filter or 20 ml for smaller) of preheated buffer in hybridization boxes. The buffer should be just enough to cover the filter.
8. Also prepare hybridization/SS DNA solution by adding 100 ml of SSDNA for every 10 ml of buffer and mix well.
9. Add the filters one by one to the hybridization/SSDNA solution, completely covering each with the solution before adding the next.
10. Close the boxes tightly to prevent evaporation and place in the 65°C incubator, shaking with around 50–60 rpm.
11. For first use of filters, pre-hybridize for at least 4–5 hrs before. After the first use, only 2 hrs prehybridization is enough.

B. *Hybridization*

1. Get the boxes containing the pre-hybridized filters from the incubator.
2. Lift off the filters using the forceps; pour the labeled probe into the appropriate box.
3. Mix the probes with the already existing buffer gently.
4. Put the filters back into the hot buffer one by one. Make sure that each filter is well coated with the liquid before adding the next. Close the boxes and then stack them in the incubator.
5. Allow the hybridization to occur at 65°C overnight (about 18–24 hrs.) with gentle shaking (50–60 rpm).

C. *Washing Filters and Autoradiography*

1. Prepare three wash solutions:

   Wash I: 2xSSC + 0.1% SDS

   Wash II: 1xSSC + 0.1% SDS

   Wash III: 0.5xSSC + 0.1% SDS
2. With the help of forceps, remove the filters from the boxes and place them in the wash I Incubate here at room temperature for 10 minutes.
3. Get the wash II heated to 65°C in micro oven. Pour the wash I in the P^{32} waste container and add the wash II to the filter and incubate at 65°C with moderate shaking (100 rpm) for 15 minutes.
4. Similarly heat the wash III to 65°C in micro oven. Pour the wash II in P^{32} waste container and add wash III and incubate at 65°C with moderate shaking for 10 minutes.
5. After post-hybridization washing, allow blots just air dry (not too dry, otherwise the hybridized probe is hard to wash away) and also keep DNA side up.
6. Sort the filters according to the probes they were probed with. Filters probed with the same probe should have similar counts.
7. Measure the strength of radioactive signal with a hand Geiger Mueller counter.

8. In a dark room (with red safety light) place the filter in a film cassette with the DNA containing surface facing upwards and place a sheet of X-ray film on top of it. Carefully close the cassette and keep at 70°C for 2–3 days.
9. Develop the film in the dark room under red safety light for 3 min in X-ray developer, rinse briefly with de-ionized water and fix for 3 min in X-ray fixer. Rinse with de-ionized water for 5 min and let the film air dry.

REMEMBER

1. Use gloves while handling gel and nylon filter.
2. Take care that no air bubbles are entrapped between gel, membrane and whatman papers.
3. Always remember that radioactive substances are hazardous to health. Hence, while handling radiolabeled probes, take all the necessary precautions such as use of gloves and protective clothing, washing only behind radioactive screen etc.
4. All the glassware/plastic ware used for work with radioisotopes should be clearly marked with a radiation symbol.
5. Dispose off radioactive wastes in approved containers in approved ways.
6. Remember not to touch the nylon filters with naked hands. Always wear gloves and use forceps while handling nylon membranes.
7. Always remember to adopt general precautions while handling nucleic acids.
8. Take all the precautions mentioned in Unit I.

❑❑❑

Part 3

RAPD (RANDOM AMPLIFIED POLYMORPHIC DNA)

Contents

INTRODUCTION

See Part 1, Unit X.

PRINCIPLE

The RAPD molecular marker is based on the principle of polymerase chain reaction (PCR). Primers used for RAPD analysis are of arbitrary or semi-arbitrary sequences which hybridize to template DNA at a number of random anonymous sites. Single primer act as the forward as well as reverse primer, when hybridize in near proximity, to define RAPD amplicons. RAPD use relatively low concentration (0.2 m mol/L) of single short oligonucleotide primers in PCR. The annealing temperature for these protocols generally ranges from 37–40°C and up to 20 markers can be simultaneously amplified and detected. The generation of RAPD fragments is primarily determined by the sequence and concentration of arbitrary primer. Amplifications, however, can accommodate extensive primer-template mismatching events depending mainly on primer length and template complexity. For hybridization of primer to perfect or partially complementary sites in the target DNA, it is required that first 5–6 nucleotide from the 3' terminus of the primer match faithfully those in the template. Actually it is the mass ratio between primers and template, which decides the overall stringency of the amplification reaction. RAPDs use primer to template ratio of less than one. Amplification products (within the 0.5–5Kb range) can be separated by gel electrophoresis and polymorphisms are detected as the presence or absence of bands of particular size.

REAGENTS/MATERIALS

1. Template DNA (25 ng/μl in TE, good quality).
2. Primers (should be diluted to 15 ng/μl).
3. PCR buffer (10X)
4. 5mM dNTPs

   Aliquot 50 μl each of dNTPs (dATP, dCTP, dCTP, dTTP) from the 100 mM stocks and add 800 μl of sterilized water.
5. Taq DNA polymerase.
6. 50 mM $MgCl_2$
7. Reagents for agarose gel electrophoresis (Unit IV)
8. Thermocycler
9. Gel electrophoretic apparatus

EXPERIMENTAL LAY-OUT

Isolation and purification of Template DNA, diluting to appropriate concentration

↓

Subject to PCR amplification using random primers, amplification conditions being guided by literature

↓

Amplified products run on agarose gel

↓

Data is scored using the suitable software

METHOD

1. Take out template DNA from – 20°C and keep it on ice.
2. Take out all the stocks from – 20°C needed for PCR, keep them on ice and allow them to thaw.
3. Mix the following in the proportion specified against each one for 15ml reaction volume:

Components	Volume in ml
Primer (15 ng/μl)	1.2
dNTPs (5 mM)	0.75
PCR buffer (10x)	1.5
$MgCl_2$ (50 mM)	0.9
Taq Polymerase	0.12
Sterile Water	9.53
DNA (25 ng/μl)	1.0
Total Volume	15 μl

4. If there are many samples to be analyzed, it is advisable to make cocktail for all the samples; cocktail can be prepared with one or two extra volumes so that shortage of reaction mix can be avoided due to pipeting error.
5. Add taq polymerase at the end. Do not spin the mixture after adding the enzyme.
6. Distribute the 14 μl of cocktail to each 0.5 ml tube.
7. Pipette out 1.0 μl of template DNA (25 ng/μl) and add to each of the tube.
8. Label the tube with marker (sample name and primer).
9. Program for RA PD:

Step	Temperature	Time	No of cycles
* Initial denaturation	94°C	2	1
* Denaturation	94°C	30 Sec.	
Annealing	37°C	30 Sec.	45
Extension	72°C	1 Min. 6 Sec.	
* Extended extension	72°C	7 Min.	1
Store	4°C	Until recovery	

10. Load the labeled tubes into the PCR machine and run the above program.
11. While the program is running, set the electrophoretic assembly and make the 1.5% agarose gel.
12. After the PCR program is over, take out the samples from PCR machine and add 3ml of loading dye to each of the sample and spin it briefly to mix the dye with the sample.

13. Load the samples and run the gel at 100 volts for 4–5 hrs. Better resolution is obtained when the loading dye migrates 10–11 cm towards the bottom edge of the gel.
14. Visualize and photograph the DNA fragments under UV light.
15. Designate each fragment and score them as presence or absence for further analysis.

REMEMBER

1. Always use purified, high molecular weight DNA.
2. All precautions should be taken to avoid contaminations because short, random primers used are able to amplify DNA fragments in a variety of organism.

❑❑❑

Part 4

AFLP (AMPLIFIED FRAGMENT LENGTH POLYMORPHISM)

Contents

INTRODUCTION

See Part 1, Unit X

PRINCIPLE

It is based on the principle of both RFLP and RAPD as this method uses primers for amplification and also involves restriction digestion of the chromosomes. AFLP analysis system includes three steps:

(*i*) Restriction enzyme digestion of genomic DNA and ligation of adapters.

(*ii*) Amplification of a subset of restriction fragments using a two-step amplification strategy with P^{33} or P^{32} labeled selective primers.

(*iii*) Denaturing polyacrylamide gel analysis of the amplification fragments and detection by autoradiography.

REAGENTS/MATERIALS

1. Genomic DNA (with concentration 100 ng/ml approx.)
2. Eco RI (rare cutting enzyme).
3. Mse I (frequent cutting enzyme).
4. 5Xenzyme reaction buffer

   50 mM Tris-HCl (pH 7.5)

   50 mM $MgCl_2$

   250 mM KCl
5. Agarose
6. 1XTBE
7. 50 pmol/µl MSe I adapter
8. 5 pmol/µl Eco RI adapter
9. Adapter reaction buffer
10. 10 mM ATP
11. T_4 DNA ligase (1U/µl)
12. Eco RI pre amplification primer
13. Mse I pre-amplification primer
14. 5 mM dNTP
15. 10xPCR buffer (minus $MgCl_2$)
16. 50 mM $Mgcl_2$
17. Taq DNA polymerase (5U/ml)
18. ^{33}PATP
19. T4 polynucleotide kinase enzyme
20. Kinase buffer
21. Reagents of denaturing polyacrylamide gel electrophoresis (Unit IV, Part 3).

EXPERIMENTAL LAY-OUT

Restriction digestion of genomic DNA with Eco RI and Mse I

↓

Ligation of Eco RI and Mse I adaptors to restriction products

↓

Pre amplification of ligated digested DNA with Eco RI Mse I Pre-amplification primers

↓

Check pre-amplification on 1.5% agarose gel

↓

Go for selective amplification with Eco RI + A and Mse I + C primers

↓

Separation of amplified fragments on denaturing polyacrylamide gels

↓

Analysis of band profile using suitable software

METHOD

I. Restriction Digestion of Genomic DNA

1. Incubate the genomic DNA with the rare cutting enzyme (Eco RI) and frequent cutting enzyme (MSe I). Reaction mix (30ml each) for digestion is prepared in following way:

   300 ng genomic DNA

   3U Eco RI

   3U Mse I

   6μl 5 X reaction buffer

   H_2O to 30 μl

2. Incubate for 2 hrs. at 37°C. For inactivating the enzymes then incubate at 70°C for 15min. at 70°C, followed by immediate cooling on ice, collect the contents by brief centrifugation.
3. Check digested DNA in 1.5% agarose gel along with molecular markers to see whether the DNAs were digested completely or not.

II. Ligation of Adapters

1. Preparation of ligation mix (10 µl of each reaction):

   4 µl H_2O

   2 µl 5 X reaction buffer

   1 µl Mse I adapter (50 pmol/µl adapter)

   1 µl Eco RI adapter (5 pmol/µl adapter)

   1 µl 10mM ATP

   1 µl T_4 DNA ligase
2. Add 10 µl ligation mixture to 20 µl digested genomic DNA, incubate for 3 hrs. at 20°C. Transfer 10 µl reaction mixture into a new tube. Add 90 µl H_2O to make a dilution of 1:10 and mix well. Store 1:10 diluted reaction mixture and undiluted portion at – 20°C.

III. Pre-amplification

1. Labeling of primers in a reaction mixture

50 ng/µl E.coRI primer or MseI primer	15 µl
10X T4 kinase buffer	5.5 µl
^{33}PATP	10 µl
T4 polynucleotide kinase	2 µl
Add water to 55 µl	

   Incubate for one hour at 37°C, then for 10 minutes at 70°C and store at -20°C.

2. Prepare pre-amplification reaction mix as follows:

	1PCR
Eco R1 pre-amplification primer (100 ng/µl)	0.5 µl
MSe I pre amplification primer (100 ng/µl)	0.5 µl
5 mM dNTP	1.3 µl
10xPCR buffer (minus $MgCl_2$)	3.0 µl
50 mM $MgCl_2$	1.2 µl
Taq DNA polymerase (5U/µl)	1.3 µl
H_2O	18.3

3. Prepare a PCR mix in a manner:

   5 µl 1:10 diluted ligated digested DNA

   25 µl pre amplification mixture

4. Run the PCR with programmes:
   94°C for 30 sec.
   50°C for 1 min.
   72°C for 1 min.
5. Check pre amplification on 1.5% agarose gel with 10 μl PCR product, light smear will be visible.
6. Dilute pre-amplification mixture in 1:50 ratio *i.e.,* 3 μl pre-amplification mixture +147 μl H_2O.
7. Keep all reaction mixtures (diluted or not) at –20°C.

IV. Selective Amplification

1. Prepare the reaction mix in following way:

Reaction mix	1 PCR
H_2O	3 μl
10xPCR buffer (minus $MgCl_2$)	1.25 μl
50mM $MgCl_2$	0.6 μl
5mM dNTP	0.6 μl
Unlabeled primer	0.5 μl
P^{33} labeled primer(Eco RI + A)	0.35 μl
Taq DNA Polymerase (5U/μl)	0.05 μl

2. Prepare the PCR mix by adding 6 μl of above reaction mix to 6 μl of 50 fold diluted pre-amplified DNA.
3. Run the PCR with following programme
   30 sec. 94°C
   1min. 65°C
   1 min. 72°C   Cycle 1
   Cycle 2–13; annealing 0.7°C
   Cycle 14–36; annealing at 56°C

   The pre amplification products are diluted and used for the selective amplification with two AFLP primers. You can use different primer combinations, such as 2 + 2, 2 + 3, 3 + 3 selective nucleotides depending on the complexity of the genome.

REMEMBER

1. Remember to use only purified, high molecular weight DNA.
2. Take care of all the precautions while handling radiolabeled probes.
3. Always remember the precautions and important points relating to restriction digestion, PCR and denaturing polyacrylamide gel electrophoresis (mentioned earlier).

Part 5 SIMPLE SEQUENCE REPEATS (SSRs) OR MICRO SATELLITES

Contents

INTRODUCTION

See Part-1, Unit X.

PRINCIPLE

It is a PCR–based molecular marker system which utilizes the conserved flanking sites of the SSR region (di or tri nucleotide repeats) in primer designing. The amplified fragments give polymorphism because of the difference in the number of repeats at a particular SSR locus possibly caused by slippage during DNA replication.

REAGENTS/MATERIALS

1. SSR Primer Pairs (forward as well as reverse primer)
2. Genomic DNA

3. 10 X PCR buffer
4. dNTPs (5 mM each)
5. Taq DNA polymerase
6. Reagents for polyacrylamide gel preparation
   (*a*) 4% acrylamide solution (1000 ml)
   420 g urea
   450 dH_2O
   100 ml 5 X TBE
   100 ml 40% acrylamide/bis-acrylamide (19:1)
   (*b*) 0.5% acetic acid in 95% ethanol
   (Add 1ml of glacial acetic acid to 199 of 95% ethanol)
   (*c*) 20% APS (ammonium persulphate)
   (Add 2.0 gm of ammonium persulphate to 10 ml of dH_2O)
   (*d*) Bind silane
   (*e*) Sigmacote
7. Polyacrylamide gel electrophoresis assembly.
8. Sharks tooth comb.
9. 10 X TBE (10 L) – 1080 g Tris base
   92 g EDTA
   552 g basic acid
   add dH_2O to final volume
10. 0.5 x TBE
    (Add 50 ml 10xTBE to 950 dH_2O)
11. 3xSTR loading solution
    0.2 ml 5M NaOH
    95 ml (95%) formamide
    50 mg bromophenol blue
    50 mg Xylene cyanol
    dH_2O to 100 ml
12. Reagents for silver staining of gel
    (*a*) Developer solution
    60 g sodium bicarbonate
    3 ml 37% formaldehyde (H_2CO)
    400 μl of 10 mg/ml sodium thiosulfate
    2000 ml dH_2O
    Prepare the fresh solution and chill to 10°C before each use

(*b*) Staining solution

2 g silver nitrate

3 ml 37% formaldehyde

2000 ml dH_2O

(*c*) Fix/stop solution (10% acetic acid)

200 ml glacial acetic acid

1800 ml dH_2O

METHOD

1. First of all go for PCR amplification of genomic DNA with SSR primer pairs. PCR conditions for micro satellites markers are variable depending on individual PCR system. Initial conditions should be based on published protocols or standard profiles followed by adjustments in annealing/denaturation temp or adjusting Mg^{2+} concentration in order to obtain best amplification. Following is the standard reaction mix for PCR.

Component	*Addition sequence*	*Volume (μl)*	*Final concentration*
Autoclaved deionized	1	13.9	
10XPCR buffer (with 15 mM $MgCl_2$)	2	2.0	1X
dNTPs (5 mM each)	3	0.8	200 μM
Forward primer (10 μM)	4	0.4	0.2 μM
Reverse primer (10 μM)	5	0.4	0.2 μM
Taq DNA polymerase (5 units/μl)	6	0.5	2.5 units
DDA template (25ng/μl)	7	2.0	(50ng/20ml)
Total Volume		20 μl	

2. Immediately put the PCR tube with above mix in Thermocycler with following programme:

Type	*Segment*	*Activity*
Time Delay	1	94°C for 5 min.
Step Cycle	1	94°C for 1 min.

	2	50°C for 1 min.
	3	72°C for 2 min.
Cycle 35 times		
Time delay	1	72°C for 7 min.
Soak	1	4°C until recovery

3. While the amplification is going on, prepare a 4% polyacrylamide gel in following manner:
   (*i*) Wash and clean the short and long glass plates with 95% ethanol.
   (*ii*) Prepare fresh binding solution, in a chemical fume hood, by adding 3 μl of Bind Silane to 1 ml of 0.5% acetic acid in 95% ethanol in a 1.5 ml micro centrifuge tube. Wipe the long glass plate with this solution using Kimwipe tissue.
   (*iii*) Wait 5 minutes for binding solution to dry then wipe the treated surface of long plate with 95% ethanol to remove excess binding silane.
   (*iv*) Apply 3 ml of sigmacote onto one side of the short glass plate. With a dry tissue spread the sigmacote over the entire surface of the plate.
   (*v*) Wait for 5 min, then remove the excess sigmacote with dry tissue.
   (*vi*) Assemble the plates with 4 mm spacers (treated surface inside).
   (*vii*) Pour 4% acrylamide solution (100 ml for one gel) into a squeeze bottle, for each 100 ml solution, add 100 μl pf TEMED (N, N, N′, N′-Tetraethyl methylenediamine) and 500 μl of 20% APS to the acrylamide solution and mix gently.
   (*viii*) Pour the contents between the plates, taking care that no bubbles are formed.
   (*ix*) Insert a multiwell sharktooth comb, straight side into the gel, between the glass plates.
   (*x*) Allow polymerization to proceed for at least 1 hour. If the gel is left overnight, seal with saran wrap. The gel may be stored overnight at room temperature if a paper towel saturated with dH_2O and plastic wrap are placed around the end of gel to prevent the gel form drying out.
4. After polymerization, shave any excess polyacrylamide away from the comb. Remove the comb and clear the plates.
5. Assemble the electrophoretic apparatus and pour in the electrophoretic buffer.
6. Clear the air bubbles and unpolymerized acrylamide form the well areas. Remove air bubbles between glass plates on the bottom of the gel also.
7. Pre-run the gel at 100 W to achieve a gel surface temperature of approximately 45–50°C.
8. Prepare the samples by loading/adding 10 μl of 3 X STR loading solution to each 0.2 ml tube containing 20ml of PCR product.
9. Prepare 10 μl molecular weight marker for each marker lane.
10. Denature the samples and markers by heating at 95°C for 5min and immediately chill on ice.

11. After the pre-run, use a 50–100 cc syringe filled with buffer to flush the urea from the well area and carefully insert the sharkstooth comb teeth between the plates. Take care that you don't pierce the gel.
12. Load 6ml of each sample and marker into the respective wells. To prevent the gel from cooling, gel loading should not exceed 20 minutes.
13. Run the gel at 100 W till the dye reaches the middle/lower side.
14. After complete running, stain the bands by DNA silver standing method in following manner:
    (*i*) Disassemble the sequencing unit and separate the two plates by using a plastic wedge or blaze.
    (*ii*) Place the gel, attached to longer plate, in a shallow plastic tray, on a shaker with gel upside.
    (*iii*) Add fixing solution in tray and let the gel fix for 30 min or until the dyes can not be visible.
    (*iv*) Wash the gel 3 times with water.
    (*v*) Stain the gel in staining solution for 30 min.
    (*vi*) Again rinse the gel (no more than 10 secs).
    (*vii*) Add the pre-chilled developer solution (4–10°C), keep for 2–5 min or until the bands are visible.
    (*viii*) Fix the gel in Fix/Stop solution using the same solution in the starting of staining procedure.
    (*ix*) Rinse gel with distilled water for 2–3 min.
    (*x*) Take the gel (long plate) out and allow it to dry overnight.
15. After drying take the photograph.

❑❑❑

Unit XI MOLECULAR MARKERS AND MAPPING

PART 1 MOLECULAR MARKERS AND MAPPING

Part 1 MOLECULAR MARKERS AND MAPPING

Contents

INTRODUCTION

Molecular markers and marker mapping are part of the latest technology that is making it's way into all areas of modern biology like genomics, breeding, transgenics, developmental biology, systematics, ecology, and perhaps especially, into plant and crop physiology. One of the difficulties is that the genetic science of molecular markers and their mapping is a complete mystery to many students. One can get lost in the language, or be confused over the genetical concepts, or end up in just mystery of the software. Not with standing these barriers, the fact remains that to put physiology on the map there is an absolute need to 'first find your gene'. The physiologist argue that, almost by definition, a gene is identified by a change in it's function, which is true, but it's also true that until we can map genes, or have clear signposts to their location in a linkage group, we can not do much with them-which is where molecular markers come in.

MOLECULAR MARKERS

See Unit X, Part 1.

MAPPING

Mapping is putting markers in order, indicating the relative genetic distance between them, and assigning them to their linkage groups on the basis of the recombination values from all their pair wise combinations. For this, it is obligatory to know the genetic concepts of segregation and recombination as illustrated in the coming part with classical Medelian markers showing full dominance. Dominant and recessive alleles are given as upper and lower case letters respectively.

Segregation and Recombination

Two alleles of a locus will segregate (separate from one another) with equal frequencies into gametes because of meiosis. *e.g.,* if a and A are two such alleles then a diploid individual heterozygous at this locus will give gametes half of which are A and half of which are a. Similarly alleles b and B at a separate locus will segregate fifty-fifty into the gametes. If a/A locus and the b/B locus are unlinked (*i.e.,* on different chromosomes) then the alleles will undergo independent segregation, giving four possible combinations in the gamates; AaBb – AB, Ab aB, ab, The simplest way to follow such events, and to introduce recombination is first to make a cross between two homozygous parents – P_1 (AABB) X P_2 (aabb) (Fig. 1.1)

Parents	AABB		X		aabb
Gametes	AB				ab
F_1			AaBb		
Test cross	AaBb		X		aabb
Gametes	AB	Ab	aB	ab	ab

Test cross	AaBb	Aabb	aaBb	aabb
Progeny	(Parental)	(Recombinant)	(Recombinant)	(Parental)

FIGURE 1.1 **A cross between two homozygous parents.**

Recombinants comprise 50% of the test cross progeny.

If the genes are linked: It genes of two character are linked *i.e.,* present on the same chromosome, following is the nature of crosses and progeny (Fig. 1.2 see on next page).

Above is the case when genes are closely placed on the chromosomes and such phenomenon of segregating together in the gamete formation is known as linkage (given by TH Morgan in 1910). It was further suggested that the strength of linkage will be determined by distance between two genes in question. The greater this distance, lower will be the linkage strength. The linkage is broken down due to the phenomenon of crossing over occurring during meiosis. Crossing over will be relatively more frequent, if distance between two genes is more than in a case where the distance between two genes is lesser. The phenomenon of crossing over involves exchange of chromosomes segment which will be clear from following illustration (Fig 1.3 see on next page).

Depending upon distance between any two genes which is inversely proportional to strength of linkage, non-crossovers will vary from 50–100% (100%–non-crossvers is a stage where no crossing over takes place). The cross over will similarly vary from 0–50% and will never exceed 50%.

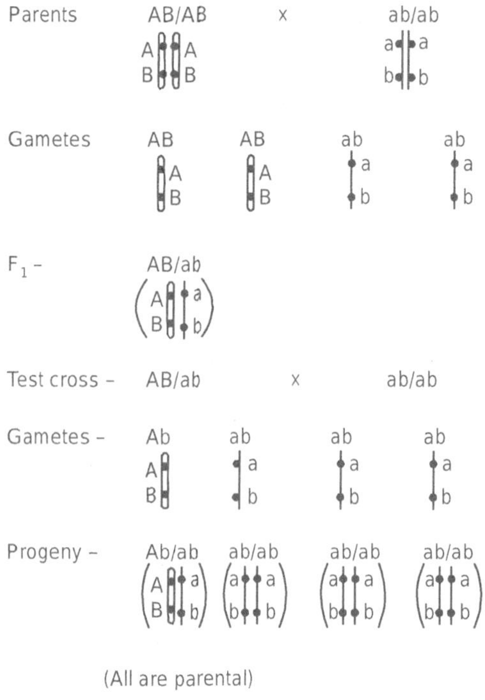

FIGURE 1.2 **A cross showing linkage of genes.**

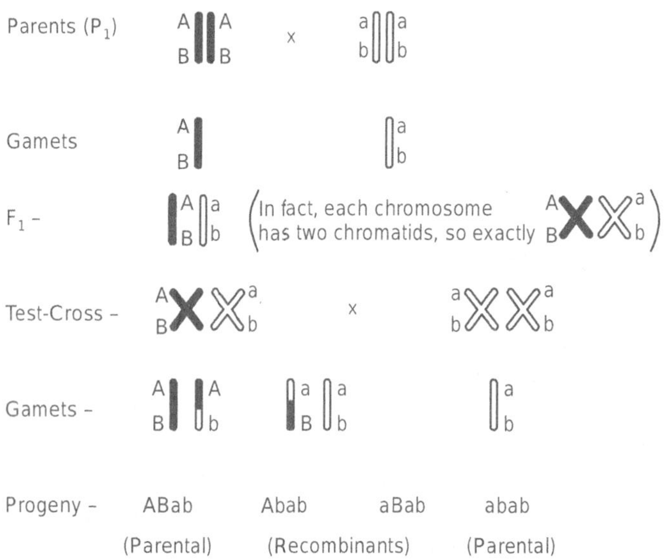

FIGURE 1.3 **A test-cross showing linkage and crossing over.**

Recombination and Linkage Maps

Since recombination frequencies are directly proportional to distances between genes in question, these values can be used in preparation of linkage maps.

Recombination Frequencies From a Test Cross

A three point test cross (involving three genes) can be taken for having a clear understanding.

Let us presume that there are three genes A, B and C present on the same chromosome (*i.e.,* these are linked). There could be three possible linear order in which genes may be present on a chromosome. These are A-B-C, A-C-B or B-A-C (In one case B is present in the middle, while in the other two cases C and A respectively are present at the middle position). In finding out the linear order, one has to really find out the gene which is present in the centre. For this a three point test-cross is made, which involves crossing of trihybrid ABC/abc (obtained from a cross ABC/ABC x abc/abc) with triple homozygous recessive abc/abc. The progeny obtained will represent the gametes formed by the hybrid. Presuming A-B-C as the order of genes, results expected can be diagrammatically represented as in Fig. 1.4.

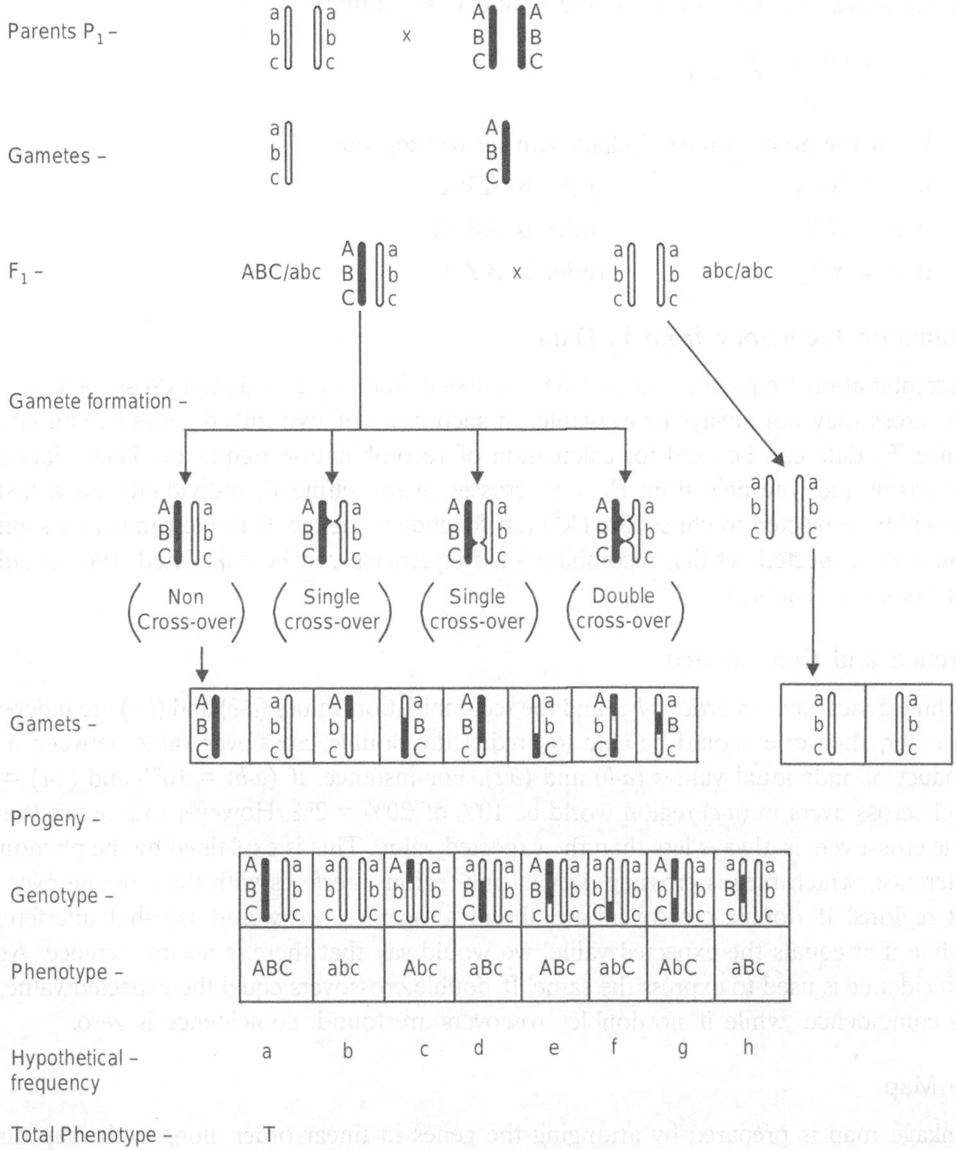

FIGURE 1.4 **A three point test-cross.**

The following procedure will be followed for calculation of recombination frequency.

If crossing over *i.e.,* recombination value (per cent) between A and B is called X, that between B and C is called Y and that between A and C is called Z , then

(*a*) Crossing over (A-B) (*i.e.,* where A and B separated)

$$x = \frac{c+d+g+h}{\text{T}} \times 100$$

(*b*) Crossing over (B-C) (*i.e.,* where B and C separated)

$$y = \frac{e+f+g+h}{\text{T}} \times 100$$

(*c*) Crossing over (A-C) (*i.e.,* where A and C separated)

$$z = \frac{c+d+e+f}{\text{T}} \times 100$$

From the above values, linkage can be worked out.

If Z = X+Y	order is A-B-C
If Z = X-Y	order is A-C-B
If Z = Y-X	order is B-A-C

Recombination Frequency from F_2 Data

Recombination frequency can also be calculated from F_2 data as test cross data of a three point test cross may not always be available. In such cases, if two linked genes exhibit complete dominance, F_2 data can be used for calculation of recombination frequency. These data can be obtained easily and routinely from $F_1 \times F_1$ crosses or by selfing F_1 individuals. As a first step, the data will be subjected to chi-square (X^2) test for independence. If the test indicates significant deviation from expected results, recombination frequencies can be calculated. (For details, see any text book on genetics).

Interference and Coincidence

If three genes have an order *a-b-c,* and the recombination values (*a-b*) and (*b-c*) are independent of each other, then one should be able to predict the double cross-over value between a and *c* as a product of individual values (*a-b*) and (*b-c*). For instance, if (*a-b*) = 10% and (*b-c*) = 20%, the double cross-overs in (*a-c*) region world be 10% of 20% = 2%. However in practice the value of double crossovers is always less than the expected value. This is explained by the phenomenon of interference, which means crossing over in one region interferes with the crossing over in the adjacent regions. If double crossovers are absent altogether, we would say that interference is 100% while if it equals the expected value, we would say that there is no interference. Another term coincidence is used to express the same. If double crossovers equal the expected value, there is 100% coincidence, while if no double crossovers are found, coincidence is zero.

Linkage Map

Linkage map is prepared by arranging the genes in linear order along with map distance between them. The map distance are given by recombination frequencies and are sometimes represented in map units (m.u.) also called centimorgan (1 Centi Morgan = 1 cM = 1% recombination). In many cases the map distance (or true distance) between two loci or a linkage map is not linearly related to recombination frequency (RF), when the latter (*i.e.,* RF) increases beyond a certain level. This is due to the presence of double cross-overs and multiple crossovers. In view of this, the recombination frequency needs a mathematical treatment to get relatively precise estimates of map distance between loci. This is achieved through the use of mapping function.

Now a day a number of computer software packages such as Mapmaker (sander et al 1987), Mapmanager (Manly and Eliot, 1991) and Joinmap (Stam, 1993) are available for computerized map construction. These programme use data obtained from the segregating populations to estimate recombination frequency between molecular markers that are then used to determine the linear arrangement of genetic markers.

Molecular Markers and Mapping

The development of molecular marker technology has caused renewed interest in genetic mapping. Molecular markers, as already explained in unit X, are alleles of loci at which there is sequence variation in DNA that is neutral in terms of phenotype. An appropriate mapping population, suitable marker system and the software for analysis of data are the key requirements for a molecular mapping and molecular breeding programme. The genetic map construction requires that the researchers:

1. Select the most appropriate mapping population.
2. Calculate pair wise recombination frequencies between molecular markers using these populations.
3. Establish linkage groups of molecular markers and estimate map distance.
4. Determine map order of molecular markers.

MAPPING POPULATIONS

A population used for gene mapping is commonly called a mapping population. To obtain it, requires choosing of parents and going for right mating scheme. It further depends on type of markers available and the objectives of experiments.

Selection of parents should by governed by the fact that sufficient variation should be present at both the DNA as well as phenotype level. If the parents are greatly different at phenotypic level for a trait, possibility of genetic variation is also there. Utility of a map is governed by marker-trait association so emphasis should be given to include as many qualitatively inherited morphological traits as possible. There are different types of mapping population available for linkage mapping.

(*i*) F_2 population

(*ii*) Back crosses

(*iii*) Doubled Haploids (DHs)

(*iv*) Recombinant Inbred Lines (RILs)

(*v*) Near isogonics lines (NILs)

(*vi*) Bulk segregant analysis.

(*i*) **F_2 population:** It is produced by selfing of F_1 population generated by crossing the selected parents. As per the Mendelian's law the ratio expected for dominant marker is 3:1 and for co dominant marker is 1:2:1. F_2 population is considered good for preliminary mapping. It can be developed with little efforts when compared with other population.

The major drawback is that F_2 populations are exhaustive so no source will be there for DNA or protein isolation after some time. Further F_2 individuals are products of single meiotic cycle. Quantitative traits cannot be precisely mapped using F_2 population.

(*ii*) **Backcross Population:** These are generated by crossing the F_1 with either of the parents, dominant or recessive (here term backcross is replaced by test cross). With respect to molecular markers, the backcross with dominant parent would segregate in a ratio 1:0 and 1:1 for dominant and co dominant markers, respectively. However testcross would segregate in a ratio of 1:1 irrespective of the nature of marker. The backcross populations can be prepared in less of time with less of labor. The disadvantage is that they get exhausted with time and give recombination information of only one parent (*i.e.*, F_1).

(*iii*) **Double haploids (DHs):** Production of double haploids involve the plant tissue culture techniques as these are developed by chromosome doubling of anther culture derived haploid plants from F_1. Because of the involvement of tissue culture, DHs are permanent mapping population and hence can be obtained anywhere, anytime. Double haploids like the F_2 populations are the products of one meiotic cycle. The expected ratio for the marker is 1:1, irrespective of genetic nature of marker (whether dominant or co-dominant). It is useful for mapping of both qualitative and quantitative characters.

Sometimes suitable methods may not be available for haploid production for a number of crops which is a major disadvantage of this population. Further, certain variations may also occur during anther culture.

(*iv*) **Recombinant Inbred Lines (RILs):** Recombinant inbred lines are developed by single seed selections from individual plants of an F_2 population. Single seed descent is repeated for several generations. At this point, all of the seed from an individual plant is bulked. In a way RILs are produced by continuous selfing or sib-mating the progeny of individual members of an F2 population until complete homozygous is achieved. RILs also equalize marker types like DHS, the genetic segregation ratio for both dominant and co-dominant marker would be 1:1. Further RILs are immortal population; they can be replicated over locations and years and therefore are of immense value in mapping QTLs.

Development of RILs may require many season to develop. In crops with high inbreeding depression, development of RILs is relatively difficult.

(*v*) **Near—Isogenic Lines (NILs):** NILs are generated either by repeated selfing or by backcrossing the F_1 plants to the recurrent parents. Expected segregation ratio of the markers is 1:1 irrespective of the nature of marker. Like DHS and RILs, NILs are also immortal mapping population and are quite useful in functional genomics.

Major drawback is that they require many generations for development.

(*vi*) **Bulk Segregant Analysis.** The core of this procedure is the formation of a bulk sample of DNA for analysis by pooling DNA from individuals with similar phenotypes, may be from segregating population like F_2 *e.g.,* if a molecular locus linked to a disease resistance locus is to be found, two bulk DNA samples are prepared, one containing DNA from plants/lines that are resistant to the disease and other bulk from plants susceptible to the disease. Each of these bulk DNA samples will contain a random sample of all the loci in the genome, expect for those that are in the region of the gene upon which the bulking occurred. Therefore, any difference in DNA marker pattern between these two bulks should be linked to the locus upon which the bulk was developed. This is a popular method for gene mapping.

ADVANTAGES OF MAP

1. Identification of Gene/QTLs

Mapping of molecular markers provide numerous extra signposts which can point to genes of interest. To put a given gene on to a molecular marker map there must be phenotypic variation for the trait controlled by that gene within the mapping population. For example, a population might include polymorphism for alleles at a particular flower colour locus. These alleles will segregate together with some particular marker let's say RFLP markers. By computing linkage values between alleles at that locus and the RFLPs, the pigmentation gene can be included in the map. Exactly the same approach can be applied to loci controlling traits such as disease resistance or morphological markers.

Major genes: Term Major gene is used to describe a gene which is inherited in a Mendelian manner and whose allelic forms give qualitatively distinct phenotypes. Mapping of such genes is a relatively simple exercise. For example, in a mapping population segregating for presence or absence of mildew resistance we may discover that the resistance locus always segregate together with a certain molecular marker. This marker than becomes gene tag for resistance. To find a truly coincident marker for a gene of interest is really fortunate.

Quantitative Trait Loci: A major gene trait is digital-in most cases, the character is either expressed or not. But many traits, particularly those of significance for crop physiology, are analogue and are observable in a segregating population as a more or less continuous range of extremes which may even lie outside the mean range of parents *e.g.,* characters like yield, stress acclimatization, size and so on. The genes that contribute to these complete phenotypes will usually be several in number (polygene) may be linked only in the physiological, but not the genetic sense. Molecular marker maps can help us to resolve these complex characters into their contributing quantitative trait loci (QTLs). QTLs can not be mapped in the standard way that we have described for major genes, because the individual loci can not be identified. The principle of QTL mapping is to associate the QTLs with molecular markers, in their inheritance and thereby to identify them by having their map locations. In reality we would have data relating the QTL effects to many molecular marker loci and gradations of effects according to the strength of linkages. Without losing the concept in complexity let us just say that we would plot the likelihood according to effect against locus positions and then map our QTL in relation to its nearest marker.

2. Marker Assisted Selection

As discussed earlier that a molecular marker closely/tightly linked to a major gene is termed as gene tag. If a disease resistant parent selected to carry a gene tag for this character were used in breeding programme, the progress of the resistance gene through the generations could be simple by tag screening of for example DNA from seedling-a far less time, space and resource demanding process than carrying out full scale disease sensitivity trials at each stage. The same approach can be applied to a QTL of interest. It can also work for negative selection, where the aim is to discriminate the undesirable trait-one that, for example, might hitch a ride with a useful character that a breeder wants to transfer from a wild species into a related crop plant. The marker assisted selection is central to the improvement strategies for the world's major crops.

3. Gene Cloning

Putting a major gene or complex trait on the molecular map also affects the prospect of isolating the corresponding DNA locus, using nearby markers as jumping off points. If a genomic library is available, it may be possible to move from the clone carrying the marker tag to the clone with the gene of interest through a series of intermediate overlapping clones by chromosome walking. If a locus is saturated with markers, we can even go straight to the gene and achieve chromosomal landing. The availability of vehicles such as YACs and BACs for cloning large fragments of genomic DNA, and the production of libraries that relate directly to fully contiguous and highly saturated molecular maps has made map-based cloning a practical proposition for gene isolation.

4. Synteny

Molecular marker mapping has strengthened the fact that in several taxonomic groups of crop plants *e.g.,* gramineae which share the same common basic chromosome number, the linkage groups and the individual chromosome maps look very similar. When we take out the repetitive DNA and compare the maps for single copy sequences, we find that they are syntenic. This means that even between crops as diverse as wheat & rice the genes we are interested in are basically the same in both species and they line up into maps which are very similar. The added value to mapping is that not only we can use the same set of probes across wide species groups, but we can transfer map information, even entire maps, from one species to another. It is possible to know the location of genes of interest in, for example, wheat by reading the markers map for rice.

❑❑❑

Unit XII DNA SEQUENCING

PART 1	A GENERAL ACCOUNT
PART 2	PERFORMING DNA SEQUENCING IN LABORATORY BY THE SANGER'S DI-DEOXY METHOD USING SEQUENASE ENZYME

Part 1 A GENERAL ACCOUNT

Contents

INTRODUCTION

DNA Sequencing is the technique by which correct and precise order of nucleotides in a piece of DNA can be determined. Acquiring knowledge of nucleic acid sequences is at the heart of molecular biology. DNA sequencing today is chiefly used to characterize newly cloned DNA, to confirm the identity of a clone or mutation, to check the fidelity of a newly created mutation, ligation, junction or product of a polymerase chain reaction (PCR) and in some cases as a screening tool to identify polymorphisms and mutations in genes of particular interest.

The history of sequencing goes back to 1964 when attempt was made by Robert Holly and group to sequence tRNA and they were quite successful in doing so for 77 nucleotide long yeast alanine tRNA. Several attempts were made to apply this method to DNA but with limited success. Within 10 years, as a consequence of repeated and focused efforts, one American group led by Maxam and Gilbert at Harvard University and one English group led by Sanger at Cambridge, independently developed two methods for the large scale sequencing of DNA.

Maxam and Gilbert used a chemical cleavage protocol while Sanger et al designed a procedure similar to the natural process of DNA replication. The techniques, though radically different, are equally valuable. Both allow DNA sequences of several Kb in length to be determined in the minimum of time. Remarkably these two different methods were published in the same year (1977) in the same journal and both the teams shared the 1980 Nobel Prize. The above two methods and also a modified version-automated DNA sequencing are briefly described below one by one:

SANGER'S DI-DEOXY SEQUENCING METHOD

Also known as chain termination sequencing technique as controlled synthesis of DNA is carried out so that fragments are generated that terminate at specific point along the target sequence. The chain termination method is based on two facts about DNA synthesis.

(*i*) When a single stranded DNA template is placed together with the 4 deoxynucleoside triphosphates (dNTPs) and a short primer that hybridizes to the beginning of the DNA template, DNA polymerase will direct the synthesis of a complementary strand (Complementary to the DNA strand to be sequenced).

(*ii*) If di deoxynucleoside triphosphates (di-dNTPs) are included, DNA strand elongation will terminate when a di-dNTP is incorporated. This is because di-dNTP lack a 3' hydroxyl group which is needed to form a bond to an adjacent nucleotide (Fig 1.1).

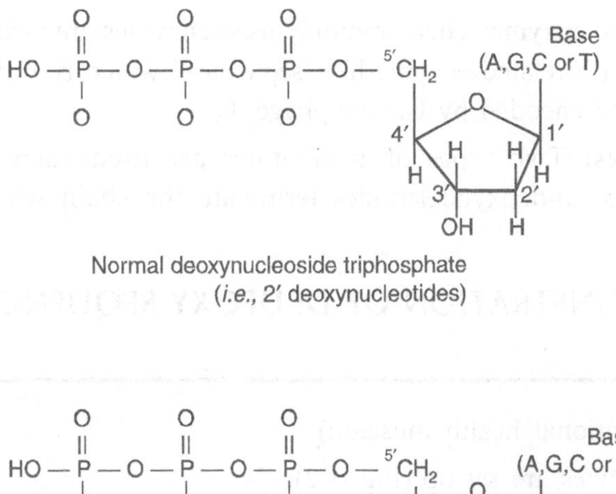

Normal deoxynucleoside triphosphate
(*i.e.*, 2′ deoxynucleotides)

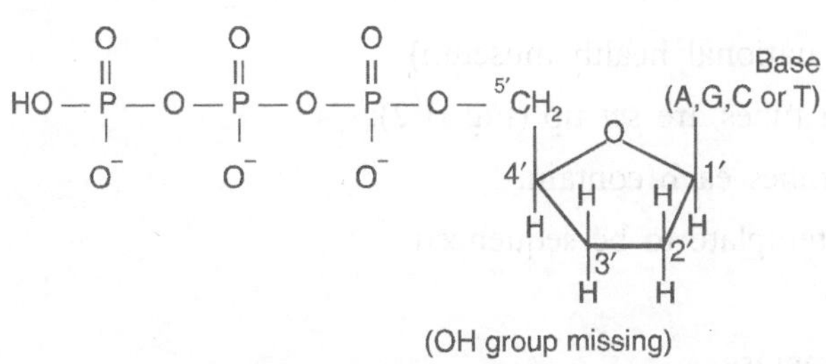

Di-deoxynucleoside triphosphate (i.e., 2′, 3′ di-deoxynucleotide)

FIGURE 1.1 **Schematic representation of deoxynucleotides and di-deoxynucleotides.**

Thus when a small amount of one of the di-dNTPs is included with the four conventional dNTPs in a reaction mixture for DNA synthesis, there is competition between two types of nucleotides for extension of the chain. If di-dNTPs succeed in this process, chain termination takes place. The products of the reaction are a population of oligonucleotide chains whose lengths are determined by the distance between the 5' terminus of the primer used to initiate DNA synthesis and the sites of chain termination. In a sequencing reaction containing di-dTTP for example, the termination points correspond to all positions normally occupied by a deoxythymine residue. By using the four different di-dNTPs in four separate enzymatic reactions, populations of oligonucleotide are generated that terminate at positions occupied by every A, C, G or T in the template strand. These populations of oligonucleotide are separated on gel by electrophoresis and location of each band can be ascertained by autoradiography or the emission of fluorescence. Following are the reaction components of a typical Sanger's di-deoxy reaction-

1. **Single stranded DNA template molecule (to be sequenced):** Single stranded DNA template can be obtained by a number of ways:
   (*i*) By cloning in M13 bacteriophage
   (*ii*) By asymmetric PCR
   (*iii*) By denaturing the double stranded PCR amplified DNA using heat or alkali.
2. **The primer:** The first step in sequencing experiment is to anneal a short oligonucleotide primer on to the single stranded DNA molecule. This primer will act as the starting point for the complementary strand synthesis reaction carried out by the polymerase enzyme.
3. **DNA polymerase enzyme**: The commonly used enzymes are klenow fragment of DNA Polymerase I , or a related enzyme like 'sequenase' which is a modified version of the DNA polymerase encoded by bacteriophage T_7.
4. **The Nucleotides:** Two types of nucleotides are used—deoxynucleotides and di-deoxynucleotides. di-deoxynucleotides terminate the chain wherever incorporated.

THE DRY-LAB DEMONSTRATION OF DI-DEOXY SEQUENCING OF CLONED DNA

(Source: www.ae@the national health museum)

The four reaction tubes are set up (Fig. 1.2):

The A, T, G, C tubes each contain:

(*i*) The DNA template to be sequenced
(*ii*) The primer
(*iii*) DNA Polymerase
(*iv*) All four dNTPs (dATP, dTTP, dCTP, dGTP), one of the dNTPs is radioactively labeled with ^{32}P.

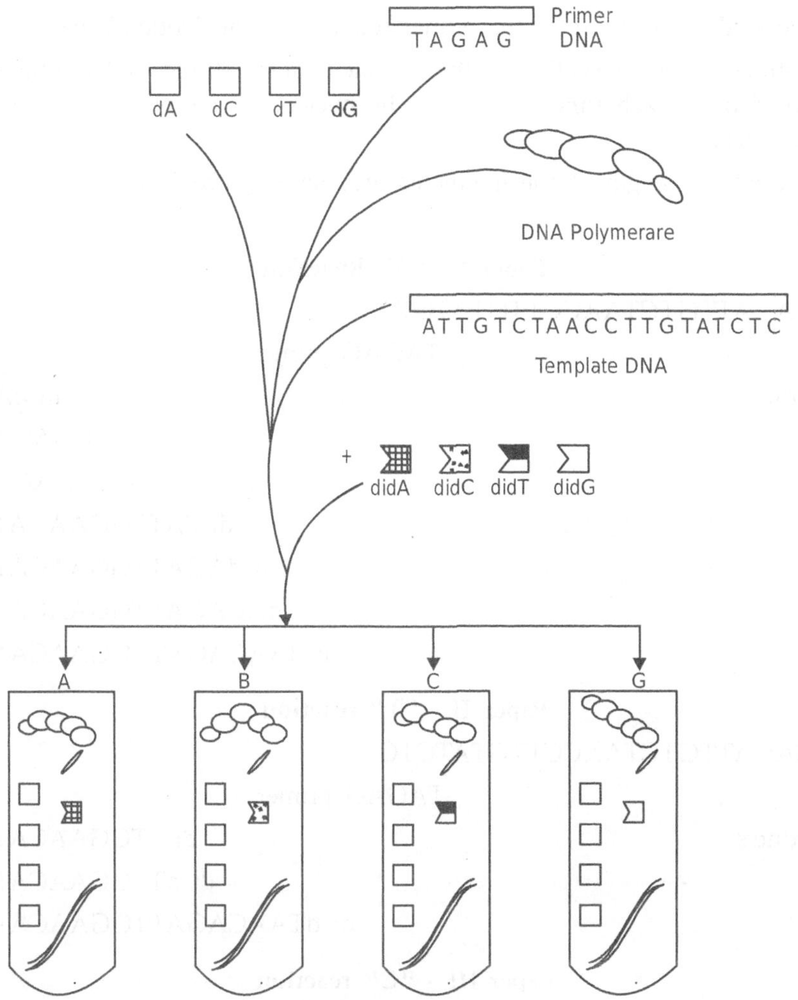

***FIGURE 1.2* Four reactions to be set up in Sanger's method of DNA sequencing.**

(*v*) One di-deoxynucleoside triphosphate per tube:

A reaction tube : di dATP
T reaction tube : di dTTP
G reaction tube : di dGTP
C reaction tube : di dCTP

The ratio of the di-deoxynucleotide to the deoxynucleotide in each tube is one to one hundred (1:100). So that once for every 100 nucleotides added, a di-deoxy form of the nucleotide is added to the elongating chain.

Activity

- Students are given 4 pages each with the sequence of DNA (which is to be determined) written at the top of each page.

- Label each page: "A Reaction", "T reaction", "G Reaction", "C Reaction".
- Ask the students to write in the primer sequence (5 nucleotides) may be.
- Then students are to synthesize the complementary strand and terminate with a di-deoxynucleotide each time they reach the nucleotide complementary to that reaction tube di dNTP.
- Then count the length of the fragments they have synthesized.

Example:

Paper I - "A" Reaction

DNA: ATTGTCTAACCTTGTATCTC

TAGAG primer

Products:

di dATAGAG (6)
di dACATAGAG (8)
di dAACATAGAG (9)
di dATTGGAACATAGAG (14)
di dAGATTGGAACATAGAG (16)
di dACAGATTGGAACATAGAG (18)
di dAACAGATTGGAACATAGAG (19)

Paper II - "T" reaction

DNA: ATTGTCTAACCTTGTATCTC

TAGAG primer

Products:

di dTGGAACATAGAG (12)
di dTTGGAACATAGAG (13)
di dTAACAGATTGGAACATAGAG (20)

Paper III - "C" reaction

DNA: ATTGTCTAACCTTGTATCTC

TAGAG primer

Products:

di dCATAGAG (7)
di dCAGATTGGAACATAGAG (17)

Paper IV - "G" reaction

DNA: ATTGTCTAACCTTGTATCTC

TAGAG primer

di dGAACATAGAG (10)
di dGGAACATAGAG (11)
di dGATTG GAACATAGAG (15)

Note that all the fragment sizes are different because each one corresponds to a different nucleotide location on the DNA to be sequenced. When all the fragment sizes are listed in order, the numbers should be sequential from 6–20 (the size of the DNA).

Ask the students to draw a gel picture and have them fill in the fragments according to size in 4 lanes corresponding to the 4 reactions (Fig 1.3).

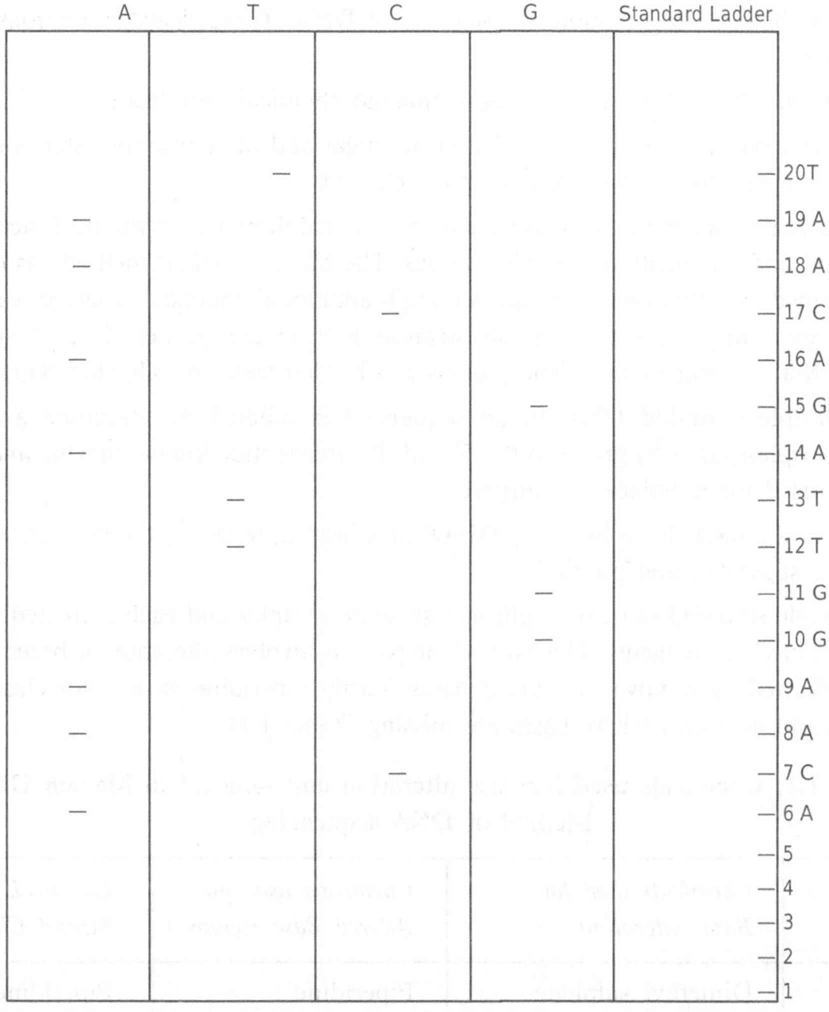

FIGURE 1.3 **Migration pattern of different fragments, obtained during Sanger's dideoxy chain termination process on polyacrlyamide gel.**

Now read the sequence of nucleotides from the bottom to up *i.e.,* on the above gel the lowest band (the smallest fragment which traveled the farthest) is A, then the next smallest is C, then A. The sequence that is read is the complement to the DNA that was sequenced.

Read up sequence **TAACAGATTGGAACA**

The DNA sequence in question- **ATTGTCTAACCTTGT**

(Complementary to the read up sequence)

MAXAM AND GILBERT'S METHOD OF DNA SEQUENCING

Unlike the chain-termination technique, which involves enzymatic synthesis, the Maxam-Gilbert method entails chemical degradation of a segment of DNA. These cleavage reactions are carried out in 2 stages:

(*i*) Specific bases (or types of bases) undergo chemical modification.

(*ii*) The modified base is removed from its sugar and the phosphodiester bonds towards the 3' end to the modified bases are cleaved.

For large scale sequencing, it is now-a-day less favored than the enzymatic di-deoxy method. However it still finds applications and illustrations. The Maxam-Gilbert method has not changed significantly since its initial development. Although additional chemical cleavage reactions have been devised by many scientists, most investigators rely on the quintet of reactions originally described by Maxam and Gilbert. The process can be described in following stages:

(*i*) Double stranded DNA to be sequenced is labeled by attaching a radioactive phosphorous (P^{32}) group to the 5' end. Polynucleotide kinase enzyme and ^{32}P-dATP is used for radiolabeling purpose.

(*ii*) Using dimethyl sulphoxide (DMSO) and heating to 90°C, the two strands of DNA are separated and purified.

(*iii*) Single stranded sample is split into separate samples and each is treated with one of the cleavage reagents. This part of the process involves alteration of bases (Fig 12.1.4) followed by removal of altered bases. Lastly piperidine is used for cleavage of the strand at points where bases are missing (Table 1.1).

Table 1.1. Chemicals used for base alteration and removal in Maxam Gilbert Method of DNA sequencing

Base Specify	*Chemicals used for Base Alteration*	*Chemicals used for Altered Base Removal*	*Chemicals used for Strand Cleavage*
G	Dimethyl sulphate	Piperidine	Piperidine
A+G	Acid	Acid	Piperidine
C+T	Hydrazine	Piperidine	Piperidine
C	Hydrazine+1.5M NaCl	Piperidine	Piperidine
A>C	1.2 N NaOH	Piperidine	Piperidine

In every case, the reactions are carried out under carefully controlled conditions to ensure that on an average only one of the target bases in each DNA molecule is modified.

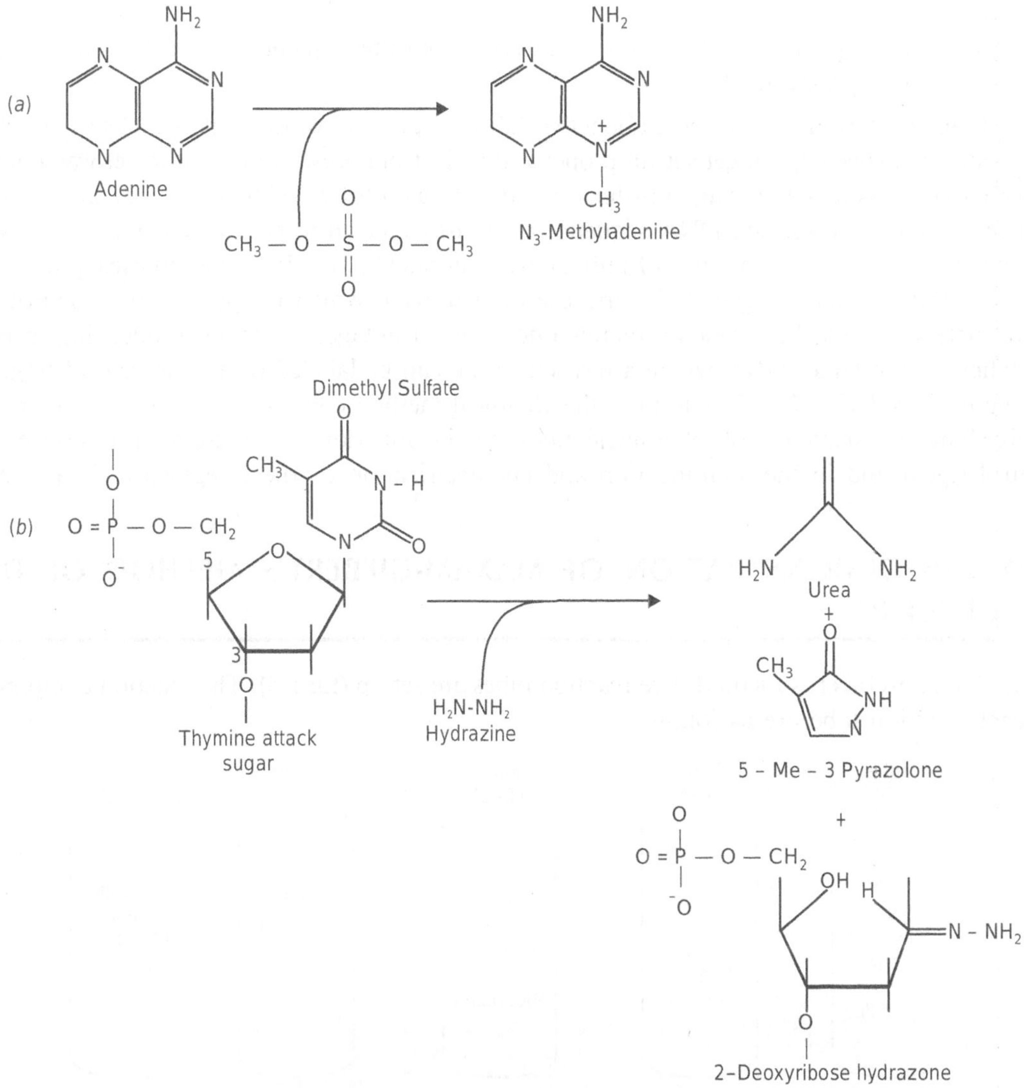

FIGURE 1.4 **(*a*) Alteration of Purines with dimethyl sulphate**
(*b*) Alteration of Pyrimidine (*e.g.,* Thymine) with hydrazine.

(*iv*) If the reactions have been arranged to give only one, or a few, cleavages per DNA molecule, a nested set of end–labeled DNA fragments of different lengths is produced.

(*v*) The samples are run together on a sequencing gel which separates the fragments by electrophoresis depending on their size. A sequencing gel is a high resolution gel designed to fractionate single stranded (denatured) DNA fragments on the basis of their length. Such gels routinely contain 6–20% polyacrylamide and 7 M Urea. The urea is a denaturant whose function is to minimize DNA secondary structure which effects electrophoretic mobility. The gel is run at sufficient power to heat up to about 70°C. This also minimizes DNA secondary structure. The labeled DNA bands obtained after such electrophoresis are revealed by autoradiography on large sheets of X-ray film.

(*vi*) The sequence can then be read directly from the sequencing ladders in the adjacent base specific tracks.

If all went well *i.e.,* if the radioactive labeling process worked, if the cleavage actions performed as expected, the gel set up properly, the electrophoresis worked, the gel was not torn or otherwise destroyed during transfer, and the X-ray film developer did not break down, 200–300 bases of confirmed DNA sequence could be expected to get every few days. However this involves–use of large amounts of radioactive material (^{35}S or ^{32}P), labor intensive preparation of paper thin acrylamide gels daily, and use of hydrazine which happens to be a neurotoxin. Nevertheless, chemical sequencing retains one clear advantage: it enables sequencing to begin anywhere in the target DNA where a restriction site can be labeled or an end-labeled fragment can be produced via PCR. In addition, the chemical method continues to have a crucial role in establishing the sequence of oligonucleotides, in the functional dissection of transcriptional control signal and in the identification and characterization of these regions in intact cells.

DRY LAB DEMONSTRATION OF MAXAM-GILBERT'S METHOD OF DNA SEQUENCING

In the Maxam-Gilbert's method, five reaction tubes are set up (fig 1.5). The reaction components of each reaction tube are as follow:

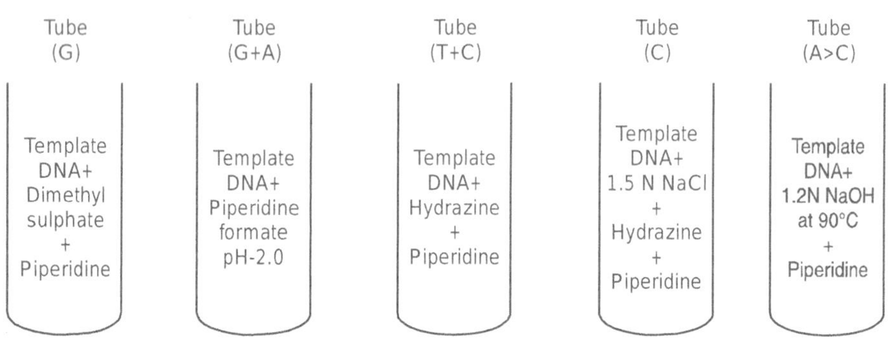

FIGURE 1.5 **The five reactions set-up in Maxam-Gilbert's Method of DNA sequencing.**

(*i*) Tube G – Radiolabeled template DNA, Dimethyl sulphate, Piperidine

(*ii*) Tube G + A – Radiolabeled template DNA, Piperidine formate (pH 2.0).

(*iii*) Tube T + C – Radiolabeled template DNA, Hydrazine, Piperidine

(*iv*) Tube C – Radiolabeled template DNA, 1.5N NaCl, Hydrazine, Piperidine

(*v*) Tube A > C – Radiolabeled template DNA, 1.2N NaOH at 90°C.

Activity

(*i*) Students are given 5 pages each with the sequence of DNA to be determined written at the top of each page.

(*ii*) Label each page "G reaction", "G + A" reaction, "T + C" reaction, "C" reaction and "A > C" reaction.

(*iii*) Ask them to make the possible fragments cleaved from the target DNA in each reaction tube (Phosphodiester towards 3' of modified base is cleaved).

Example:

DNA to be sequenced:

5' ^{32}P-GATCCAAGT$_3$

Page I - G reaction

DNA: 5 '^{32}P-G A T C C A A G T- 3'

Products: ^{32} P-G A T C C A A G

^{32}P-G

Page II - "G + A" reaction

DNA: 5' ^{32}P-G A T C C A A G T 3'

Products: 5'^{32}P-G

5'^{32}P-G A T C C A A G

5' ^{32}P-G A

5' ^{32}P-G A T C C A

5' ^{32} P-G A T C C A A

Page III - "T + C" reaction

DNA: 5'^{32}P-G A T C C A A G T

Products: 5'^{32}P-G A T

5'^{32}P-G A T C C A A G T

5'^{32}P-G A T C

5' ^{32}P- G A T C C

Page IV - "C" Reaction

DNA: 5' ^{32}P G A T C C A A G T$^{3'}$

Products: 5' ^{32}P-G A T C

5' ^{32}P-G A T C C

Page V- "A > C" Reaction

DNA: 5' ^{32}P G A T C C A A G T$^{3'}$

Products: 5' ^{32}P G A

5' ^{32}P G A T C C A

5' ^{32}P G A T C C A A

5' ^{32}P G A T C (Less frequent)

5' ^{32}P G A T C C (Less frequent)

(*iv*) Now give each student a piece of paper which represents a polyacrylamide gel. Have them draw in the fragments according to size in 5 lanes corresponding to 5 reactions (Fig 1.6).

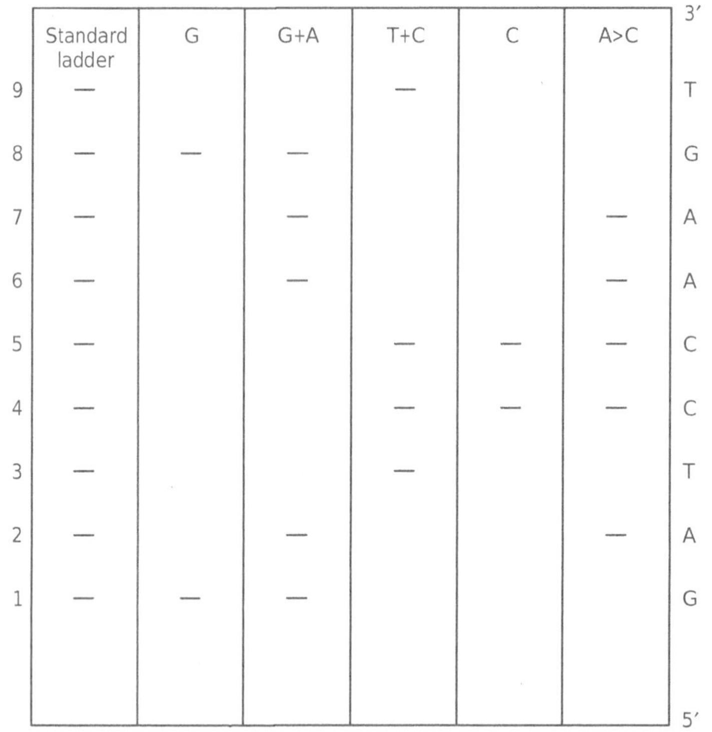

FIGURE 1.6 **Migration pattern of fragments obtained by cleavage in five reaction tubes during Maxam-Gilbert's method of DNA sequencing.**

(*v*) Read the sequence from bottom to top which will directly give the sequence from 5' end to 3' end.

AUTOMATED DNA SEQUENCING

Automated DNA sequencing utilizes a fluorescent dye to label the nucleotides instead of a radioactive isotope. The fluorescent dye is not an environmentally hazardous chemical and has no special handling or disposal requirement. Instead of using X-ray film to read the sequence, a laser is used to stimulate the fluorescent dye. The fluorescent emissions are collected on a charge coupled device that is able to determine the wavelength. Some systems like perkin-Elmer applied Biosystems (ABI), DNA sequencers are designed to discriminate all four fluorescent dye wavelengths simultaneously, which allows for complete DNA sequencing in one lane on the gel.

Varying degrees of automation are also available. For full automation, all which is required is to load a sample tray with template DNA; the equipment performs the labeling and analysis. The other option is to perform the labeling reactions with fluorescent dyes, load the samples on to a gel, and place the gel into the DNA sequencer. The equipment performs the separation and analysis. The system automatically identifies the nucleotide sequence and saves the information on the computer. Thus, only a review of the data is necessary to ensure no anomalies were misidentified by the computer.

The greatest obstacle to researchers when converting from manual to automatic DNA sequencing is being required to learn the use of computer software necessary to interpret the results. Automated DNA sequencing provides more reliable research results than manual DNA sequencing, thus maintaining the integrity of the research (For more detail see Molecular cloning-A Laboratory Manual by Sambrook and Russell).

❑❑❑

Part 2

PERFORMING DNA SEQUENCING IN LABORATORY BY THE SANGER'S DIDEOXY METHOD USING SEQUENASE ENZYME

Contents

INTRODUCTION

The objective of this chapter is to perform DNA sequencing by the Sanger di-deoxy method using Sequenase enzyme, a modified version of bacteriophage T_7 DNA polymerase. The reaction is carried out in following three steps:

1. Annealing of primer to SS DNA.
2. A brief polymerization reaction in which the primer is elongated by 20-100 nucleotides. This step is performed when DNA polymerase is present in excess and the

concentrations of the four dNTPs are limiting. A radio labeled dNTP is included in the reaction so that the short elongation products become radio labeled at multiple sites. During this preliminary polymerization reaction most of the available dNTPs, including the radio labeled dNTP, are incorporated into DNA.

3. Extension and termination of the elongated radio labeled primers. This step is carried out in four reactions that contain high concentrations of unlabeled dNTPs and a single unlabeled ddNTP (ddA, ddG, ddC or ddT).

PRINCIPLE

Di-deoxy DNA sequencing is based on the principle that DNA polymerization (replication) on a template strand can be interrupted by the presence of dideoxyribonucleoside triphosphates (di-dNTP) in the reaction mixture (explained in detail in Part 1 of the same unit).

Several vector systems have been designed which are of the upmost importance in facilitating the development of sequencing technology. A series of M13 derived vectors constitutes one family which takes advantage of the relative ease of preparation of SS DNA from viral vectors.

M13 phage vectors: The important feature of this system is that the DNA produced is single stranded and is extruded into the medium without lysis of the host. In this system, as opposed to purely plasmid vectors, the direct production of SS DNA simplifies the work because there is no necessity to separate complementary strands prior to sequencing. SS DNA can be conveniently harvested by centrifugation of cellular material. The supernatant containing the SS DNA is phenol extracted to remove viral proteins followed by precipitation of the SS DNA with polyethylene glycol (PEG).Separation of the DNA from the PEG affords single stranded DNA for sequencing.

REAGENTS/MATERIALS

1. Single stranded DNA (to be sequenced) (1 μg) for each set of four sequencing reactions.
2. Deionized distilled H_20 (ice cold)
3. Dithiothreitol (100mM)
4. 5 X labeling mixture (contains unlabeled dATP, dCTP, dTTP, each at a concentration of 7.5mM).
5. ddNTP extension/termination mixtures (contain all four dNTPs at a concentration of 160μm and a single ddNTP at a concentration of 0.16 μM).
6. Formamide-loading buffer.
7. Sequenase dilution buffer.

   10 mM Tris-Cl (pH 7.5)

   5 mM dithiothreitol

   0.50 mg/ml bovine serum albumin
8. 5X sequenase reaction buffer with $MgCl_2$

   200 mM Tris-Cl (pH 7.5)

   100 mM $MgCl_2$

125 mM NaCl

Store at – 20°C

9. Sequenase enzyme
10. Oligonucleotide primer
11. 5'^{32}P dATP
12. TE buffer
13. Microfuge tubes
14. Microfuge rotor
15. Water bath

EXPERIMENTAL LAY-OUT

To 0.5 ml microfuge tube, add SS DNA, primer, sequenase reaction buffer and water

↓

Incubate at 65°C for 2 min (Annealing reaction)

↓

Remove and cool

Remove and cool + ^{32}P dATP + Diluted sequenase enzyme + Diluted labeling buffer →

Labeling Reaction

↓

Transfer to pre-warmed microfuge having ddNTP extension/termination mixtures

↓

Centrifuge for few seconds

↓

Incubate in water bath at 37°C for 3-5 minutes

↓

Add 4 μl of formamide loading buffer

↓

Load on sequencing gel

↓

Run the gel and analyze the results

METHOD

1. To a 0.5 ml microfuge tube, add:

   Single Stranded template DNA (1μg/μl) – 1 μl.

   Oligonucleotide primer (1ng/μl) – 3 μl

   5X sequenase reaction buffer with $MgCl_2$ – 2 μl.

   Add H_2O to 10 μl.
2. Incubate the above tube for 2 minutes at 65°C in water bath. Remove the tube and allow it to cool.
3. Meanwhile thaw the 5X labeling and ddNTP extension/termination mixtures and radiolabeled dATP. Store the thawed solutions on ice.
4. Transfer 2.5 ml of each ddNTP extension/termination mixture into separate 0.5 ml microfuge tube, labeled C, T, A and G.
5. Make a five fold dilution of the 5X labeling mixture in ice-cold H_2O (A Volume of 2 μl of the diluted labeling mixture is required for each template sequenced).
6. Dilute sequence in ice-cold sequenase dilution buffer.
7. To carry out the labeling reaction, add the following to the 10 μl annealing reaction of step2:

   Diluted labeling mixture (from step 5) - 2 μl

   0.1 M dithiotreitol - 1 μl

   ^{32}P dATP – 0.5 μl

   Diluted sequenase (1.6 units/μl) – 2.0 μl
8. Warm the labeled microfuge of step 4. Then transfer the 3.5 μl of the labeling reaction (step7) to the pre-warmed labeled microfuge tubes.
9. Place the microfuge tubes in a microfuge at room temperature and centrifuge for a few seconds at 2000 rpm to mix the components of the reaction. Immediately transfer the reaction to water bath for 3–5 minutes at 37°C.
10. Stop the reaction by adding 4 ml of formamide loading buffer.
11. The reaction may be stored up to 5 days at –20°C or analyzed directly by denaturing gel electrophoresis. After heat denaturation and quick cooling on ice, load 3 ml of each of the C, T, A and G reactions into individual wells of a sequencing gel.

REMEMBER

1. Amount and ratio of primer and template DNA in the reaction must be optimized. Because if this is not optimum *i.e.,* the number of primer: template complex in the reaction is too low, this result in synthesis of a labeling reaction, rather than many short products. Consequently bands in the gel will appear faint at the bottom.

2. This protocol is suitable for obtaining the sequence of DNA lying up to 300-400 nucleotides from the 5' end of the primer.
3. Pyrophosphatase can be mixed with sequenase at a ratio of 1:3. It is used to prevent reduction in band intensity during prolonged incubation of sequencing reactions.
4. Do not incubate the labeling reaction at temperature >20°C for longer than 5 minutes.

❑❑❑

Unit **XIII**

IN VITRO MUTAGENESIS

Part 1 A GENERAL ACCOUNT

Contents

INTRODUCTION

The term in vitro mutagenesis means a process of introducing mutations in DNA by artificial means. Mutations so induced can be expressive (*i.e.,* phenotypic change is there) or silent (*i.e.,* phenotype remains unaltered) depending upon which sequence has been mutated. Mutagenesis, particularly site directed mutagenesis, is very informative in genomics (studying DNA structure and function) as well as in proteomics (for producing desirable changes in proteins).

DIFFERENT METHODS

1. By the use of chemicals (known as chemical mutagenesis)

A variety of techniques are available for introducing mutations in vitro like:

A large number of chemicals are available which are capable of introducing mutations in single as well as double stranded DNA. Some of these are listed below:

(*i*) Hydroxylamine (converts G-C basepairs to A-T basepairs and vice versa).

(*ii*) Acridines (causes addition as well deletion of one to at least twenty bases).

(*iii*) Nitrous acid (converts G-C basepairs to A-T basepairs and vice versa).

(*iv*) Hydrazine (causes all the possible basepair changes).

(*v*) Sodium bisulphate (deaminates cytosine resulting in conversion of cytosine into uracil. It results in incorporation of a complementary A residue during DNA synthesis and ultimately G-C is replaced by A-T).

Of the above sodium bisulphate is the most popular chemical mutagen. It can mutagenise DNA quite efficiently, but can produce only transition mutations. Further these mutations often result in conservative amino acid substitutions which limit the use of sodium bisulphate for inducing mutations. Other mutagens find a very limiting space in mutagenesis experiments because of the very low frequency of mutant's recovery.

2. By the use of mutated oligonucleotide (known as oligonucleotide directed mutagenesis)

This is the most common method used now-a-day for site directed mutagenesis. As the name indicates, here a short synthetic oligonucleotide is used for introducing the desired mutation. The basic plan (Fig 1.1) can be discussed in following points:

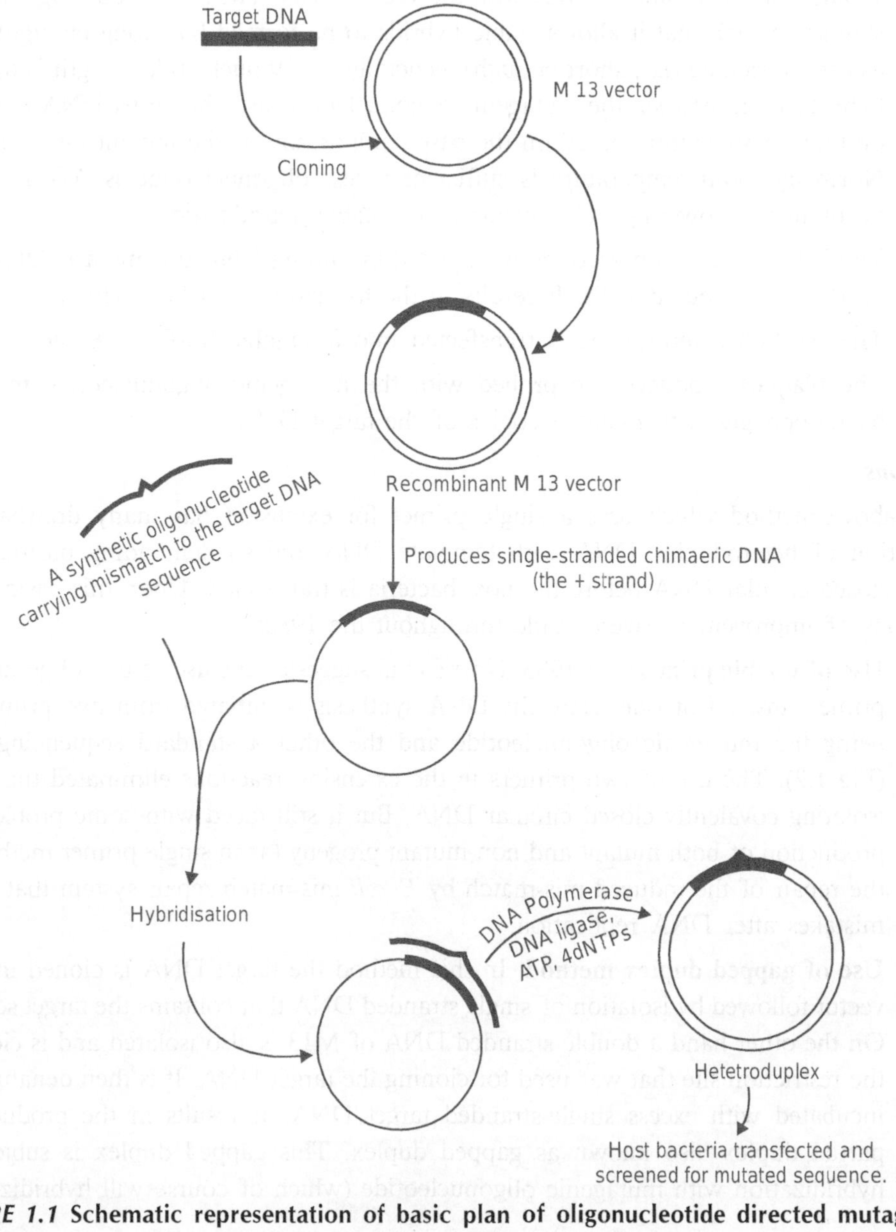

FIGURE 1.1 **Schematic representation of basic plan of oligonucleotide directed mutagenesis.**

(*i*) To clone the target DNA (where mutation is to be introduced) into a vector capable of producing single stranded DNA like bacteriophage M13 vectors. The size of the target DNA in the M13 or equivalent vector should be kept to a minimum because large DNA fragments are unstable in single stranded vectors and can undergo spontaneous deletions.

(*ii*) A synthetic oligonucleotide, complementary to the target DNA, is prepared in such a way that it contains a mis-match base at a desired site. The mutagenic oligonucleotide should be free from features that lead to formation of secondary structure as such structures decrease the efficiency of the process. Further mutagenic oligonucleotide should not form stable hybrids with the vector DNA. The size of the oligonucleotide should be such that it allows stable hybrids to be formed (*i.e.,* long enough) as well as easy screening (*i.e.,* short enough); generally 17–19 nucleotides length is used. The hybridization between the mutagenic oligonucleotide and the target DNA requires an optimum temperature based on the base composition of the mutant oligonucleotide. Normally room temperature is sufficient unless oligonucleotide is A-T rich which needs further lowering of temperature for stable hybridization.

(*iii*) The hybridized oligonucleotide is extended by adding DNA polymerase, DNA ligase, ATP and the four dNTPs. It results in the formation of a heteroduplex.

(*iv*) The resulted heteroduplex is transfected into a suitable host say *E.coli.*

(*v*) The plaques produced are probed with the mutagenic oligonucleotide to identify bacteriophages with mutated copies of the target DNA.

Modifications

The above method which uses a single primer for extension has many drawbacks like contamination of heteroduplex DNA with template DNA and so demanding purification of covalently closed circular DNA before the host bacteria is transfected. Upon this basic strategy a successions of improvements were made throughout the 1980s:

(*i*) **Use of double primers**: In 1983, Norris et al suggested the use of two oligonucleotide primers instead of one. Here the DNA synthesis is initiated from two primers, one being the mutagenic oligonucleotide and the other a standard sequencing primer (Fig 1.2). The use of two primers in the extension reactions eliminated the need of isolating covalently closed circular DNA. But it still faced with some problems like production of both mutant and non-mutant progeny (as in single primer method) and the repair of the induced mis-match by *E. coli* mis-match repair system that corrects mistakes after DNA replication.

(*ii*) **Use of gapped duplex method**: In this method the target DNA is cloned in a M13 vector followed by isolation of single stranded DNA that contains the target sequence. On the other hand a double stranded DNA of M13 is also isolated and is cleaved at the restriction site that was used for cloning the target DNA. It is then denatured and incubated with excess single-stranded target DNA. It results in the production of partial duplex also known as gapped duplex. This gapped duplex is subjected to hybridization with mutagenic oligonucleotide (which of course will hybridize to the

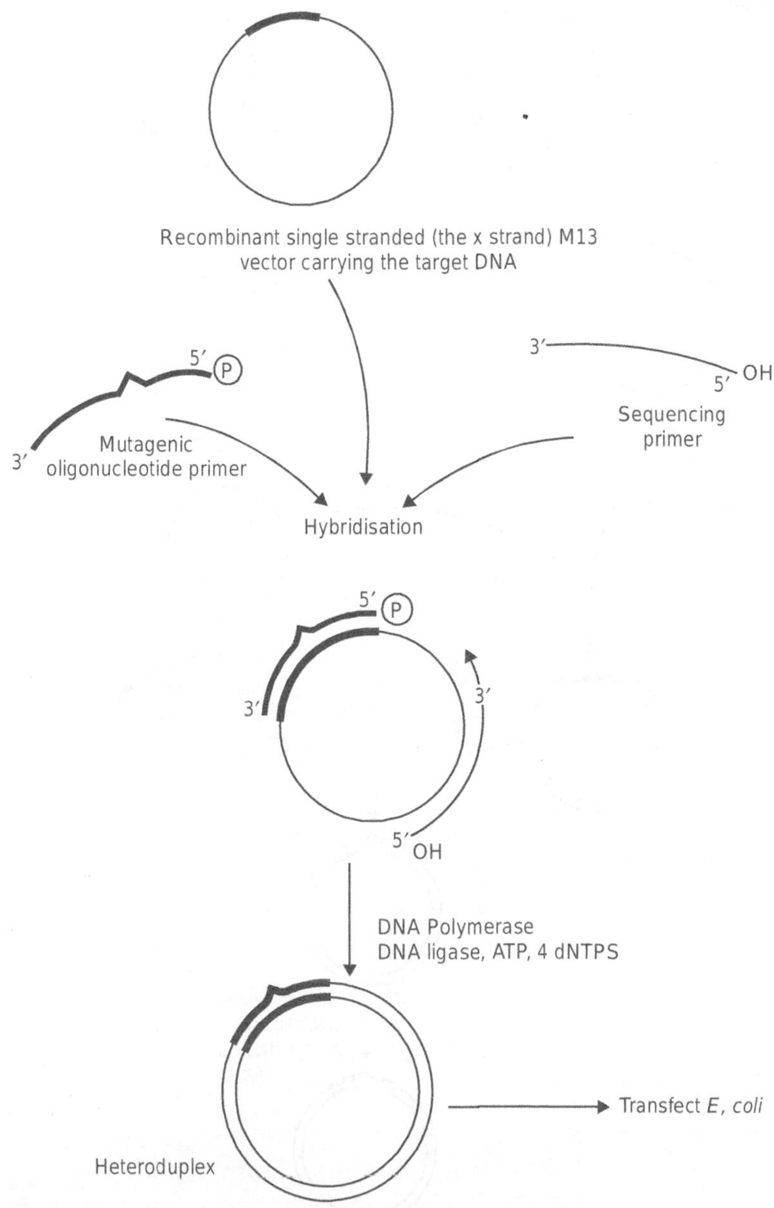

FIGURE 1.2 **Oligonucleotide—directed mutagenesis using double primer method.**

single stranded gap which is the target DNA).The gap is then filled in by the activity of Klenow enzyme and DNA ligase (Fig 1.3). After this the *E.coli* is then transfected with the recombinant vector. The method is highly efficient as compared to the previously discussed methods.

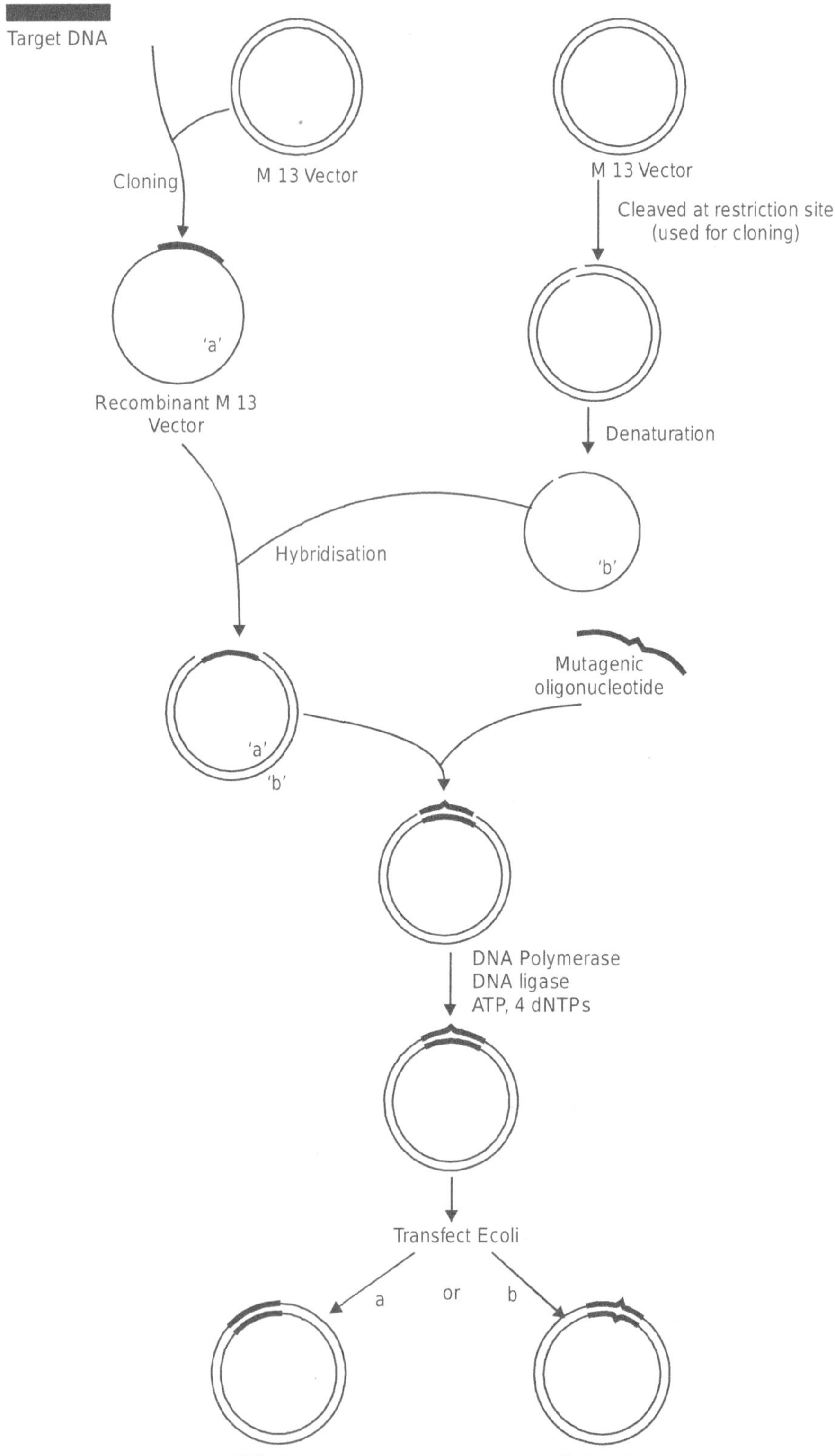

FIGURE 1.3 **Schematic representation of gapped duplex method of mutagenesis.**

(*iii*) **Use of coupled priming method**. This is based on the use of a pair of reciprocating selectable markers which enable a second selectable mutation to be introduced as the first is removed. So repeated round of mutagenesis can be carried out without the need of re-cloning. This is also a very efficient modification.

(*iv*) **The kunkel method**. This method depends on the use of a specialized host which is deficient in two enzymes, namely dUTPase (dut) and Uracil glycosylase (ung), for the production of single stranded DNA prior to mutagenesis (Fig 1.4). This will

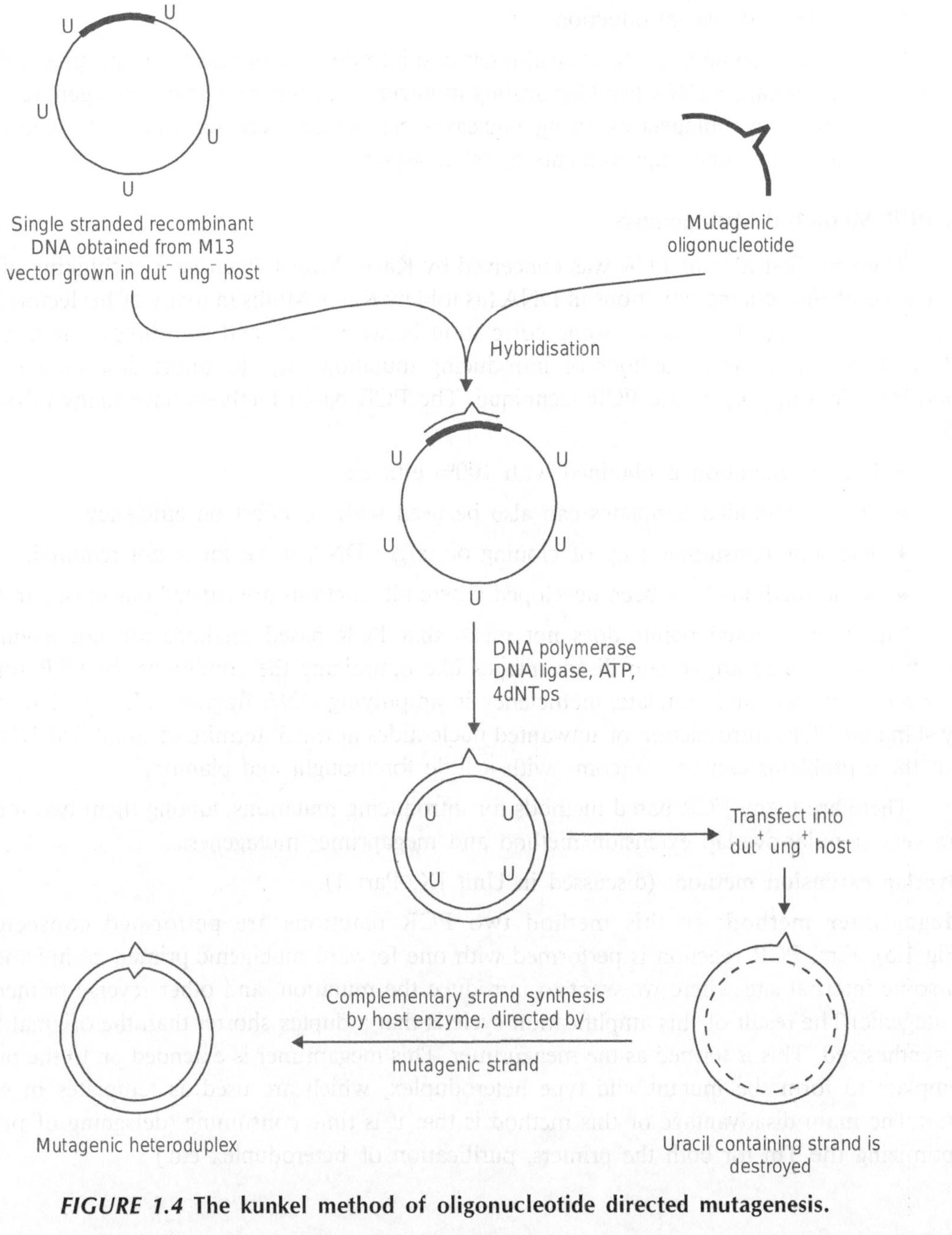

FIGURE 1.4 **The kunkel method of oligonucleotide directed mutagenesis.**

increase the intracellular pool of uracil and so in the subsequent cycles of DNA replication uracil will replace the thymine. This replacement can not be repaired because the host is ung-. If everything goes as expected on an average 20–30 uracil residues are incorporated in the genome of bacteriophage M13. Now if this bacteriophage is transfected into a ung+ host the uracil will be removed rapidly resulting in cleavage of phosphodiester backbone at the resulting apyrimidinic site. This strand is then subjected to destruction and a new strand complementary to the mutagenic strand is synthesized. This is a very popular method and has changed very little since its introduction.

Besides some methods were also given to introduce small insertions and small deletions in the target DNA like Gap sealing mutagenesis, Linker insertion mutagenesis, Linker scanning mutagenesis, using nucleases etc. which were successful in some aspects needing some improvements in other aspects.

3. PCR Mediated Mutagenesis

The very first idea of PCR was conceived by Karry Mullis when he was thinking of some new ways of introducing mutations in DNA (as told by Karry Mullis in many of his lectures).This is sufficient enough to think of some correlation between PCR and site directed mutagenesis. Most of the PCR based methods of introducing mutations are the direct descendants of the thermostable Taq polymerase PCR technique. The PCR based methods have many advantages like:

- Desired mutation is obtained with 100% efficiency.
- Double stranded templates can also be used with no effect on efficiency.
- The time consuming step of cloning of target DNA in vector is not required.
- Some methods have been developed where all reactions are carried out in one test tube.

The above quoted points does not mean that PCR based methods are not having any disadvantage. There are certain disadvantages like optimizing the conditions for PCR for each new set of primers and template, inefficiency in amplifying DNA fragments longer than 2–3 kb by standard PCR, introduction of unwanted nucleotides at the 3' termini of amplified DNA etc. But these problems can be overcome with a little forethought and planning.

There are many PCR based methods for introducing mutations, among them two methods are very popular-overlap extension method and megaprimer mutagenesis.

Overlap extension method: (discussed in Unit IX, Part 1).

Megaprimer method: In this method two PCR reactions are performed consecutively (Fig 1.5). First PCR reaction is performed with one forward mutagenic primer, complementary to some internal site where we want to introduce the mutation, and other reverse primer (non mutagenic).The result of this amplification cycle is that a duplex shorter than the original length is synthesized. This is termed as the megaprimer. This megaprimer is extended on to the original template to form the mutant:wild type heteroduplex, which are used as templates in second PCR.The main disadvantage of this method is that it is time consuming (deisgning of primers, optimizing the Tm for both the primers, purification of heteroduplex etc.)

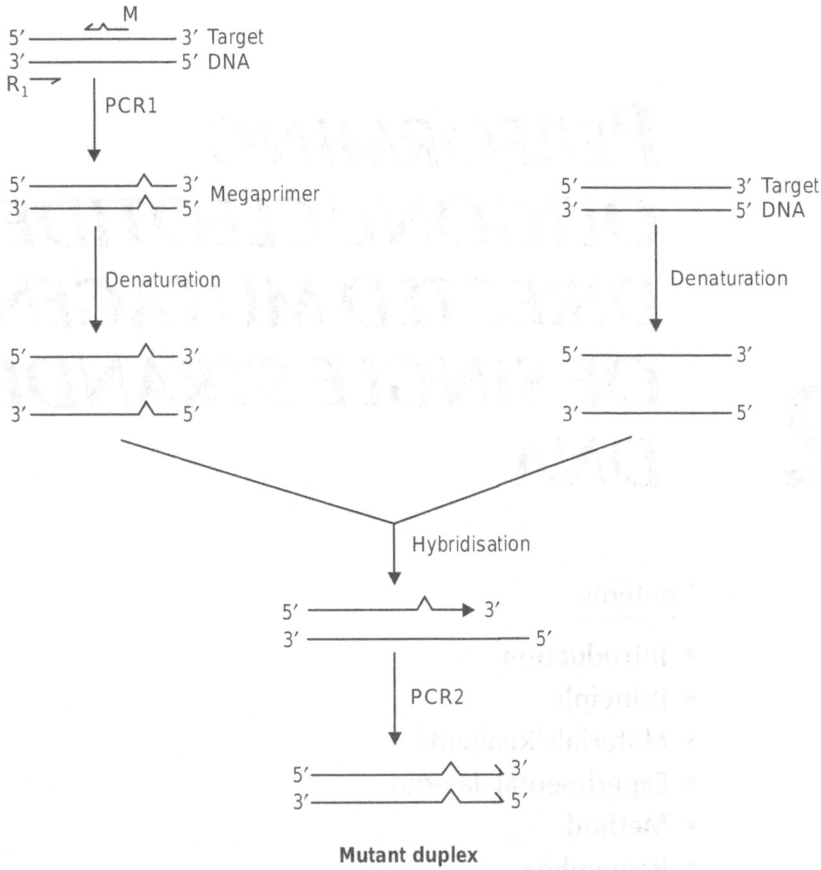

FIGURE 1.5 **The megaprimer method of mutagenesis.**

4. Commercial Kits for Mutagenesis

A wide variety of commercial kits are also available for introducing mutation at a desired site like:

Supplier	***Mail Address***
1. Bio-Rad	www.bio-rad.com
2. Life Technologies	www.lifetech.com
3. Promega	www.promega.com
4. Amersham-Pharmacia Biotech	www.apbiotech.com
5. Stratagene	www.stratagene.com

❑❑❑

Part 2

PERFORMING OLIGONUCLEOTIDE DIRECTED MUTAGENESIS OF SINGLE STRANDED DNA

Contents

INTRODUCTION

See Part 1, Unit XIII.

PRINCIPLE

This is based on the principle of double primer method and the Kunkel method. Latter offers great advantage of the strong selection against uracil substituted DNA exhibited by strains of E. coli that express uracil DNA glycosylase. Template DNA is first prepared in ung$^-$ dut$^-$ host by cloning in M 13 bacteriophage. The resulting uracil (instead of thymine) containing DNA is used as a template which result in the formation of a heteroduplex DNA with uracil in the template strand and thymine in the newly *in vitro* synthesized strand.

MATERIALS/REAGENTS

1. Ethanol
2. NaCl
3. Phenol
4. Chloroform
5. Sodium acetate
6. TE
7. Agarose
8. YT medium
9. ATP (10 mM)
10. Buffer I (10X)
    200 mM Tris-Cl
    100 mM $MgCl_2$
    500 mM NaCl
    10 mM dithiothreitol
11. Buffer II (10X)
    200 mM Tris-Cl
    100 mM $MgCl_2$
    100 mM dithiothreitol
12. E. coli dut^- ung^- strain
13. M13 vector
14. DNA ligase
15. DNA polymerase
16. T4 polynucleotide kinase
17. dNTPs mix
18. M13 universal primer
19. Mutagenic oligonucleotide primer
20. Mutagenic M13 SS DNA template
21. Target DNA
22. Marker DNA

EXPERIMENTAL LAY-OUT

Cloning the target DNA into M13 vector

↓

Harvest the sufficient recombinant bacteriophage plaques

↓

Transfect the E.coli strain CJ 236 (dut^- ung^- F') with the recombinant bacteriophage

↓

Isolate the bacteriophage pellet

↓

Extract single stranded uracil containing M13 bacteriophage DNA using phenol

↓

Purify the above template by spun column chromatography

↓

Anneal the mutagenic oligonucleotide primer and universal primer to the above template

↓

Formation of Double stranded DNA by adding suitable DNA polymerase, DNA ligase in the presence of dNTPs mix

↓

Transfect competent E.coli strain TGI

↓

Screen the plaques either by sequencing or by hybridization with radiolabeled probe.

METHOD

1. Target DNA is cloned into the M13 vector following the appropriate strategy.
2. Transfect the normal E. coli host with this recombinant bacteriophage and allow the plaques to grow.
3. Transfer a single plaque produced by the recombinant bacteriophage M13 into a microfuge tube containing 1 ml of YT medium.
4. Incubate the above microfuge at 65°C for 5 minutes to kill the bacterial cells.
5. Go for vortexing of the tube to release the bacteriophages trapped in top agar.
6. Centrifuge the tube for 30 sec to remove dead bacterial cells which form a pellet.
7. Take approximately 40 µl of the supernatant to a 500 ml flask containing 40 ml of 2X YT medium with 0.20 µg/ml uridine.
8. Add 5 ml of mid log phase culture of *E.coli* strainCJ236 (dut$^-$ ung$^-$ F').
9. Incubate the culture with constant shaking for 5–7 hrs at 37°C.
10. Centrifuge the above culture at 6000 rpm for 25–30 min at 4°C to pellet the cells.
11. Transfer the supernatant to a fresh centrifuge bottle.
12. Determine the relative titre of the bacteriophage suspension on *E.coli* dut$^-$ ung$^-$ strain (say CJ236) and a normal strain (such as JM109). The titre on strain CJ236 should be four to five orders of magnitude greater than that on the JM109 strain.
13. Add 0.25 volumes of 2.5 M of NaCl containing 15% (w/v) PEG. Mix the contents and incubate on ice for 1 hour.
14. Centrifuge at 5000rpm for 20 min at 4°C to pellet down the bacteriophages.
15. Transfer the bacteriophage pellet to a centrifuge tube containing 4 ml of TE. Vortex it vigorously and incubate on ice for 1 hour.
16. Centrifuge at 6000 rpm for 20 minutes at 4°C to pellet the bacterial debris.
17. Transfer the supernatant to a 15 ml polypropylene tube. Extract the suspension twice with phenol and once with phenol: chloroform.
18. Centrifuge at 5500 rpm for 5 minutes to separate the phases.
19. Transfer the aqueous phase form the final extraction to a glass centrifuge tube
20. Add 0.2 volume of sodium acetate followed by 2 volumes of ethanol at 0°C.
21. Incubate on ice for 30 minutes and recover the DNA pellet by centrifugation.
22. Dissolve the DNA in approximately 150 µl of TE.
23. Purify the uracil containing template by spun-column chromatography.
24. Phosphorylate the mutagenic oligonucleotide and the universal primer with T4 polynucleotide kinase in two different tubes. In each tube, add the respective oligonucleotide, kinase enzyme, buffer, ATP and water for phosphorylation to occur.
25. Anneal the phosphorylated mutagenic oligonucleotide and universal sequencing primer to the single stranded M13 containing the target sequence in the presence of Buffer I.

26. After cooling, mix the following in a fresh microfuge tube:

10X buffer II	1.0 μl
2 mM dNTP solution	1.0 μl
10 mM ATP	1.0 μl
DNA ligase	5 units
DNA polymerase	2.5 units
H_2O	to 10 μl

27. Add 10 μl of the solution in step 25 to the reaction mixture of step 26.
28. Incubate this final reaction mixture at 15°C for 5-15 hours.
29. Make a series of dilutions of the above mix (1:10, 1:100, and 1:500) in 10mM Tris-Cl.
30. Into different ice chilled tubes, add 1 μl of the original reaction buffer and 5 μl of the dilution. Add 200 μl of competent JM109 cells to each tube.
31. Incubate on ice for 30 minutes.
32. Add 2.5 ml of YT top agar (melted and cooled) to each tube and plate on the YT agar plate.
33. Incubate for 12–16 hours at 37°C to allow plaques to form.
34. Screen the recombinant plaque either by hybridization with a radiolabeled probe or by sequencing for the confirmation of mutation.

REMEMBER

1. The mutated DNA can be introduced into E.coli using electroporation also but frequency of this is generally low as compared with the transfection method.
2. Always purify the mutagenic oligonucleotide by column chromatography unless it is more than 30 nucleotides in length needing polyacrylamide gel electrophoresis for purification, before carrying out mutagenesis.
3. The molar ratio of primer to template should be between 10:1 and 50:1.
4. The reaction conditions should be selected keeping in view the type of polymerase being used.

❑❑❑

MOLECULAR BIOLOGY GLOSSARY

A

A—Adenine residue, in either DNA or RNA.

Acridine dyes—A class of positively charged polycyclic molecules that intercalate into DNA and induce frame shift mutations.

Acrylamide gels—See polyacrylamide gels.

Adaptor—A synthetic double-stranded oligonucleotide that has one end blunt, while the other end has a nucleotide extension that can base pair with a cohesive end created by a specific type II restriction endonuclease

Adenine ($C_5H_5N_5$; f.w. 135.14) (symbol: A)—A white crystalline purine base, a constituent of DNA, RNA. Also, a B-group vitamin (B_4), generally available as $C_5H_5N_5.3H_2O$, with molecular weight 189.13. It is added to some tissue culture media, as adenine sulphate, to promote shoot formation and for its weak cytokinin effect. It is present in plant tissues combined with aminoamide, phosphoric acids and D-ribose.

Adenosine disphosphate (ADP)—*See* ATP.

Adenosine monophosphate (AMP)—*See* ATP.

Adenosine triphospate (ATP)—*See* ATP.

AFLP—*See* amplified fragment length polymorphism.

Agar (Malay, *agar-agar*)—A polysaccharide solidifying agent used in nutrient media preparations and obtained from certain types of red algae (*Rhodophyta*). Both the type of agar and its concentration can affect the growth and appearance of cultured explants.

Agarose—The main constituent of agar.

Agarose gel electrophoresis—A process in which a matrix composed of a highly purified form of agar is used to separate larger DNA and RNA molecules. *See* electrophoresis.

Aggregate—1. A clump or mass formed by gathering or collecting units.

2. A body of loosely associated cells, such as a friable callus or cell suspension.

3. A serological reaction (aggregation) in which the antibody and antigen react and precipitate out of solution.

Agrobacterium—A genus of bacteria that includes several plant pathogenic species, causing tumour-like symptoms. *See Agrobacterium tumefaciens;* crown gall; hairy root culture; *Ri* plasmid; *Ti* plasmid.

Agrobacterium tumefaciens—*A* bacterium that causes crown gall disease in some plants. The bacterium infects a wound, and injects a short stretch of DNA into some of the cells around the wound.

Agrobacterium tumefaciens-mediated transformation—A naturally occurring process of DNA transfer from the bacterium *A. tumefaciens* to plants.

Alkylating agents—Chemicals that transfer alkyl (methyl, ethyl, etc.) groups to the bases in DNA.

Allele (Gr. *allelon*, of one another, mutually each other); **allelomorph** (adj: allelic, allelomorphic): One of a pair, or series, of variant forms of a gene that occur at a given locus in a chromosome.

Allele frequency—The number of copies of an allele in a population, expressed as a proportion of the total number of copies of all alleles at a locus in a population.

Allele-specific amplification (ASA)—The use of polymerase chain reaction (PCR) at a sufficiently high stringency so that only a primer with exactly the same sequence as the target DNA will be amplified.

Alu sequences—A family of 300-bp sequences occurring nearly a million times in the human genome.

Ambient temperature—Air temperature at a given time and place; not radiant temperature.

Amino acid (Gr. *Ammon*, from the Egyptian sun god, in M. L. used in connection with ammonium salts): An acid containing the amine group, NH_2. In particular, any of 20 basic building blocks of proteins with a free amino (NH_2) and a free carboxyl (COOH) group, and having the basic formula NH_2: C_R - COOH is the amino acid.

AMP (adenosine monophosphate)—*See* ATP.

Ampicillin (â-lactamase)—A penicillin-derived antibiotic that prevents bacterial growth by interfering with synthesis of the cell wall.

Amplification—1. Treatment (*e.g.,* use of chloramphenicol) designed to increase the proportion of plasmid DNA relative to that of bacterial (host) DNA.

2. Replication of a gene library in bulk.

3. Duplication of gene(s) within a chromosomal segment.

4. Creation of many copies of a segment of DNA by the polymerase chain reaction (PCR)

Amplified fragment length polymorphism (AFLP)—A type of DNA marker, generated by digestion of genomic DNA with two restriction enzymes to create many DNA fragments, ligation of specific sequences of DNA (called adaptors) to the ends of these fragments, amplification of the fragments via PCR (using a set of primers with sequences corresponding to the adapters, plus various random combinations of three additional bases at the end), and visualization of fragments via gel electrophoresis.

Anion—A negatively charged ion; opposite to cation.

Anneal—The pairing of complementary DNA or RNA sequences, via hydrogen bonding, to form a double-stranded polynucleotide. Most often used to describe the binding of a short primer or probe.

Annealing—The process of heating (de-naturing step) and slowly cooling (re-naturing step) double-stranded DNA to allow the formation of hybrid DNA or complementary strands of DNA or of DNA and RNA.

Anonymous DNA marker—A DNA marker (q.v.) detectable by virtue of variation in its sequence, irrespective of whether or not it actually occurs in or near a coding sequence. Microsatellites are typical anonymous DNA markers.

Antibiotic—A class of natural and synthetic compounds that inhibit the growth of or kill some micro-organisms.

Antibiotic resistance—The ability of a micro-organism to produce a protein that disables an antibiotic or prevents transport of the antibiotic into the cell.

ARS (autonomous replicating sequence)—Any eukaryotic DNA sequence that initiates and supports chromosomal replication; they have been isolated in yeast cells. Also called as autonomous(ly) replicating segment.

Artificial medium—*See* culture medium.

Assay—1. To test or evaluate.

2. The procedure for measuring the quantity of a given substance in a sample (chemically or by other means).

3. The substance to be analysed.

ATP (adenosine triphosphate)—A nucleotide of fundamental importance as a carrier of chemical energy in all living organisms. It consists of adenosine with three phosphate groups, linked together linearly. The phosphates are attached to adenosine through its ribose (sugar) portion. Upon hydrolysis, bonds yield either one molecule of ADP (adenosine diphosphate) and an inorganic phosphate, or one molecule of AMP (adenosine monophosphate) and pyrophosphate; in both cases releasing energy that is used to power biological processes. ATP is regenerated by rephosphorylation of AMP and ADP, using chemical energy derived from the oxidation of food.

Autoclave—1. An enclosed chamber in which substances are heated under pressure to sterilize utensils, liquids, glassware, etc., using steam. The routine method uses steam pressure of 103.4×10^3 Pa at 121°C for 15 minutes, or longer to allow large volumes to reach the critical temperature.

2. A pressure cooker used to sterilize growth medium and instruments for tissue culture work.

B

BAC (bacterial artificial chromosome)—A cloning vector constructed from bacterial fertility (F) factors; like YAC vectors, they accept large inserts of size 200 to 500 kb. *See* cloning vector; YAC.

Bacillus thuringiensis (Bt)—A bacterium that kills insects; a major component of the microbial pesticide industry.

Backcross—Crossing an organism with one of its parents or with the genetically equivalent organism. The offspring of such a cross are referred to as the backcross generation or backcross progeny. *See* testcross.

Bacteria—Plural of bacterium, q.v.

Bacteriophage—A virus that infects bacteria. Also called as bacteriophage. Altered forms are used in DNA cloning work, where they are convenient vectors.

Bacterium (Gr. *bakterion*, a stick; pl: bacteria)—Common name for the class Schizomycetes : minute (0.5-5 m), unicellular organisms, without a distinct nucleus. Bacteria are prokaryotes, and most of them are identified by means of Gram staining (q.v.).

Base analogues—Unnatural purine or pyrimidine bases that differ slightly in structure from the normal bases, but can be incorporated into nucleic acids. They are often mutagenic.

Base pair (bp)—The two strands that constitute DNA are held together by specific hydrogen bonding between purines and pyrimidines (A pairs with T; and G pairs with C). The size of a nucleic acid molecule is often described in terms of the number of base pairs (symbol : bp) or thousand base pairs (kilobase pairs; symbol: kb; a more convenient unit) it contains.

Base substitution—Replacement of one base by another in a DNA molecule. *See* transition; transversion.

Bifunctional vector—*See* shuttle vector.

Binary vector system—A two-plasmid system in *Agrobacterium tumefaciens* for transferring into plant cells a segment of T-DNA that carries cloned genes. One plasmid contains the virulence gene (responsible for transfer of the T-DNA), and another plasmid contains the T-DNA borders, the selectable marker and the DNA to be transferred.

Biodiversity —1. The variety of species (species diversity) or other taxa of animals, micro-organisms and plants in a natural community or habitat, or of communities in a particular environment (ecological diversity), or of genetic variation in a species (genetic diversity, q.v.). The maintenance of a high level of biodiversity is important for the stability of ecosystems.

2. The variety of life in all its forms, levels and combinations, encompassing genetic diversity, species diversity and ecosystem diversity. cf agrobiodiversity (Source: FAO, 1999)

Bio-informatics—The use and organization of information of biological interest. In particular, it is concerned with organizing bio-molecular databases, in getting useful information out of such databases, in utilizing powerful computers for analysing such information, and in integrating information from disparate biological sources.

Biological containment—Restricting the movement of (genetically engineered) organisms by arranging barriers to prevent them from growing outside the laboratory. Biological containment can take two forms: making the organism unable to survive in the outside environment, or making the outside environment inhospitable to the organism. The latter is rarely suitable for bacteria, which, in principle, could survive almost anywhere. Thus for bacteria and yeasts, the favoured approach is to mutate the genes in the organism so that they require a supply of a specific nutrient that is usually available only in the laboratory. If they get out, they then cannot grow. Making the environment unfriendly to the organism is partly a biological control, partly a physical one. Thus, some of the first genetically engineered rice strains were developed in England (which is too cold for rice to grow) and tried in the field in Arizona (where it is too dry). Biological containment may also involve the use of vector molecules and host organisms which have been genetically disabled such that they can survive only in the peculiar conditions provided by the experimenter and which are unavailable outside the laboratory.

Biotechnology—1. The use of biological processes or organisms for the production of materials and services of benefit to humankind. Biotechnology includes the use of techniques for the improvement of the characteristics of economically important plants and animals and for the development of micro-organisms to act on the environment.

2. The scientific manipulation of living organisms, especially at the molecular genetic level, to produce new products, such as hormones, vaccines or monoclonal antibodies.

Blot—1. As a verb, this means to transfer DNA, RNA or protein to an immobilizing matrix.

2. As a noun, it usually refers to the autoradiograph produced during the Southern or northern blotting procedures. The variations on this theme depend on the molecules:

- **Southern blot:** the molecules transferred are DNA molecules, and the probe (q.v.) is DNA.
- **Northern blot:** the molecules transferred are RNA, and the probe is DNA.
- **Western blot:** the molecules transferred are protein, and the probe is labelled antibody.
- **Southwestern blot:** the molecules transferred are protein, and the probe is DNA.
- **Dot blot:** DNA, RNA or protein are dotted directly onto the membrane support, so that they form discrete spots.
- **Colony blot:** the molecules (usually DNA) are from colonies of bacteria or yeast growing on a bacteriological plate.

Blunt end—The end of a DNA duplex molecule in which neither strand extends beyond the other.

Blunt-end cut—To cleave phosphodiester bonds in the backbone of duplex DNA between the corresponding nucleotide pairs on opposite strands. This cleavage process results in both strands finishing at the same residue, *i.e.,* there are no nucleotide extensions on either strand.

Blunt-end ligation— Joining (ligation) of the nucleotides that are at the ends of two blunt-ended DNA duplex molecules.

Broad-host-range plasmid—A plasmid that can replicate in a number of different bacterial species.

C

C—Cytosine residue in either DNA or RNA.

cDNA; complementary DNA—The double-stranded DNA complement of an mRNA sequence; synthesized *in vitro* from a mature RNA template using reverse transcriptase (to create a single strand of DNA from the RNA template) and DNA polymerase (to create the double-stranded DNA). **cDNA.**

Clone—A double-stranded DNA molecule that is carried in a vector and was synthesized *in vitro* from an mRNA sequence by using reverse transcriptase and DNA polymerase.

cDNA cloning—A method of cloning the coding sequence of a gene, starting with its mRNA transcript. It is normally used to clone a DNA copy of a eukaryotic mRNA. The cDNA copy, being a copy of a mature messenger molecule, will not contain any intron sequences and may be readily expressed in any host organism if attached to a suitable promoter sequence within the cloning vector.

cDNA library—A collection of cDNA clones that were generated *in vitro* from the mRNA sequences isolated from an organism or a specific tissue or cell type or population of an organism. cf library.

centiMorgan (cM)—One percent recombination between two loci.

Centrifugation—Separating molecules by size or density using centrifugal forces generated by a spinning rotor. G-forces of several hundred thousand times gravity are generated in ultracentrifugation. *See* density gradient centrifugation.

Centrifuge—A device in which solid or liquid particles of different densities are separated by rotating them in a tube in a horizontal circle. The denser particles tend to move along the length of the tube to a greater radius of rotation, displacing the lighter particles to the other end.

Chemically-defined medium—When all of the chemical components of a plant tissue culture medium are fully known and defined. cf undefined; organic complex.

Chemical mutagen—A chemical or product capable of causing genetic mutation in living organisms exposed to it.

Chimeric DNA—A recombinant DNA molecule containing unrelated genes.

Chimeric gene—A semi-synthetic gene, consisting of the coding sequence from one organism, fused to promoter and other sequences derived from a different gene. Most genes used in transformation are chimeric.

Chromosome mutation—A change in the gross structure of a chromosome, usually causing severely deleterious effects in the organism. They are often due to an error in pairing during the crossing-over stage of meiosis. The main types of chromosome mutation are translocation, duplication, deletion and inversion.

Chromosome jumping—A technique that allows two segments of duplex DNA that are separated by thousands of base pairs (about 200 kb) to be cloned together. After sub-cloning, each segment can be used as a probe to identify cloned DNA sequences that, at the chromosome level, are roughly 200 kb apart.

Chromosome walking—A technique that identifies overlapping cloned DNA fragments that form one continuous segment of a chromosome. These fragments can be generated either by random shearing or by partial digestion with a four-base-pair cutter such as *Sau*3A. A series of colony hybridizations is then carried out, starting with some cloned fragment which has already been identified and which is known to be in the region encompassed by the overlapping clones. This identified fragment is used as a probe to pick out clones containing adjacent sequences. These are then used as probes themselves to identify clones carrying sequences adjacent to them and so on. At each round of hybridization one "walks" further along the chromosome from the initial fragment. *See* positional cloning.

Cloning—1. The mitotic division of a progenitor cell to give rise to a population of identical daughter cells or clones.

2. Incorporation of a DNA molecule into a chromosomal site or a cloning vector.

3. Animal cloning: the creation of a whole animal by mitotic divisions from a single diploid somatic cell , typically by the process of nuclear transfer (q.v.).

Cloning site—*See* insertion site.

Cloning vector—A small, self-replicating DNA molecule: usually a plasmid or viral DNA chromosome: into which foreign DNA is inserted in the process of cloning genes or other DNA sequences of interest. It can carry inserted DNA and be perpetuated in a host cell. Also called a cloning vehicle, vector, or vehicle.. *See* vector.

Cloning vehicle—*See* cloning vector.

Co-culture—The joint culture of two or more types of cells, such as a plant cell and a micro-organism, or two types of plant cells. Used in various dual-culture systems or in nurse-culture, q.v.

Coding—The specification of a peptide sequence, by the code contained in DNA molecules.

Co-dominance—The situation in which both alleles in a heterozygous individual are expressed, so that the phenotype of heterozygotes incorporates the phenotypic effect of each allele.

Co-dominant alleles—Alleles that produce independent effects when in the heterozygous condition.

Coefficient—A number expressing the amount of some change or effect under certain conditions (*e.g.,* the coefficient of inbreeding).

Co-enzyme—An organic molecule of low molecular weight and usually non-protein, such as a vitamin, that binds to an enzyme and promotes its catalytic activity.

Cohesive ends—Double-stranded DNA molecules with single-stranded ends which are complementary to each other, enabling the different molecules to join each other.

Coincidence—The ratio of the observed frequency of double crossings-over to the expected frequency, where the expected frequency is calculated by assuming that the two crossing-over events occur independently of each other.

Co-integrate vector system—A two-plasmid system for transferring cloned genes to plant cells.

Co-integrate A DNA molecule formed by the fusion of two different DNA molecules, usually mediated by a transposable element.

Comparative gene mapping: The comparison of map locations of genes between species. The results of these comparisons indicate substantial conservation of blocks of genes and even large segments of chromosomes between species. Great use can be made of this conservation of map position.

Competence—Ability of a bacterial cell to take up DNA molecules and become genetically transformed.

Complementary homopolymeric tailing—The process of adding complementary nucleotide extensions to different DNA molecules, *e.g.,* dG (deoxyguanosine) to the 3′-hydroxyl ends of one DNA molecule and dC (deoxycytidine) to the 3′-hydroxyl ends of another DNA molecule to facilitate, after mixing, the joining of the two DNA molecules by base pairing between the complementary extensions. Also called dG:dC tailing, dA:dT tailing.

Contaminant—Bacterial, fungal or algal micro-organism accidentally introduced into a culture or culture medium. Contaminant may overgrow the plant cells and consequently inhibit their growth. Working under aseptic conditions with a rigorous exclusion of potential contaminants must be practised in plant tissue culture.

Control—1. Noun: Unchanged (standard) protocol or treatment for comparison with the experimental treatment. The term is commonly used for untreated organisms.

2. Verb: To direct or regulate cultures with addition of plant growth regulators.

Cosmid—A plasmid vector which contains the two cos (cohesive) ends of phage lambda (?) and one or more selectable markers such as an antibiotic resistance gene.

Cot curve—When duplex DNA is heated, it dissociates into single strands. When the temperature is lowered, complementary strands tend to anneal or re-nature. The extent of re-naturation depends on the product of DNA concentration in moles of nucleotides per litre, and time in seconds. A graph showing the proportion of re-natured DNA against cot is known as a cot curve. The cot at which half the DNA has re-natured is the half-cot, a parameter indicating the degree of complexity of the DNA.

Co-transformation—In genetic engineering experiments, it is often necessary to transform with a plasmid for which there is no selectable phenotype and then screen for the presence of that plasmid within the host cell. Co-transformation is a technique in which host cells are incubated with two types of plasmid, one of which is selectable and the other not. Cells which have been transformed with the first plasmid are then selected. If transformation has been carried out at high DNA concentration, then it is probable that these cells will also have been transformed with the second (non-selectable) plasmid. The technique is frequently used in experiments with mammalian cells.

Covalently closed circle (CCC)—A double-stranded DNA molecule with no free ends. The two strands are interlinked and will remain together even after denaturation. In its native form, a CCC will adopt a supercoiled configuration.

Cross—In genetic studies, the mating of two individuals or populations. Also called mating.

Cross-breeding—Mating between members of different populations (lines, breeds, races or species).

Cross hybridization—The hydrogen bonding of a single-stranded DNA sequence that is partially but not entirely complementary to a single-stranded substrate. Often, this involves hybridizing a DNA probe for a specific DNA sequence to the homologous sequences of different species.

Crown gall (A.S. *gealla,* gall)—A bulbous growth that occurs at the base of certain plants as result of infection, especially by *Agrobacterium tumefaciens*; a bacterial gene carried by the Ti plasmid is transferred by the bacteria into a higher plant cell, where it causes a tumour-like growth. a.k.a. crown gall tumour. *See Agrobacterium;* hairy root disease.

Cultivar (from **culti**vated + **var**iety) (abbr: cv.)—A category of plants that are, firstly, below the level of a sub-species taxonomically, and, secondly, found only in cultivation. It is an international term denoting certain cultivated plants that are clearly distinguishable from others by stated characteristics and that retain their distinguishing characters when reproduced under specific conditions.

Culture—A population of plant or animal cells or micro-organisms that is grown under controlled conditions.

Culture medium—Any nutrient system for the cultivation of cells of plants, bacteria or other organisms; usually a complex mixture of organic and inorganic nutrients.

Culture room—Room for maintaining cultures and often in a controlled environment of light, temperature and humidity. *cf* growth cabinet; incubator.

D

dA: dT tailing—*See* complementary homopolymeric tailing.

ddNTP: *See* di-deoxynucleotide.

De-ionized water—Water which is free of most inorganic (not completely free, since Na is present in ample quantities) and most organic compounds.

Deletion—A mutation involving the removal of one or more base pairs in DNA sequence. Large deletions are visible as the lack of chromosomal segments.

Denaturated DNA—Duplex DNA that has been converted to single strands by breaking the hydrogen bonds of complementary nucleotide pairs. Usually achieved by heating.

Denaturation—Loss of native configuration of a macro-molecule (protein or nucleic acid) by physical or chemical means, usually accompanied by loss of biological activity. Denatured proteins often unfold their polypeptide chains and express changed properties of solubility. Most commonly used by genetic engineers to describe the destruction of hydrogen bonds maintaining the double-stranded nature of all or part of a DNA molecule.

de novo (L. "from the beginning, anew")—Arising, anew, afresh, once more. Also *ex novo*.

Density gradient centrifugation—High-speed centrifugation in which molecules "float" at a point where their density equals that in a gradient of caesium chloride or sucrose. The density gradient may either be formed before centrifugation by mixing two solutions of different density (as in sucrose density gradients) or it can be formed by the process of centrifugation itself (as in CsCl and Cs_2SO_4 density gradients). *See* centrifugation.

Deoxyribonucleic acid—*See* DNA.

Deoxyribonuclease (DNase)—Any enzyme that hydrolyses DNA.

Detergent—Substance which lowers the surface tension of a solution, improving its cleaning properties (*e.g.,* Tween-20TM, a surfactant and wetting agent). *See* surfactant; wetting agent.

Di-deoxynucleotide (ddNTP)—A deoxynucleotide that lacks a 3′—hydroxyl group, and is thus unable to form the 3′-5′ phospho-diester bond necessary for chain elongation.

Dimethyl sulphoxide; dimethyl sulfoxide (C_2H_6OS; m.w. 78.13)—A highly hygroscopic liquid and powerful solvent with little odour or colour. It is an organic co-solvent used in small quantities to dissolve neutral organic substances in tissue culture media preparation. DMSO also has uses as a cryoprotectant.

Diploid (Gr. *diploos*, double + *oides*, like)—1. The status of having two complete sets of chromosomes, most commonly one set of paternal origin and the other of maternal origin.

2. An organism or cell with a double set (2n) of chromosomes (most commonly one of paternal origin, and the other of maternal origin), or referring to an individual containing a double set of chromosomes per cell. Somatic tissues of higher plants and animals are ordinarily diploid in chromosome constitution, in contrast with the haploid gametes.

Direct repeat—Two or more stretches of DNA within a single molecule which has the same nucleotide sequence in the same orientation. Direct repeats may be either adjacent to one another or far apart on the same molecule.

Directed mutagenesis—The process of generation of nucleotide changes in cloned genes by any one of several procedures, including site-specific and random mutagenesis. Also called as *in vitro* mutagenesis.

Directional cloning—The technique by which DNA insert and vector molecules are digested with two different restriction enzymes to create non-complementary sticky ends at either end of each restriction fragment, so allowing the insert to be ligated to the vector in a specific orientation and preventing the vector from re-circularizing. *See* cloning.

Disarm—To delete from a plasmid or virus genes that are cytotoxic or tumour inducing.

DMSO—*See* dimethyl sulphoxide.

DNA (deoxyribonucleic acid; formerly spelt desoxyribonucleic acid)—The long chain of molecules in most cells that carries the genetic message and controls all cellular functions in most forms of life. DNA is a macro-molecule composed of a long chain of deoxyribonucleotides joined by phospho-diester linkages. Each deoxyribonucleotide contains a phosphate group, the five-carbon sugar 2-deoxribose, and a nitrogen-containing base.

DNase (deoxyribonuclease)—An enzyme that catalyses the cleavage of DNA. DNase I is a digestive enzyme secreted by the pancreas, that degrades DNA into shorter nucleotide fragments. Many other endonucleases and exonucleases are involved in DNA repair and replication. *cf* nuclease.

DNA fingerprint—The unique pattern of DNA fragments identified originally by Southern hybridization (using a probe that binds to a polymorphic region of DNA) or now by polymerase chain reaction (PCR) (using primers flanking the polymorphic region). *See* genetic fingerprinting.

DNA ligase—An enzyme that catalyses a reaction that links two DNA molecules via the formation of a phospho-diester bond between the 3′ hydroxyl and 5′ phosphate of adjacent nucleotides.

DNA micro-array—A small glass surface to which has been fixed an array of DNA fragments, each with a defined location. A typical DNA chip would contain 10 000 discrete spots (each containing a different DNA fragment) in an area of just a few square centimetres. When a solution of fluorescently labelled DNA fragments is hybridized to the chip, spots to which hybridization occurs are visible as fluorescence. If the spots on the chip are genes (expressed sequence tags, q.v.), hybridization with cDNA from a particular tissue shows which genes are expressed in that tissue. If the spots are short, synthesized oligonucleotides (approximately 25 bases) corresponding to that part of a gene containing a single nucleotide polymorphisms (SNP)(q.v.), with a separate spot for each of the 4 possible bases at that site, hybridization with genomic DNA from an individual plant or animal enables that individual to be genotyped at as many SNP loci as are represented on the chip. The big advantage of DNA chips is the extent to which the process of genotyping can be automated, thereby enabling huge numbers of plants or animals to be genotyped for a huge number of loci.

DNA polymerase—An enzyme that catalyses the synthesis of double-stranded DNA, using single-stranded DNA as a template. *See* polymerase.

DNA polymorphism—The existence of two or more alternative forms (alleles) of a chromosomal locus that differ in nucleotide sequence or have variable numbers of repeated nucleotide units. *See* allele.

DNA probe—A labeled (tagged) segment of DNA that is able, after a DNA hybridization reaction, to detect a specific DNA sequence in a mixture of sequences. If the tagged sequence is complementary to any one in the mixture, the two sequences will form a double helix. This will be identified thanks to its label (either by radioactivity or fluorescence).

DNA repair—A variety of mechanisms that repair errors that occur during DNA replication.

DNA repair enzymes—Enzymes that catalyse the repair of DNA.

DNA replication—The process whereby DNA makes exact copies itself, under the action of and control of DNA polymerase.

DNAase—*See* DNase.

DNA sequencing—Procedures for determining the nucleotide sequence of a DNA fragment.

DNA transformation—*See* transfection; transformation.

DNA delivery system—A generic term for any procedure that transports DNA into a recipient cell.

Dominance—The type of gene action exhibited by a dominant allele (q.v.).

Dominant—1. Describing an allele whose effect with respect to a particular trait is the same in heterozygotes as in homozygotes. The opposite is *recessive*.

2. Describing the most conspicuously abundant and characteristic species of a community.

3. Describing an animal that is allowed priority in access to food, mates, etc., by others of its species because of its success in previous aggressive encounters.

Dominant marker selection—Selection of cells via a gene encoding a product that enables only the cells that carry the gene to grow under particular conditions. For example, plant and animal cells that express the introduced Neo^r gene are resistant to the compound G418, while cells that do not carry the Neo^r gene are killed by G418, also called as positive selection; *see also* positive selectable marker.

Double crossing-over—Two simultaneous reciprocal breakage and reunion events between the same two chromatids.

Double helix—Describes the coiling of the antiparallel strands of the DNA molecule, resembling a spiral staircase in which the paired bases form the steps and the sugar-phosphate backbones form the rails.

E

EDTA—*See* ethylene-diamine tetra-acetic acid.

Electrophoresis—A technique that separates charged molecules: such as DNA, RNA or protein: on the basis of relative migration in an appropriate matrix (such as agarose gel or polyacrylamide gel) subjected to an electric field. *See* agarose gel electrophoresis; polyacrylamide gel electrophoresis (PAGE); pulsed-field gel electrophoresis (PFGE).

Electroporation—1. An electrical treatment of cells that induces transient pores, through which DNA can enter the cell.

2. The introduction of DNA or RNA into protoplasts or other cells by the momentary disruption of the cell membrane through exposure to an intense electric field.

End-labelling —The introduction of a radioactive atom at the end of a DNA or RNA molecule. A commonly used method is to use T4 polynucleotide kinase to introduce a ^{32}P atom onto the end of a DNA molecule.

Escherichia coli—A commensal bacterium inhabiting the colon of many species. *E. coli* is widely used in biology, both as a simple model of cell biochemical function and as a host for molecular cloning experiments. In environmental studies, it is a key indicator of water pollution due to human sewage effluent.

Ethidium bromide—A fluorescent dye used to stain DNA and RNA. The dye fluoresces when exposed to UV light.

Ethylenediamine tetraacetic acid (EDTA)—A chelating compound. In tissue culture it is used to keep nutrients, such as iron, bound in a form that leaves them still available to the plant but which prevents them from precipitating out.

Eukaryote (Gr. *eu*, true + *karyon*, true nucleus)—Any organism characterized by having the nucleus enclosed by a membrane. Eukaryotic organisms include animals, plants, fungi and some algae. They also possess membrane-bound functional organelles, such as mitochondria and chloroplasts, in the cytoplasm of their cells. *cf* prokaryote.

Excision—1. The natural or *in vitro* enzymatic release (removal) of a DNA segment from a chromosome or cloning vector.

2. Cutting out and preparing a tissue, organ, etc., for culture.

3. Removing adventitious shoots from callus tissue.

Exonuclease III—An *E. coli* enzyme that removes nucleotides from the 3′-hydroxyl ends of double stranded DNA. a.k.a. exo III; exodeoxyribonuclease III.

Explant—A plant part aseptically excised and prepared for culture in a nutrient medium. *cf* donor plant.

3′-Extension—A short single-stranded nucleotide sequence on the 3′-hydroxyl end of a double-stranded DNA molecule. a.k.a. 3′ protruding end; 3′ sticky end; 3′ overhang.

5′-Extension—A short single-stranded nucleotide sequence on the 5′-hydroxyl end of a double-stranded DNA molecule. a.k.a. 5′ protruding end; 5′ sticky end; 5′ overhang.

F

F factor—A bacterial episome that confers the ability to function as a genetic donor in conjugation; the fertility factor in bacteria.

Fluorescent probe—Probe whose response is based on the fluorescence intensity of individual cells or cell components.

Forced cloning—The insertion of foreign DNA into a cloning vector in a predetermined orientation.

Foreign DNA—A DNA molecule that is incorporated into either a cloning vector or a chromosomal site. *See* exogenous DNA.

Four-base cutter; four-base-pair-cutter; four-cutter—A type II restriction endonuclease that binds (and subsequently cleaves) DNA at sites that contain a sequence of four nucleotide pairs that is uniquely recognised by that enzyme. Because any sequence of four bases occurs more frequently by chance than any sequence of six bases, four-base cutters cleave more frequently than do six-base cutters. Thus, four-base cutters create smaller fragments than six-base cutters.

G

G—Guanine residue in either DNA or RNA.

Gall—A tumorous growth in plants.

Gapped DNA—A duplex DNA molecule with one or more internal single-stranded regions.

GDP—Guanosine diphosphate, *See* guanosine.

Gel—A lyophilic colloid that has coagulated to a rigid or jelly-like solid. It is used for the electrophoretic separation of nucleic acids or proteins, and for encapsulation. *See* encapsulating agent; gel electrophoresis.

Gel electrophoresis—An analytical method for separating molecules according to their size. Samples are put at one end of a slab of polymer gel; an electric field across the gel pulls the molecules through it; the smaller molecules pass more easily and so move towards the other end faster; the various sizes of molecules end up at different positions according their size.

Gelatin—A glutinous, proteinaceous gelling and solidifying agent. Gelatin is produced by boiling animal connective tissues, which partially hydrolyses the collagen. Gelatin is used to gel or solidify nutrient solutions for tissue culture. *cf* agar.

Gene (Gr. *gen*, race, offspring)—Conceptually, the unit of heredity transmitted from generation to generation during sexual or asexual reproduction. More generally, the term is used in relation to the transmission and inheritance of particular identifiable traits. Since the molecular revolution, it is now known that a gene is a segment of nucleic acid that encodes peptide or RNA.

Gene pool—1. The total genetic information in all the genes in a breeding population at a given time.

Gene probe—A single-stranded DNA or RNA fragment used in genetic engineering to search for a particular gene or other DNA sequence. The probe has a base sequence complementary to the target sequence and will thus attach to it by basepairing. By labelling the probe, it can be identified after subsequent separation and purification.

Genetic map—The linear array of genes on a chromosome, based on recombination frequencies (linkage map) or physical location (physical or chromosomal map). *See* mapping.

Genetic mapping—Determining the linear order of genes and/or DNA markers along a chromosome. a.k.a. mapping.

Genetic marker—A DNA sequence used to "mark" or track a particular location (locus) on a particular chromosome. *cf* marker gene.

Genetic polymorphism—*See* polymorphism.

Genetic transformation—The transfer of extracellular DNA among and between species by using bacterial or viral vectors.

Genotype (from gene + type)—1. The genetic constitution (gene makeup) of an organism.

2. The pair of alleles at a particular locus, *e.g., Aa* or *aa*.

3. The sum total of all pairs of alleles at all loci that contribute to the expression of a quantitative trait.

GTP—Guanosine triphosphate. *See* guanosine.

H

Helper plasmid—A plasmid that provides a function or functions to another plasmid in the same cell.

Herbicide—Any substance that is toxic to plants; usually applied to agrochemicals intended to kill specific unwanted plants, such as weeds.

Herbicide resistance—The ability of a plant to withstand herbicide.

Heredity—Resemblance among individuals related by descent; transmission of traits from parents to offspring.

Heteroduplex—A double-stranded DNA molecule or DNA-RNA hybrid, where each strand is of a different origin, and consequently containing one or more mismatched (non-complementary) base pairs. A DNA duplex is prepared by the hybridization of single-stranded DNA molecules derived from two different sources. Where the two DNAs have identical or very similar sequences, a double-stranded molecule will be established, whereas where the two DNAs differ in sequence, single-stranded regions will remain. A heteroduplex will be revealed as single-strand "bushes" when DNA is observed electron microscopically. A map of homologous and non-homologous regions of the two molecules may thereby be constructed. This process is known as heteroduplex mapping.

Heterologous—From a different source, as in heterologous DNA.

Heterologous probe—A DNA probe that is derived from one species and used to screen for a similar DNA sequence from another species.

Hybrid (L. *hybrida*)—1. The offspring of two parents that are genetically different. A cross between two genetically unlike individuals.

2. A heteroduplex DNA or DNA-RNA molecule.

I

Inbred line—The product of inbreeding, *i.e.,* the mating of individuals that have ancestors in common; in plants and laboratory animals, it refers to populations resulting from at least 6 generations of selfing or 20 generations of brother-sister mating, that are for all practical purposes, completely homozygous. In farm animals, the term is sometimes used to describe populations that have resulted from several generations of the mating of close relatives, without having reached complete homozygosity.

Inbreeding—Matings between individuals that have one or more ancestors in common, the extreme condition being self-fertilization, which occurs in many plants and some primitive animals.

Incompatibility—1. Selectively-restricted mating competence, which limits fertilization, such as lack of proper functions by otherwise normal pollen grains or certain pistils, a condition that may be caused by a variety of factors.

2. A physiological interaction resulting in graft rejection or failure.

3. A function of a related group of plasmids. Plasmids which are closely related share similar replication functions and this leads to the exclusion of one or the other plasmid if they are present in the same cell; thus such plasmids are incompatible. Plasmids are placed in incompatibility groups by this simple reaction and, in general, plasmids belonging to one incompatibility group (q.v.) are very closely related.

Incompatibility group—A classification scheme indicating which plasmids can co-exist within a single cell. Plasmids must belong to different incompatibility groups to co-exist within the same cell. Plasmids that belong to the same incompatibility group are unstable when placed in the same cell. A plasmid cloning vector should always belong to an incompatibility group different from that of the host bacterium's endogenous plasmids.

Incomplete digest—*See* partial digest.

Inheritance—The transmission of particular characteristics and/or genes from generation to generation.

Insecticide—A substance that kills insects.

Insert—A DNA molecule that is incorporated into a cloning vector.

Insertion site; cloning site—A unique restriction site in a vector DNA molecule into which foreign DNA can be inserted. The term is also used to describe the position of integration of a transposon or insertion site element.

in situ (Latin for "in place")—Meaning in the natural place or in the original place.

1. Experimental treatments performed on cells or tissue rather than on extracts from them.

2. Assays or manipulations performed with intact tissues.

***in situ* colony; *in situ* plaque hybridization**—A procedure for screening colonies or plaques growing on plates or membranes for the presence of specific DNA sequences by the hybridization of nucleic acid probes to the DNA molecules present in these colonies or plaques.

Integrating vector—A vector that is designed to integrate cloned DNA into the host cell chromosomal DNA.

Integration—The recombination process which inserts a small DNA molecule (usually by homologous recombination) into a larger one. If the molecules are circular, integration involves only a single crossing-over; if linear, then two crossings-over are required. A well known example is the integration of phage ? (lambda) DNA into the *E. coli* genome.

Integration-excision (I/E) **region**—The portion of bacteriophage lambda (?) DNA that enables bacteriophage lambda (?) DNA to be inserted into a specific site in the *E. coli* chromosome and excised from this site.

Intercalating agent—A chemical capable of inserting between adjacent base pairs in a DNA molecule.

Interference—Crossing over at one point that alters the chance of another crossing-over nearby; detected by studying the pattern of crossings-over with three or more linked genes. Interference is positive or negative depending on whether the chance of another crossing-over nearby is reduced or increased, respectively.

Intraspecific—Within a species or its populations, including subspecies, such as an intraspecific cross, or variation.

Introgression—The introduction of new gene(s) into a population by crossing between two populations, followed by repeated backcrossing to that population while retaining the new gene(s).

Inversion—A chromosome re-arrangement that reverses the order of a linear array of genes in it.

Inverted repeat—Two regions of a nucleic acid molecule which have the same nucleotide sequence but in an inverted orientation, such as

5′ GCACTTG... ...CAAGTGC 3′
3′ CGTGAAC... ...GTTCACG 5′

Because they contain exactly the same message when read in either direction, inverted repeats are said to be palindromes.

in vitro (L. for "in glass")—Living in test tubes, outside the organism or in an artificial environment, typically in glass vessels in which cultured cells, tissues, organs or whole plants may reside.

J

Joining segment (J)—A small DNA segment that links genes in order to yield a functional gene encoding an immunoglobulin.

Joule (SI symbol: J)—The amount of energy needed to apply a force of 1 newton over a distance of 1 metre.

Jumping genes—*See* transposable elements.

K

Kanamycin—An antibiotic of the aminoglycoside family that poisons translation by binding to the ribosomes.

Kan^r—Kanamycin-resistance gene. *See* selectable marker.

Kb—Kilobase pairs. *See* base pair; kilobase.

Kilobase (kb)—A length unit equal to 1 000 base pairs of a double-stranded nucleic acid molecule. One kilobase of double-stranded DNA has a mass of about 660 kilodalton.

Kilobase pairs (kb)—Used interchangeably with kilobase.

Kinase—An enzyme that can transfer a phosphate from a high energy phosphate such as ATP, to an organic molecule.

Klenow fragment—A product of proteolytic digestion of the DNA polymerase I from *E. coli*; it has both polymerase and 3′-exonuclease activities but not 5′-exonuclease activity.

K_m—A dissociation constant that characterizes the binding of an enzyme to a substrate. The smaller the value of K_m, the tighter the binding of the enzyme to the substrate. Also called the Michaelis constant.

L

Label—A compound or atom that is either attached to or incorporated into a macromolecule and is used to detect the presence of a compound, substance, or macromolecule in a sample. a.k.a. tag.

Labeling—The process of replacing a stable atom in a compound with a radioactive isotope of the same element to enable it to be detected by autoradiography or other techniques. Increasingly, radioactive labelling is being replaced by fluorescent labelling. The method is used to trace the path of the labelled compound through a biological or chemical system.

Lac repressor-*lac* promoter system—*See* IPTG.

Lag phase—1. A state of apparent inactivity preceding a response; called also a latent phase.

2. The initial growth phase, during which cell number remains relatively constant, prior to rapid growth.

3. The first of five growth phases of most batch-propagated cell suspension cultures, being the phase in which inoculated cells in fresh medium adapt to the new environment and prepare to divide. *See* growth phases.

Laminar air-flow cabinet; laminar air-flow hood—Cabinet for inoculation of cultures. The working area is kept sterile by a continuous, non-turbulent flow of sterilized air through a HEPA filter. *See* HEPA filter; prefilter.

Lawn—In biotechnology: a uniform and uninterrupted layer of bacterial growth, in which individual colonies cannot be observed.

LCR—*See* ligase chain reaction.

Ligase chain reaction (LCR)—A technique for determining the presence or absence of a specific nucleotide pair within a target gene.

Ligation—The joining of two linear nucleic acid molecules by the formation of phosphodiester bonds. In cloning experiments, a restriction fragment is often ligated to a linearized vector molecule using T4 DNA ligase.

Linearized vector—*See* ligation.

Linkage—The tendency of non-allelic genes to be inherited together more than would be expected if they were assorting independently. Linkage exists between two loci when they are located sufficiently close on the same chromosome that some gametes are produced without crossing-over occurring between the two loci.

Linkage disequilibrium—*See* gametic (phase) disequilibrium.

Linkage equilibrium—*See* gametic phase equilibrium.

Linkage map—A linear or circular diagram that shows the relative positions of genes on a chromosome as determined by recombination fraction q.v.. *See* genetic map.

Linked genes; **linked markers**—Genes or markers that show linkage q.v..

Linker—A synthetic double-stranded oligonucleotide that carries the sequence for one or more restriction endonuclease sites.

Liquid nitrogen—Nitrogen gas condensed to a liquid with a boiling point of about -196°C. Very commonly the medium in which containers of genetic material are stored. *See* cryopreservation.

Locus (pl:loci)—A site on a chromosome.

Long terminal repeat (LTR)—A string of bases that occurs at each end of the "genome" of a retrovirus that has become integrated into the host genome. Involved in the integration process.

M

M13—A single-stranded DNA bacteriophage used as a vector for DNA sequencing.

M13 strand—The single-stranded DNA molecule that is present in the infective form of bacteriophage M13.

Map—1. Verb: To determine the relative positions of loci on a DNA molecule. Linkage mapping is done by estimating the recombination fraction between loci, from the genotypes of offspring of particular matings. The further apart two loci are on a chromosome, the greater will be the frequency of recombination between them up to a maximum of 50%, the situation observed when they are sufficiently far apart on a chromosome that recombinant gametes are as frequent as non-recombinant gametes, or when they are on different chromosomes. Physical mapping is usually performed by the use of *in situ* hybridisation of cloned DNA fragments to metaphase chromosomes.

2. Noun: A diagram showing the relative positions of, and distances between, loci.

Map distance—The standard measure of distance between loci, expressed in centiMorgans (cM). Estimated from recombination fraction via a mapping function (q.v.). For small recombination fractions, map distance equals the percentage of recombination (recombination frequency) betwen two genes. 1% recombination = 1 cM. Sometimes called a map unit.

Mapping—Determining the location of a locus (gene or genetic marker) on a chromosome. *See* continuous map; linkage map; physical map.

Mapping function: A mathematical expression relating observed recombination fraction (q.v.) to map distance expressed in centiMorgans. Two common mapping functions are those developed by Haldane (1919; *J. Genet.,* **8**:299-309) and Kosambi (1944; *Ann. Eugen.*,

12: 172-175). In both functions, the relationship between recombination fraction and map distance is approximately linear for recombination fractions less than 10%; as recombination fraction increases above 10% (up to its maximum of 50%), map distance is increasingly greater than recombination fraction.

Map unit—*See* map distance; crossing-over unit.

Marker—An identifiable DNA sequence that facilitates the study of inheritance of a trait or a gene.

Marker-assisted introgression—The use of DNA markers to increase the speed and efficiency of introgression (q.v.) of a new gene or genes into a population. The markers will be closely linked to the gene(s) in question.

Marker-assisted selection (MAS)—The use of DNA markers to increase the response to selection in a population. The markers will be closely linked to one or more quantitative trait loci (q.v.).

Marker gene—A gene of known function and known location on the chromosome. *cf* genetic marker.

Marker peptide—A portion of fusion protein that facilitates its identification or purification.

MAS—*See* marker-assisted selection.

MCS—*See* polylinker.

Melting temperature (abbr: T_m)—The temperature at which a double-stranded DNA or RNA molecule denatures into separate single strands. The T_m is characteristic of each DNA species and gives an indication of its base composition. DNAs rich in G:C base pairs are more resistant to thermal denaturation than A:T rich DNA since three hydrogen bonds are formed between G and C, but only two between A and T.

Mendelian population—A natural, interbreeding unit of sexually reproducing plants or animals sharing a common gene pool.

Mendelism—The theory of heredity that forms the basis of classical genetics, proposed by Gregor Mendel in 1866 and formulated in two laws (*see* Mendel's Laws).

Mendel's Laws—Two laws summarizing Gregor Mendel's theory of inheritance. The Law of Segregation states that each hereditary characteristic is controlled by two 'factors' (now called alleles), which segregate and pass into separate germ cells. The Law of Independent Assortment states that pairs of 'factors' segregate independently of each other when germ cells are formed. *See* independent assortment; linkage.

Mcr.—Monomer Repeat Units.

Microsatellite—A form of VNTR (q.v.). Specifically, a segment of DNA characterized by the occurrence of a variable number of copies (from a few up to 30 or so) of a sequence of around 5 or fewer bases (called a repeat unit, q.v.).

Molality—The number of moles of solute per litre of solvent.

Molarity—The number of moles of a substance contained in a kilogram of solution. *See* mole.

Mole (symbol: M) Amount of substance that has a weight in grams numerically equal to the molecular weight of the substance. Also called gram molecular weight. A mole contains 6.023×10^{23} molecules or atoms of a substance.

Molecular biology—The area of knowledge concerned with the molecular aspects of organisms and their cells.

Molecular cloning—The biological amplification of a specific DNA sequence through mitotic division of a host cell into which it has been transformed or transfected. *See* cloning.

Molecular genetics—The area of knowledge concerned with the genetic aspects of molecular biology, especially with DNA, RNA and protein molecules.

Molecule (L. diminutive of *moles,* a little mass)—A unit of matter, the smallest portion of an element or a compound that retains chemical identity with the substance in mass. The molecule usually consist of a union of two or more atoms; some organic molecules containing a very large number of atoms.

Multi-copy—Describing plasmids which replicate to produce many plasmid molecules per host genome, *e.g.,* pBR322 is a multi-copy plasmid, there are usually 50 pBR322 molecules (or copies) per *E. coli* genome.

Multigene family—A group of genes that are similar in nucleotide sequence or that produce polypeptides with similar amino acid sequences.

Multigenic—Controlled by several genes, as opposed to monogenic.

Multi-locus probe—A probe that hybridizes to a number of different sites in the genome of an organism. *See* probe.

Mutagen—An agent or process which is capable of inducing a mutation, such as UV light. *cf* mutation.

Mutagenesis—Change(s) in the genetic constitution of a cell through alterations to its DNA.

Mutant An organism or an allele that differs from the wild type because it carries one or more genetic changes in its DNA. A mutant organism may carry mutated gene(s) (= gene mutation); mutated chromosome(s) (= chromosome mutation); or mutated genome(s) (= genome mutation). a.k.a. a variant.

Mutation (L. *mutare,* to change)—A sudden, heritable change appearing in an individual as the result of a change in the structure of a gene (= gene mutation); changes in the structure of chromosomes (= chromosome mutation); or in the number of chromosomes (= genome mutation).

N

Nanometre—1×10^{-9} m. One millionth of a millimetre, a.k.a. a millimicron; equals ten angstroms.

Narrow-host-range plasmid—A plasmid that can replicate in one, or at most a few, different bacterial species.

Negative control system—A mechanism in which the regulatory protein(s) is required to turn off gene expression.

Negative selection—Selection of individuals that do not possess a certain character. A method by which growing cells that do not carry a DNA insert integrated at a specific chromosomal location are selected. *See* positive selection.

Nick—Verb: To break a phospho-diester bond in the backbone of one of the strands of a duplex DNA molecule. *cf* cleave; cut.

Nicked circle; **relaxed circle**—During extraction of plasmid DNA from the bacterial cell, one strand of the DNA becomes nicked. This relaxes the torsional strain needed to maintain supercoiling, producing the familiar form of plasmid.

Nick translation—A procedure for labelling DNA. A DNA fragment is treated with DNase to produce single-stranded nicks. The nick is moved along the DNA molecule in the presence of labelled deoxyribonucleoside triphosphates by the concerted action of the 5′-> 3′ exonuclease and 5′-> 3′ polymerase activities of *E. coli* DNA polymerase I.

Northern blot—A cellulose or nylon membrane to which RNA molecules have been attached by capillary action. The transferred RNA is hybridized to single-stranded DNA probes. northern blot technique is often used to measure expression (transcription) of a gene for which a specific cDNA is available for use as a probe. *See* Southern blot, western blot.

Northern blotting—Similar to Southern blotting, except that RNA is transferred onto a matrix and the presence of a specific RNA molecule is detected by DNA-RNA hybridization. *See* northern hybridization.

Northern hybridization—Hybridization of a labelled DNA probe to RNA fragments that have been transferred from an agarose gel to a nitrocellulose filter. *See* northern blotting; hybridization.

Nucleoside—A base (purine or pyrimidine) that is covalently linked to a 5-carbon (pentose) sugar. When the sugar is ribose, the nucleoside is a ribonucleoside; when it is deoxyribose, the nucleoside is a deoxyribonucleoside. Adenine, guanine and cytosine occur in both DNA and RNA; thymine occurs in DNA; and uracil in RNA. They are the building blocks of DNA and RNA. *See* nucleoside analogue.

Nucleoside analogue—A synthetic molecule that resembles a naturally occurring nucleoside, but that lacks the bond site needed to link it to an adjacent nucleotide. *See* nucleoside.

Nucleotide—A nucleoside with one or more phosphate groups linked to the 5′ carbon of the pentose sugar. Ribose-containing nucleosides include ribonucleoside monophosphate (NMP), ribonucleoside diphosphate (NDP), and ribonucleoside triphosphate (NTP). When the nucleoside contains the sugar deoxyribose, the nucleotides are called deoxyribonucleoside mono-, di-, or tri-phosphates (dNMP, dNDP, or dNTP). A building block of DNA and RNA.

O

Oligonucleotide—A short molecule (usually 6 to 100 nucleotides) of single-stranded DNA.

Oligonucleotide ligation assay (OLA)—A diagnostic technique for determining the presence or absence of a specific nucleotide pair within a target gene, often indicating whether the gene is wild type (normal) or mutant (defective).

Oligonucleotide-directed mutagenesis; oligonucleotide-directed site-specific mutagenesis.—*See* site-specific mutagenesis.

Opine—The condensation product of an amino acid with either a keto-acid or a sugar, produced by the crown gall tissues. Opine synthesis is a unique characteristic of tumour cells.

P

Palindrome (Gr. *palindromos*, running back again)—Originally a word, sentence or verse that reads the same from right to left as it does from left to right. In biotechnology, the term is applied to a segment of DNA in which the base-pair sequence reads the same in both directions from a central point of symmetry. *See* the entry for *inverted repeat* for an example of a palindrome. *See also* palindromic sequence.

Palindromic sequence—A segment of duplex DNA whose 5′-to-3′ sequence is identical on each DNA strand. The sequence is the same when one strand is read left-to-right and the other strand is read right-to-left. Typically, recognition sites for type II restriction endonucleases are palindromes. *See* the entry for *inverted repeat* for an example of a palindrome.

pBR322—One of the first plasmid vectors widely used; especially used for cloning DNA in *E. coli*. *See* plasmid.

PCR—*See* polymerase chain reaction.

Physical map—A map showing physical locations on a DNA sequence, such as restriction sites and sequence-tagged sites. Also a diagram of a chromosome or a karyotype, showing the location of loci (genes and markers) *See* mapping.

Plaque—A clear spot on an otherwise opaque culture plate of bacteria or cultured bacteria cells, showing where cells have been lysed by viral infection.

Plasmid—An extrachromosomal, autonomous circular DNA molecule found in certain bacteria, capable of autonomous replication. Plasmids can transfer genes between bacteria and are important tools of transformation in genetic engineering. They exist in an autonomous state and are transferred independently of chromosomes. *See* nicked circle; pAMP; relaxed plasmid; stringent plasmid; supercoiled plasmid.

Poly (A) polymerase—Enzyme that catalyses the addition of adenine residues to the 3′ end of pre-mRNAs to form the poly-(A) tail. *See* polyadenylation; polymerase.

Poly-(A) tail—*See* polyadenylation.

Polyacrylamide gel electrophoresis (PAGE)—A method for separating nucleic acid or protein molecules according to their molecular size.

Polyacrylamide gels—Often referred to incorrectly as acrylamide gels. These gels are made by cross-linking acrylamide with *N,N′*-methylene-*bis*-acrylamide. Polyacrylamide gels are used for the electrophoretic separation of proteins, DNA and RNA molecules. Polyacrylamide beads are also used as molecular sieves in gel chromatography, marketed as Bio-gelTM.

Polyadenylation—Post-transcriptional addition of a polyadenylic acid tail to the 3′ end of eukaryotic mRNAs. Also called poly-(A) tailing. The adenine-rich 3′ terminal segments is called a poly (A) tail.

Polygene—One of many genes of small effect influencing the development of a quantitative trait; results in continuous variation and in quantitative inheritance. *See* gene.

Polygenic—Controlled by many genes of small effect.

Polylinker—A segment of DNA that contains a number of different restriction endonuclease sites. a.k.a. multiple cloning site (MCS).

Polymerase chain reaction (PCR)—A procedure that amplifies a particular DNA sequence. It involves multiple cycles of ***denaturation***, ***annealing*** to oligonucleotide primers, and ***extension*** (polynucleotide synthesis), using a thermostable DNA polymerase, deoxyribonucleotides, and primer sequences in multiple cycles of denaturation-renaturation-DNA synthesis. *See* polymerase.

Polynucleotide—A chain of nucleotides in which each nucleotide is linked by a single phospho-diester bond to the next nucleotide in the chain. They can be double—or single-stranded. The term is used to describe DNA or RNA. *See* nucleotide.

Primer—A short DNA or RNA fragment annealed to a template of single-stranded DNA, providing a 3′ hydroxyl end from which DNA polymerase extends a new DNA strand to produce a duplex molecule.

Primer DNA polymerase—DNA polymerase that provides primers for the DNA polymerization. Unlike RNA polymerase, DNA polymerase is unable to initiate the *de novo* synthesis of a polynucleotide chain. DNA polymerase can only add nucleotides to a free 3′ hydroxyl group at the end of a pre-existing chain. A short oligonucleotide, known as a primer, is therefore needed to supply such a hydroxyl group for the initiation of DNA synthesis.

Probe—1. For diagnostic tests, the agent that is used to detect the presence of a molecule in a sample.

2. A DNA or RNA sequence labelled or marked with a radioactive isotope or that is used to detect the presence of a complementary sequence by hybridization with a nucleic acid sample.

Prokaryote (L. *pro*, before + Gr. *karyon*, a nut, referring in modern biology to the nucleus)—A member of a large group of organisms, including bacteria and blue-green algae, which do not have the DNA separated from the cytoplasm by a membrane in their cells. The DNA is usually in one long strand. Prokaryotes do not undergo meiosis and do not have functional organelles such as mitochondria and chloroplasts. *cf* eukaryote.

PUC—A widely used expression plasmid, containing a galactosidase gene. *See* plasmid.

Pulsed-field gel electrophoresis (PFGE)—A procedure used to separate very.

Pyrimidine—A single-ring, nitrogen-containing base present in nucleic acids; cytosine (C) and thymine (T) are commonly present in DNA, whereas uracil usually replaces thymine in RNA.

Q

Quantitative inheritance—Inheritance of measurable traits (height, weight, colour intensity, etc.) that depend on the cumulative action of many genes.

Quantitative trait—A measurable trait that shows continuous variation; a trait that can not be classified into a few discrete classes.

Quantitative trait locus (QTL)—A locus that affects a quantitative trait. The plural form (quantitative trait loci) is also abbreviated as QTL.

R

Random amplified polymorphic DNA (RAPD; pronounced 'rapid')—A technique using single, short (usually 10-mer) synthetic oligonucleotide primers for PCR. The primer, whose sequence has been chosen at random, initiates replication at its complementary sites on the DNA, producing fragments up to about 2 kb long, which can be separated by electrophoresis and stained with ethidium bromide.

Random primer method—A protocol for labelling DNA *in vitro*. A sample of random oligonucleotides (6 or 14 nucleotides long) containing all possible combinations of nucleotide sequences is hybridized to a DNA probe. Then, in the presence of a DNA polymerase and the four deoxyribonucleotides: one of which is labelled: the 3′ hydroxy ends of the hybridized oligonucleotides provide initiation sites for DNA synthesis that uses the separated strands of the probe DNA as a template. This reaction produces labelled copies of portions of the probe DNA.

Recombinant—A term used in both classical and molecular genetics.

1. In classical genetics: An organism or cell that is the result of recombination (crossing-over), *e.g.,* Parents: *AB/ab* and *ab/ab*; recombinant offspring: *Ab/ab.*

2. In molecular genetics: A molecule containing DNA from different sources. The word is typically used as an adjective, *e.g.,* recombinant DNA.

Recombinant DNA—The result of combining DNA fragments from different sources.

Recombinant DNA technology—A set of techniques which enable one to manipulate DNA. One of the main techniques is DNA cloning (because it produces an unlimited number of copies of a particular DNA segment), and the result is sometimes called a DNA clone or gene clone (if the segment is a gene), or simply a clone. An organism manipulated using recombinant DNA techniques is called a genetically modified organism (GMO).

Re-naturation—The re-association of two nucleic acid strands after denaturation. The restoration of a molecule to its native form. In nucleic acid biochemistry, this term usually refers to the formation of a double-stranded helix from complementary single-stranded molecules. Some simple proteins can also be re-natured and regain their function.

Replacement; gene replacement—A method of substituting a cloned gene, or part of a gene, which may have been mutated *in vitro*, for the wild-type copy of the gene within the host's chromosome. *See* homogenotization.

Replicative form (RF) —The molecular configuration of viral nucleic acid that is the template for replication in the host cell.

Restriction endonuclease [enzyme]—A class of endonucleases that cleaves DNA after recognizing a specific sequence, *e.g., Bam*H1 (5′GGATCC3′), *Eco*RI (5′GAATTC3′), and *Hin*dIII (5′AAGCTT3′). There are three types of restriction endonuclease enzymes:

Restriction enzyme—*See* restriction nuclease.

Restriction fragment—A fragment of DNA produced by cleaving (digesting, cutting) a DNA molecule with one or more restriction endonucleases.

Restriction fragment length polymorphism (RFLP)—The occurrence of variation in the length of DNA fragments that are produced after cleavage with a type II restriction endonuclease. The differences in DNA lengths are due to the presence or absence of recognition site(s) for that particular restriction enzyme.

Restriction map—The linear array of restriction endonuclease sites on a DNA molecule. *See* mapping.

Restriction site—The specific nucleotide sequence in DNA that is recognized by a type II restriction endonuclease and within which it makes a double-stranded cut.

Rhizobacterium—A micro-organism whose natural habitat is near, on or in plant roots.

Rhizobium (pl: rhizobia)—Prokaryote able to establish symbiotic relationship with leguminous plants, as a result of which elemental nitrogen is fixed or converted to ammonia. *See* nitrogen fixation.

Ri plasmid—A class of large conjugative plasmids found in the soil bacterium *Agrobacterium rhizogenes*. Ri plasmids are responsible for hairy root disease of certain plants. A segment of the Ri plasmid is found in the genome of tumour tissue from plants with hairy root disease.

RNase—Ribonuclease. A group of enzymes that catalyse the cleavage of nucleotides in RNA.

S

Satellite DNA—That portion of the DNA in plant and animal cells consisting of highly repetitious sequences (millions of copies) typically in the range from 5 to 500 bases. Thousands of copies occur tandemly (end-on-end) at each of many sites. It can be isolated from the rest of the DNA by density gradient centrifugation.

SDS—Sodium dodecyl sulphate. *See* gel electrophoresis.

Selectable marker—A gene whose expression allows the identification of:

1. A specific trait or gene in an organism.
2. Cells that have been transformed or transfected with a vector containing the marker gene. *See* -lactamase, kanr.

Sequence-tagged site (STS)—Short, unique DNA sequence (usually 200 to 500 bp) that, by being able to be amplified by PCR, is uniquely "tagged" to the site on the chromosome from which it was amplified.

Sequencing—The determination of the order of nucleotides in a DNA or RNA molecule, or that of amino acids in a polypeptide chain. *See* DNA sequencing.

Simple sequence repeat (SSR)—*See* microsatellite.

Single-nucleotide polymorphism (SNP; pronounced "snip")—A polymorphism (q.v.) at a particular base site in a coding sequence, *e.g.,* at base 306 in a particular gene, one individual could be heterozygous for A and G: the maternal allele could have an A at this site, while the paternal allele has a G at this site. This type of polymorphism is extensive throughout the genome, and has the great advantage of being detectable without the need for gel electrophoresis, which opens the way for large-scale automation of genotyping.

Single-stranded—A term used to describe nucleic acid molecules consisting of only one polynucleotide chain. The genomes of certain phages, *e.g.,* MI3, are single-stranded DNA molecules; rRNA, mRNA and tRNA are all single-stranded nucleic acids, but they all contain double-stranded regions formed by the intra-strand base-pairing of self-complementary sequences.

Site-directed mutagenesis—The introduction of base changes: mutations: into a piece of DNA at a specific site, using recombinant DNA methods.

Site-specific—A term used to describe any process or enzyme which acts at a defined sequence within a DNA or RNA molecule. Type II restriction enzymes are site-specific endonucleases and the recombination systems encoded by some transposons are site-specific, such as is the integration of phage into the *E. coli* chromosome.

Site-specific mutagenesis—A technique to change one or more specific nucleotides within a cloned gene in order to create an altered form of a protein with one or more specific amino acid changes. a.k.a. oligonucleotide-directed mutagenesis; oligonucleotide-directed site-specific mutagenesis.

Six-base cutter—A type II restriction endonuclease that binds (and subsequently cleaves) DNA at sites that contain a sequence of six nucleotide pairs that is uniquely recognized by that enzyme. Because any sequence of six bases occurs less frequently by chance than any sequence of four bases, six-base cutters cleave less frequently than do four-base cutters. Thus, six-base cutters create larger fragments than four-base cutters.

Southern blot—A cellulose or nylon membrane to which DNA fragments previously separated by gel electrophoresis have been transferred by capillary action.

Southern blotting—A technique for transferring denatured DNA molecules that have been separated electrophoretically, from a gel to a matrix (such as a nitrocellulose membrane) on which a hybridization assay can be performed.

Southern hybridization—A procedure in which a cloned, labelled segment of DNA is hybridized to DNA restriction fragments on a Southern blot (q.v.).

Staggered cuts—Symmetrically cleaved phospho-diester bonds that lie on both strands of duplex DNA but are not opposite one another.

Sterilize—1. The process of elimination of micro-organisms, such as by chemicals, heat, irradiation or filtration.

2. The operation of making an animal incapable of producing offspring.

Sterilization—The act of sterilizing.

Sticky ends; cohesive ends—The single-stranded nucleotide sequence left on a restriction fragment by type II restriction enzymes that cut each strand at a separate location. These unpaired regions are available for hybridization with complementary ends on other fragments during the creation of recombinant DNA. a.k.a. protruding end; overhang; cohesive end. *See* cohesive ends.

Stock solutions—Pre-prepared solutions of individual components and used to prepare many different types of media. Certain substances, including Ca and Mg sulphates and phosphates must not be combined until actual medium preparation because insoluble combinations are formed and precipitate.

Strain—A group of individuals from a common origin within a species.

Strain isolation—Isolation of any bacterium, animal or plant from the outside world.

Stringent plasmid—A plasmid that only replicates along with the main bacterial chromosome and is present as a single copy, or at most several copies, per cell. *See* plasmid.

STS—*See* sequence-tagged site.

Sub-cloning—1. Splicing part of a cloned DNA molecule into a different cloning vector.

2. The process of transferring a cloned DNA fragment from one vector to another. *See* cloning.

Supercoil—A DNA molecule that contains extra twists as a result of overwinding (positive supercoils) or underwinding (negative supercoils).

Supercoiled plasmid—The predominant *in vivo* form of plasmid, in which the plasmid is coiled around histone-like proteins. Supporting proteins are stripped away during extraction from the bacterial cell, causing the plasmid molecule to supercoil around itself *in vitro*. *See* Plasmid.

T

T—The thymine residue in DNA.

Tag—*See* label.

Tailing—The *in vitro* addition of the same nucleotide by the enzyme terminal transferase, to the 3′-hydroxyl ends of a duplex DNA molecule. a.k.a. homopolymeric trailing.

Tandem array—The existence of two or more identical DNA sequences in series, *i.e.,* end -to-end.

Taq polymerase—A heat-stable DNA polymerase isolated from the thermophilic bacterium *Thermus aquaticus,* and used in PCR. *See* polymerase.

Target—For diagnostic tests, the molecule or nucleic acid sequence that is being sought in a sample.

Target site duplication—A sequence of DNA that is duplicated when a transposable element inserts; usually found at each end the insertion.

Targeting vector—A cloning vector carrying a DNA sequence capable of participating in a crossing-over event at a specified chromosomal location in the host cell.

T-DNA—The segment of DNA in the Ti plasmid of *Agrobacterium tumefaciens* that is transferred to plant cells and inserted into the chromosomes of the plant.

T4 DNA ligase—An enzyme from bacteriophage-T4-infected cells, and that catalyses the joining of duplex DNA molecules and repairs nicks in DNA molecules. The enzyme requires that one of the DNA molecules has a 5′-phosphate group and that the other has a free 3′-hydroxyl group.

Template—An RNA or single-stranded DNA molecule upon which a complementary nucleotide strand is synthesized. A pattern or mould. DNA stores coded information and acts as a model or template from which information is copied into complementary strands of DNA or transcribed into mRNA.

Template strand—The polynucleotide strand that a polymerase uses for determining the sequence of nucleotides during the synthesis of a new nucleic acid strand.

Terminal transferase—An enzyme that adds nucleotides to the 3′ terminus of DNA molecules.

Termination codon —*See* stop codon.

Testcross—Backcross to the recessive parental type, or a cross between genetically unknown individuals and a fully recessive tester to determine whether an individual in question is heterozygous or homozygous for a certain allele. It is also used as a test for linkage, *i.e.,* to estimate recombination fraction.

Tetracycline—An antibiotic that interferes with protein synthesis in prokaryotes.

Thymidine—A nucleoside present in DNA but absent in RNA.

Thymidine kinase (tk)—An enzyme that allows a cell to utilize an alternate metabolic pathway for incorporating thymidine into DNA. Used as a selectable marker to identify transfected eukaryotic cells.

Thymine—A pyrimidine base found in DNA. The other three organic bases: adenine, cytosine, and guanine: are found in both RNA and DNA; in RNA, thymine is replaced by uracil.

Ti plasmid—Tumour-inducing plasmid. A giant plasmid of *Agrobacterium tumefaciens* that is responsible for the induction of tumours in infected plants. Ti plasmids are used as vectors to introduce foreign DNA into plant cells. *See* vector.

Tissue—A group of cells of similar structure which sometimes performs a special function.

Tissue culture—A general term used to describe the culture of cells, tissues or organs in a nutrient medium under sterile conditions.

Transformation—1. The uptake and establishment of DNA in a bacterium or yeast cell, in which the introduced DNA often changes the phenotype of the recipient organism.

2. Conversion by various means of animal cells in tissue culture from controlled to uncontrolled cell growth. Typically through infection by a tumour virus or transfection with an oncogene. *See* transformant; transformation efficiency.

Transformation efficiency—The number of cells that take up foreign DNA as a function of the amount of added DNA; expressed as transformants per microgram of added DNA. *See* transformation.

Transformation frequency—The fraction of a cell population that takes up foreign DNA; expressed as the number of transformed cells divided by the total number of cells in a population.

Transgene—A gene from one genome that has been incorporated into the genome of another organism. Often refers to a gene that has been introduced into a multicellular organism.

Transgenic—An organism in which a foreign gene (a transgene) is incorporated into its genome. The transgene is present in both somatic and germ cells, is expressed in one or more tissues, and is inherited by offspring in a Mendelian fashion. *See* transgenic animal; transgenic plant.

Transient—Of short duration.

Tumour-inducing plasmid—*See* Ti plasmid.

U

U—The uracil residue in RNA.

Ultraviolet light; ultraviolet radiation (UV)—The portion of the electromagnetic spectrum with wavelengths from about 100 to 400 nm; between ionizing radiation (X-rays) and visible light. UV is absorbed by DNA and is highly mutagenic to unicellular organisms and to the epidermal cells of multicellular organisms. UV light is used in tissue culture for its mutagenic and bactericidal properties.

Uracil A pyrimidine base found in RNA but not in DNA. In DNA, uracil is replaced by thymine.

V

V/v—On a volume per volume basis; the percent of the volume of a constituent in 100 units of volume, *e.g.,* (ml/100 ml) × 100.

Variable number tandem repeat—*See* VNTR.

Variance—In statistics: The sum of the squared deviations, divided by one less than the number of observations. A statistical measure of variation in a population.

Variant—An organism that is genetically different from the wild type organism. a.k.a. mutant.

Variation—Differences between individuals within a population or among populations.

Variety—A naturally occurring subdivision of a species, with distinct morphological characters and given a Latin name according to the rules of the International Code of Nomenclature. A taxonomic variety is known by the first validly published name applied to it so that nomenclature tends to be stable.

Vector (L. *vehere,* to carry)—1. An organism, usually an insect, that carries and transmits disease-causing organisms.

2. A plasmid or phage that is used to deliver selected foreign DNA for cloning and in gene transfer. *See* Ti plasmid.

Vehicle—The host organism used for the replication or expression of a cloned gene or other sequence The term is little used and is often confused with vector. *See* vector.

Vir genes—A set of genes on a Ti plasmid that prepare the T-DNA segment for transfer into a plant cell.

Virulent phage—A phage (virus) that destroys the host (bacterium).

Virus (L. *virus,* a poisonous or slimy liquid)—An infectious particle composed of a protein capsule and a nucleic acid core (DNA or RNA), which is dependent on a host organism for replication. The DNA or a double-stranded DNA copy of an RNA virus genome is integrated into the host chromosome during lysogenic infection or replicated during the cystic cycle. *See* coat protein; DNA; genome; host; nucleic acid; prion; RNA; tumour virus; viroid.

W

W/V—Weight per volume; the weight of a constituent in 100 cm^3 of solution, expressed as a percentage.

Walking—A method for cloning large regions of a chromosome. Starting from a known site, a gene library is screened for clones that hybridize to DNA probes taken from the ends of the first clone. These clones are then isolated, and their ends used to screen the library again. These clones are then isolated and their ends used, and so on.

Wavelength—The distance between two corresponding points on any two consecutive waves. For visible light it is very small and is generally measured in nanometres.

Western blot—A technique in which protein is transferred from an electrophoretic gel to a cellulose or nylon support membrane following electrophoresis. A particular protein molecule can then be identified by probing the blot with a radiolabelled antibody which binds only the specific protein to which the antibody was prepared. Useful, for example, for measurement of levels of production of a specific protein in a particular tissue or at particular developmental stage. *cf* blot; northern blot; Southern blot.

Wild type—An organism as found in nature; the dominant allele usually found in nature, and from which mutations produce other dominants or recessives alleles.

X

X—The basic number of chromosomes in a polyploid series, monoploid = x; diploid = 2x; triploid = 3x; etc.

X-linked—The presence of a gene on the X chromosome. a.k.a. X-linkage.

Y

Y-chromosome—The partner of the X-chromosome in the male of many animal species.

YAC—Yeast artificial chromosome, used as a vector system for cloning DNA fragments that can be hundreds of kilobases long. Linear cloning vectors constructed from essential elements of yeast chromosomes. They can accommodate foreign DNA inserts of 200 to 500 kb.

Yeast artificial chromosome—*See* YAC.

Yeast episomal vector (YEp)—A cloning vector for the yeast *Saccaromyces cerevisiae* that uses 2 ?m plasmid as origin of replication and is maintained as an extrachromosomal nuclear DNA molecule.

Yeast extract—A mixture of substances from yeast. *cf* organic complex; undefined.

Yeast—A unicellular ascomycete fungus, commonly found as a contaminant in plant tissue culture.

Yeast cloning vectors—The yeasts, and especially *Saccaromyces cerevisiae*, are favourite organisms in which to clone and express DNA. They are eukaryotes, and so can splice out introns, the non-coding sequences in the middle of many eukaryotic genes.

Z

Z-DNA; zig-zag DNA—A form of DNA duplex in which the double helix is wound in a left-hand, instead of a right-hand, manner. DNA adopts the Z configuration when purines and pyrimidines alternate on a single strand.

❑❑❑

SOURCES AND SUGGESTED READING

Akagi H, Yokozeki Y, Inagaki A, Fujimura T (1996) *Theo. Appl. Genet.* **93:** 1071

Alphey L, (1997) DNA sequencing. Bios scientific, Oxford.

Amersham protocols, version 2.0.

D.T. Plummer, An Introduction to practical Biochemistry (1977), Tata McGraw Hill, Bombay

.

Analytical Biochemistry, Holme D.J. & Peck H. (1998), Longman Scientific & Technical, England, III edition.

Andrews AT (1986). Electrophoresis: Theory, Techniques and Biochemical and Clinical Applications. Oxford University Press, Oxford.

Avise, J.C. (1994). Molecular Markers, Natural History and Evolution. Chapman & Hall.

Ayres NM, McClung AM, Larkin PD, Bligh HFJ, Jones CA, Park WD (1997) *Theo. Appl. Genet.* **94:** 61

Barcaccia G, Albertini, Rosellini D, Tavoletti S, Veroneri F (2000) *Genome* **43:** 528

Berto MP, Ruiz C, Pina JA, Asins MJ (2001) *Mol. Phylogenet. Evol.* **21:** 285

Beyon, R.J and esterby, J.S. (1996) Buffer solutions the basics. IRL Press, Oxford,

Bhuttramakki D, *et al.* in Plant and Animal Genome VIII Conference, 9-12 Jan 2000, San Diego, USA (html://www.intlpag.org/pag/8/abstracts/)

Birnboim, H.C. and Doly, J. (1979) A rapid alkaline extraction procedure for screening recombinant plasmid DNA a method for preparing plasmid DNA. Nucleic Acids Research, **7:** 1513-232.

Biruboim, (1983) Methods in Enzymology **100:** 243-255

Brar DS, Dhaliwal HS (1997) In: *Proceedings of third Agricultural Science Congress*, March

Breslauer, Kenneth J., Ronald Frank, Helmut Blacker, and Luis A. Marky (1986). Predicting, DNA duplex stability from the base sequence. *Proc. Natl. Acad. Sci. USA* **83:** 3746-3750.

Brock TD: Life at high temperatures. Bacteriology 303: Procaryotic Microbiology, 1994.

Brown T.A., Gene doing (third edition) John Wiley Press.

Brown TA, (1997), Genetics: A molecular approach, 3rd edn, Chapman & Hall, London.

Brown, T.A. (1994) DNA Sequencing: The Basics. Oxford University Press, Oxford – everything you need to know about DNA sequencing.

Brown, T.A. and Brown, K.A. (1994) Ancient DNA: using molecular biology to study the past. BioEssays, **16**, 719-26.

Bryan GJ, Collins AJ, Stephenson P, Orry A, Smith JB, Gale MD (1997) *Theo. App. Genet.* **94:** 557

Burr, B., Burr, F.A. 1991. Recombinant inbred lines for molecular mapping in maize. Theor Appl. Genet. 85:55-60.

CDC. Hantavirus infection—southwestern United States: interim recommendations for risk reduction (1993). MMWR; **42**(No. RR-11).

CDC. Hantavirus pulmonary syndrome—United States, (1993). MMWR 1994; **43:** 45-8.

CDC. Update: outbreak of hantavirus infection—southwestern United States, (1993). MMWR 1993; **42:** 495-6.

CDC/National Institutes of Health. Biosafety in microbiological and biomedical laboratories. 3rd ed. Washington, DC: US Department of Health and Human Services, Public Health Service, (1993). HHS publication no. (CDC) **93-**8395.

Chabra A.K. & RK Behl, (2006), Manual on Molecular markers in crop improvement, Department of Plant Breeding, CCS Haryana Agricultural University, Hisar 125004.

Childs JE, Ksiazek TG, Spiropoulou CF, et al. Serologic and genetic identification of Peromyscus maniculatus as the primary rodent reservoir for a new hantavirus in the southwestern United States. J Infect Dis (in press).

Chin EC, Senior ML, Shu H, Smith JSC (1996) *Genome* **38:** 866

Chowdhury. V.K., Santosh Dhillon and P.K. Kharb (2005). Manual on introductory plant biotechnology, Deptt. of BMB, CCSHAU, Hisar.

Christian, G D. & *'0'* Reilly, J. E. (1986) (eds), Instrumental Analysis. 2nd ed. Allyn & Bacon. Inc. Boston, USA.

Cooper, T. C."Tobls of'Biochemistry, (1997) John Wiley, New York USA.

Cregan PB, Jervik T, Bush AL, Shoemaker RC, Lark KG, Kahler AL, Kaya N, VonToai TT, 10. Lohnes DG, Chung J, Specht JE (1999) *Crop Sci.* **39:** 146

Crivello, J.F. (1997). In "The PCR Technique: Quantitative PCR" (J.W. Larrick, ed.), pp. 170-172. Eaton Publishing, Mountain View, CA.

Desmyter J, LeDuc JW, Johnson KM, Brasseur F, Deckers C, van Ypersele de Strihou C. Laboratory rat associated outbreak of haemorrhagic fever with renal syndrome due to Hantaan-like virus in Belgium (1983). Lancet; **11:**1,445-8.

Dournon E, Moriniere B, Matheron S, et al. HFRS after a wild rodent bite in the Haute-Savoie—and risk of exposure to Hantaan-like virus in a Paris laboratory, (1984) Lancet; **1:**676-7.

Dunn, M.J., (1993), Gel Electrophoresis: Proteins, Bios Scientific, Oxford.

ed. The C. Y.Mosby Company, St. Louis, USA, (1983).

Elliott LH, Ksiazek TG, Rollin PE, et al. Isolation of Muerto Canyon virus, causative agent of hantavirus pulmonary syndrome. Am J Trop Med Hyg (in press).

Estridge, B., Reynolds, A., and Walters, N. *Basic Medical Laboratory Techniques.* (2000) Delmar, a division of Thomson Learning

Feinberg & Vogelstein, (1984), Anal Biochem, **137:** 266-267.

Garland S, Lewin L, Blakeney A, Reinke R, Henry R (2000) Theo. Appl. Genet. **101:** 364

Gilliland, Gary, Steven Perrin, Kerry Blanchard, and H. FranklIn Bunn (1990). Analysis of cytokine mRNA and DNA: detection and quantitation by competitive polymerase chain reaction. *Proc. Natl. Acad. Sci. USA* 87:2725-2729.

Gupta PK, Roy JK, Prasad M (2001) Curr. Sci. **80:** 524

Gupta, P.K.(2004). Biotechnology and Genomics. Rastogi Publication, Meerut, India.

Harwood A, (1996) Basic DNA and RNA Protocols Human Press, Totowa. NY.

Hawcroft, D. M. Electrophoresis the Basics (1997) IRL Press, Oxford, UK.

Hawcroft, D.M (1996). Electrophoresis: The Basics. IR & Press, Oxford.

Hayashi, Kenshi (1994). Manipulation of DNA by PCR. In *The Polymerase Chain Reaction* (K. Mullis, F. Ferre. R.A. Gibbs, eds.). Birkhauser, Boston, pp. 3-13.

Hayashi, Kenshi (1994). Manipulation of DNA by PCR. In *The Polymerase Chain Reaction* (K. Mullis, F. Ferre. R.A. Gibbs, eds.). Birkhauser, Boston, pp. 3-13.

Helentjaris, T., G. King, M. Slocum, C. Siedestrang and S. Wegman. (1985). Restriction fragment polymorphisms as probes for plant diversity and their development as tools for applied plant breeding. Plant Mol. Biol. 5: 109-118.

Hjelle B, Jenison S, Torrez-Martinez N, et al. A novel hantavirus associated with an outbreak of fatal respiratory disease in the southwestern United States: evolutionary relationships to known hantaviruses. (1994) J Virol **68:**592-6.

ICRISAT Manual on Molecular Markers from http://www.icrisat.org.

Innis et al (eds), (1990), PCR protocols: A guide to methods and applications, san diego, CA, Academic Press, Inc.

Jain N (2003) PhD Thesis submitted to CCS Haryana Agricultural University, Hisar (Hr) INDIA

Jarvis P, Lister C, Szabo V, Dean C (1994) *Plant Mol. Biol.* **24**: 685

Jodhav, H.V.and chitins M.A. encyclopaedia of hazardous chemicals and medical Management (1993). Himalayee Publishing House, Mumbai

Johnston, M. et al. (1994) Complete nucleotide sequence of Saccharomyces erevisiae chromosome VIII. Science, 265, 2077-82.

Joshi SP, Gupta VS, Aggarwal RK, Ranjekar PK, Brar DS (2000) *Theo. Appl. Genet.* **100:** 1311

Joshi SP, Ranjekar PK, Gupta VS (1999) Curr. *Sci.* **77:** 230

Karl, S.A., B.W. Bowen, and J.C. Avise. 1992. Global population genetic structure and malemediated gene flow in the green turtle (Chelonia mydas): RFLP analyses of anonymous nuclear loci. Genetics **131:** 163-173.

Kawamata J, Yamanouchi T, Dohmae K, et al. Control of laboratory acquired hemorrhagic fever with renal syndrome (HFRS) in Japan (1987). Lab Anim Sci **37;** 431-6.

Keith Wilson and John Walker (eds), Practical Biochemistry Principles and techniques (fifth edition) Combridge University Press.

Kimpton, C.P., P.Gill, A. Walton, A. Urquhart, E. S.Millican, and M. Adams. (1993) – Automated DNA profiling employing multiplex amplification of short tandem repeat loci. PCR Methods and Applications **3:** 13-22.

Konieczyn A and Ausubel FM (1993) *Plant J.* **4:** 403

Laboratory Manual in Biochemistry, J. Jayaraman., Wiley Eastern Limited, New Delhi (1968). 2. Biochemical Calculations, Irwin H. Segel., John Wiley and Sons, New York (1976).

Lavi U, Cregan P, Schaap T, Hillel J (1994) *Plant Breed. Rev.* **12:** 195

LeDuc JW. Epidemiology of Hantaan and related viruses (1987). Lab Anim Sci **37**: 413-8.

Lee HW, Baek LJ, Seong IW, Gwag TR. Physico-chemical properties of Hantaan virus, II: the effect of temperature and pH on infectivity of Hantaan virus (1983). J Korean Soc Virol **13:** 23.

Lee HW, Johnson KM. Laboratory-acquired infections with Hantaan virus, the etiologic agent of Korean hemorrhagic fever. J Infect Dis 1982;146:645-51.

Lee M (1999) *Adv. Agro.* **55:** 265

Lee, L. W. & Schmidt,.L M. Elementary Principles of Laboratory Instruments. 5'h

Lefebvre, V., Palloix, A. Caranata, C., Pochard, E. 995 Construction of an intraspecific linkage map of pepper using molecular markers and doubled-haploid propenies. Genome **38**:112-121.

Litt M, Luty JA (1989) *Amer. J. Human Genet.* **4:** 397

Lloyd G, Bowen ETW, Jones N, *et al.* (1984) HFRS outbreak associated with laboratory rats in UK. Lancet i:1,175-6.

Lobban, P. and Kaiser, A.D. (1973) Enzymatic end-to-end joining of DNA molecules. *Journal of Molecular Biology*, **79**, 453-71-ligation.

Lunn, E.G.and Sandstone, F. (1987) ethidium bromide: destruction and decontamination of solutions anal Biochem, **162:** 453-458.

Maekawa, M., K. Sudo, and T. Kanno. (1993). Search for improved electrophoretic conditions for PCR-Single – Strand Conformation Polymorphism analysis: is an SDS buffer condition useful ? PCR Methods and Applications **3**:130-132.

Mahbubani. M. H., & Bej, A.K. (1994). Applications of polymerase chain reaction methodology in clinical diagnostics. In H.G. Griffin & A.M. Griffin (Eds.).(1994). PCR technology: Current innovations (pp. 307-326). Ann Arbor, MI: CRC Press, Inc.

Marchuk, D., Drumm, M., Saulino, A. and Collins, F.S. (1991) Construction of T-vectors, rapid and general system for direct cloning of unmodified PCR products. Nucleic Acids Research, **19:** 1154.

Marmur, J. (1961) A procedure for the isolation of deoxyribonucleic acid from micro-organisms. *Journal of Molecular Biology*, **3**, 208-18- total cell DNA preparation.

Maxam AM, and W Gilbert 1977 A new method for sequencing DNA. Proc Natl Acad Sci USA **74**: 560-564.

Maxam, A. & Gilbert, W. (1977). A new method of sequencing DNA. Proceedings of the National Academy of Sciences, USA, **74:** 560-4.

McDonell, M.W., Simon, M.N. and Studier, F.W. (1977) Analysis of restriction fragments of T7 DNA and determination of molecular weights by electrophoresis in neutral and alkaline gels. *Journal of Molecular Biology*, 110-119-**46:** an early example of the use of agarose gel electrophoresis in the analysis of restriction fragment sizes.

McKee KT Jr, LeDuc JW, Peters CJ. Hantaviruses (1991). In: Belshe RB, ed. Textbook of human virology. 2nd ed. St. Louis: Mosby Year Book, 615-32.

Meyer, E. chemistry of HAZARDOUS MATerials (1988). (3rd ed.) Brody Prentice – Hall, Inc,NJ,USA.

Miller JC and Tanksley SD (1990) *Theo. Appl. Genet.* **80**: 385

Mohan M, Nair S, Bhagwat A, Krishna TG, Yano M, Bhatia CR, Sasaki T (1997) *Mol. Breed.* **3:** 87

Molecular Markers, Theory and Practicals- a laboratory manual, NDRI, Karnal.

Mullis KB, Faloona FA: Scharf, S., Saiki, R., Horn, G., and Erlich, H. 1986. Specific enzymatic amplification of DNA in vitro: the polymerase chain reaction. Cold Spring Harb Symp Quant Biol **51**, 263-73.

Mullis, K.B. and Faloona, F.A. (1987). Methods Enzymol. **155:** 835-850.

Nelson OL and Cox M.M.; lehninger (2001) Peinciple of Biochemestry, 3rd Ed. Macmillian Press Ltd. U.K. worth Publishers USA, reprinted version.

Nichol ST, Spiropoulou CF, Morzunov S, *et al.* (1993) Genetic identification of a hantavirus associated with an outbreak of acute respiratory illness. Science **262**:914-7.

Nuzum EO, Rossi CA, Stephenson EH, LeDuc JW. (1988) Aerosol transmission of Hantaan and related viruses to laboratory rats. Am J Trop Med Hyg **38**:636-40.

Old RW and Primrose SB, (1994), Principles of gene manipulation (fifth edition). Blackwell science, Oxford.

Oliver, S.G. (1994) Back to bases in biology. Nature, **368**, 14-15 – a discussion of the status and potential importance of genome sequencing projects.

Parsons JB, Newbusy HT, Jackson MT, Ford-Lloyd BV (1997) Mol. Breed. **3:** 115

Paterson AH, Tanksley SD, Sorells ME (1991) *Adv. Agro.* **46:** 39

Piatak, M., Jr., Ka-Cheung Luk, Bill Williams, and Jeffrey D. Lifson (1993). Quantitative competitive polymerase chain reaction for accurate quantitation of mv DNA and RNA species. *BioTechniques* **14:**70-80

Powell, W., M. Morgante, C. Andre, M. Hanafey, J.Vogel, S.S. Tingey and J.A. Rafalski. (1996). The comparison of RFLP, RAPD, AFLP and SSR (microsatellite) markers for germplasm analysis. Mol. Breed. **2:**225-238.

Prince HN, Prince DL, Prince RN. (1991) Principles of viral control and transmission. In: Block SS, ed. Disinfection, sterilization, and preservation. 4th ed. Philadelphia: Lea & Febiger, **41**1-44.

ProberJM, GL Trainor, RJ Dam, FW Hobbs, CW Robertson, RJ Zagursky, AJ Cocuzza, MA Jensen, and K Baumeister (1987) A system for rapid DNA sequencing with fluorescent chain-terminating dideoxynucleotides. Science **238:** 336-341.

Protein Methods D.M. Boolag and SJ. Edelstein (1991) Wiley - Liss, New york.

Ramesh VJ, Bennetzen EJ, Zehr BEL (1995) Town and Country Conference Center, San Diego, CA.

Rapley R and Walker JM, (1998), Molecular Biomethos Handbook, Humana Press, Totowa NY.

Robert, H. Jerald, R. S. Mary, A. S. & McWhorter, C. A., (eds). (1974) Laboratory Instrumentation. Harper & Row Publishers, Maryland, USA.

Roberts, R.J. and Macelis, D. (1993) Restriction enzymes and methylases. Nucleic Acids Research, **21** 3125-37 – comprehensive list of all the known restriction endonucleases and their recognition sequences.

Robinson, R. A., and Stokes, R. H., (1968) Electrolyte solutions, the measurement and interpretation of conductance, chemical potential, and diffusion in solutions of simple electrolytes, 2nd ed., rev. London, Butter worths.

Roder MS, Korzun V, Wendehake K, Plaschke J, Tixier MH, Leroy P, Ganal MW (1998) *Genetics* **149:** 2007

Rogers, S.O. and Bendich, A.J. (1985) Extraction of DNA from milligram amounts of fresh, herbarium and mummified plant tissues. *Plant Molecular Biology*, **5:** 69-97 – the CTAB method.

Rothstein, R. J. *et. al.* (1979) Synthetic adaptors for cloning DNA, Methods in Enzymology, **68:** 98-109.

Ruiz-Martinez MC, J Berka, A Belenkii, F Foret, AW Miller, and BL Karger (1993) DNA sequencing by capillary electrophoresis with replaceable liner polyacrylamide and laser-induced fluorescence detection. Anal Chem **65:** 2851-2858.

S. Ignacimuthu, (2003), Methods in Biotechnology Phoenix Publishing house Pvt. Ltd.

Sahgai – Maroof *et al,* (1984), PNAS **81:** 8014-18.

Saiki, R.K., Gelfand, D.H., Stoffel, S., Scharf, S.J., Higuchi, R., Horn, G.T., Mullis, K.B. and Erlich, H.A. (1988) Primer-directed enzymatic amplification of DNA with a thermostable DNA polymerase. Science, **239:** 487-91 – the first description of PCR with Taq polymerase.

Salimath SS, De-Oliviera AC, Godwin ID, Bennetzen JL (1995) *Genome* **38:** 757

Sambrook, Fritsch and Maniatis (1987). Molecular Cloning, a laboratory manual.

Sanger F, S Micklen, and AR Coulson (1977) DNA sequencing and chain-terminating inhibitors. Proc Natl Acad Sci USA **74:** 5463-5467.

Sanger, F., Nicklen, S. & Coulson, A.R.(1977). DNA sequencing with chain-terminating inhibitors. Proceedings of the National Academy of Sciences, USA, **74:** 5463-7.

Sanghi-Maroof, M.A; Biyashav, R.M; Yang, G.P; Zhang, Q. and Allard, R.W. (1994) Proc. Natl. Acad. Sci. USA **91:** 5466-5470.

Shiyuza, H., Birren, B., Kim, U.J., Mancino, V., Slepak, T., Tachiiri, Y. and Simon, M. (1992) Cloning and stable maintenance of 300-kilobase-pair fragments of human DNA in Escherichia coli using an F-factor –based vector. Proceedings of the National Academy of Sciences, USA, **89**, 8794-7.

Singh, B.D. (2005). Biotechnology: Expanding Horizons, Kalyani Publication, New Delhi.

Smith, H.O. and Wilcox, K.W. (1970) A restriction enzyme from Haemophilus influenzae. *Journal of Molecular Biology,* **51:** 379-91-one of the first full descriptions of a restriction endonuclease.

Smith, LM, JZ Sanders, RJ Kaiser, P Hughes, C Dodd, CR Connell, C Heiner, SBH Kent, and LE Hood 1986 Fluorescence detection in automated DNA sequence analysis. Nature **321:** 674-679.

Smith, M.L., Bruhn, J.N. and Anderson, J.B. (1992) The fungus Armillaria bulbosa is among the largest and oldest living organisms. Nature, **356,** 428-31- an example of the use of RAPD analysis.

Sninsky, OOs. Academic Press, San Diego, CA, pp. 69-83. Quartin, Robin *S.,* and James G. Wetmur (1989). Effect of ionic strength on the hybridization of oligodeoxynucleotides with reduced charge due to methylphosphonate linkages to unmodified oligodeoxynucleotides containing the complementary sequence. *Biochemistry* **28:** 1040-1047.

Southern, E.M. (1975) Detection of specific sequences among DNA fragments separated by gel electrophoresis. Journal of Molecular Biology, **98:** 503-7-Southern hybridization.

Spiropoulou CF, Morzunov S, Feldmann H, Sanchez A, Peters CJ, Nichol ST. Genome structure and variability of a virus causing hantavirus pulmonary syndrome. Virology (in press).

Staub JE, Serquen FC & Gupta M (1996). Genetic markers, map construction, and their application in plant breeding. HortScience, **31:** 729-741.

Staub JE, Serquen FC, Gupta M (1996) *Hort. Sci.* **31:** 729

Steere, N, (ed), (1988) handbook of laboratory safety, 2nd ed. CRC Press, Boca Raton, Fl, USA

Swerdlow H, JZ Zhang, DY Chen, HR Harke, R Grey, SL Wu, NJ Dovichi, and C Fuller (1991) Three DNA sequencing methods using capillary gel electrophoresis and laser-induced fluorescence. Anal Chem **63:** 2835-2841.

Tai TH and Tanksley SD (1990), Plant Mol Biol Reporter **8 (4):** 297 – 303.

Tanksley SD (1993) *Ann. Rev. Genet.* **27:** 205

Temnykh S, Park WD, Ayres N, Cartinhour S, Hauck N, Lipovich L, Cho YG, Ishii T, 33. 31. McCouch SR (2000) *Theo. Appl. Genet.* **100:** 697

Thompson, J.D., D.G. Higgins and J.T. Gibson. (1994). CLUSTAL W: improving the sensitivity of progressive multiple alignment through sequence weighting, position-specific gap penalties and weight matrix choice Nucleic Acids Res. **22**: 4673-4680.

Towards fully automated genome-wide polymorphism screening. Net Genet 9, **341**-2.

Tsai TF. (1987) Hemorrhagic fever with renal syndrome: mode of transmission to humans. Lab Anim Sci; **37:** 428-30.

Umenai T, Lee HW, Lee PW, *et al.* Korean haemorrhagic fever in staff in an animal laboratory. Lancet (1979);**i**:1,314-6.

University Press, U. K. (1994).

Vos P, Hogers R, Bleeker M, Reijans M, Lee TVD, Hornes M, Frijters A, Pot J, Peleman J, 33. Kuiper M, Zabeau M (1995). *Nucleic Acids Res.* **23:** 4407

Vos, P., R. Hogers, M. Bleeker, M. Rijans, T. Van de Lee, M. Hornes, A. Frijters, J. Pot, J. Peleman, M. Kuiper and M. Zabeau. (1995). AFLP: a new technique for DNA fingerprinting. Nucleic Acids Rs. **23:** 4407-4414.

W.Rychlik, W.J. Spencer and RE Rhoads, (1990), Optimization of the annealing temperature for DNA amplification in vitro, Nucleic Acids Research, Vol **18**, No **21**, 6409-6413.

Walker, J.M. (1996) The Protein Protocols Handbook. Humana Press, Totowa, NJ.

Walters P, Powell W, Lagudah E, Snape J, Henderson K in Plant and Animal Genome VIII Conference, 9-12 Jan, 2000, San Diego, USA (http://www.intl pag.org/pag/8/abstracts)

Weber JL, May PE (1989) *Amer. J. Human Genet.* **44:** 386

Welsh, J. and M. McClelland. (1990). Fingerprinting genomes using PCR with arbitrary primers. Nucleic Acids Res. **18:** 7213-7218.

Wetmur, James G., and John 1. Sninsky (1995). Nucleic acid hybridization and unconventional bases. In *PCR Strategies* Innis, M.A., D.H. Gelfand, and **1.1.**

Wilson. K. & Walker, J. (1995) Princjples apd Techniques of Practical Biochemistry. 4'10 ed. Cambridge University Press. OK.

Wu YT, Zhang TZ, Yin JM (2001) *Yi Chuan Xue Bao* **28:** 1040

Xu F, Sun M (2001) *Mol. Phylogenet. Evol.* **21:** 372

Yanagihara R. Hantavirus infection in the United States: epizootiology and epidemiology. Rev Infect Dis (1990); **12:** 449-57.

Zhang JZ, Y Fang, JY Hou, HJ Ren, R Jiang, P Roos, and NJ Dovichi (1995) Use of non-cross-linked polyacrylamide for four-color DNA sequencing by capillary electrophoresis separation of fragments up to 640 bases in length in two hours. Anal Chem **67:** 4589-4593.

Zhang Q, Arbuckle J, Wessler SR (2000) *Proc. Natl. Acad. Sci. USA* **97:** 1160

Zietkiewichz E, Rafalski A & Labuda D (1994). Genome fingerprinting by simple sequence repeat (SSR)-anchored polymerase chain reaction amplification. Genomics, **20:** 176-183.

Zietkiewicz E, Rafalski A, Labuda D (1994) *Genomics* **20:** 176

❑❑❑

IMPORTANT WEBSITES

1. http://www.cfsan.fda.gov
2. http://wwww.cgn.wagennigen-ur.nl/pgr/research/molgenbibliogr.htm
3. http://www.web-books.com/MoBio/Free/Ch9D.htm
4. http://www.pyrosequencing.com/DynPage.aspre ?
5. http://web.uct.ac.Za/microbiology
6. http://www.mcb.uct.ac.Za/pcroptine.htm
7. http://www.1.qiagen.com/resources/Info/guidelines_f
8. http://www.roche-applied-science.com/pcr/application
9. http://www.crcpress.com
10. http://www.embo.org/scisoc/tw02pcr.pdf.
11. http://www.mgh.harvard.edu/gastro/pcr2006
12. http://www.mun.ca/biology/scarr/4241sq_Sanger_dideoxy_sequencing
13. http://www.csun.edu/~hcbio027/biotechnology/lec3/sanger.html
14. http://www.statwww.berkeley.edu/users/terry/Classes/s260.1998/Week8b/week8b/node9.html
15. http://www.zoo.utoronto.ca/able/volumes/vol-24/mini.2.kosinksi.pdf
16. http://www.plattsburgh.edu/acadvp/artsci/biology/bio401/DNASeq.html
17. http://www.idtdna.com/support/technical/technicalbulletinPDF
18. http://oak.cats.ohiou.edu/~ballardh/molsyst/issrs.doc
19. http://ncbi.nih.gov/books/bv.fcgi?call=bv.View.ShowSection&rid=handbook.glossary.1237
20. http://pest.cabveb.org/PDF/BER/Ber88-6/Ber88577.pdf.
21. http://www.asicoaquaticummarkers.com/Glossary.htm
22. http://upload.wikwmedia.org/wikipedia/en/b/b8/DNA-structure-and-bases.png
23. http://wwwbiowire.com/nucleus/nucleus_1_1.jsp
24. http://www.karymullis.com
25. http://www.le.ac.uk/biology/phh4/addprot.htm
26. http://www.mrcindia.org/mol-ent.htm
27. http://www.ncbi.nlm.nih.gov/entrez/query.fcgi

28. http://www.nhm.ac.uk/hosted_sites/acarology/saas/saa/pdf07/003-014.pdf
29. http://www.nobel.se/chemistry/laureates/1993/mullis-lecture.html
30. http://www.oznet.ksu.edu/biotech/glossary.htm
31. http://www.phschool.com/science/biology_place/biocoach/dnarep/i ntro.html
32. http://www.scipub.net/botany/molecular-markers-plant-genetics-biotechnology.html
33. http://www.brenda.uni.koeln.de
34. http://www.omega.cc.umb.edu
35. http://www.chemfinder.cambridgesoft.com
36. http://www.biosupplynetcom
37. http://www.instructcitcornell.edu
38. http://www.biochemistcom
39. http://www.worthington-biochem.com
40. http://www.ultranetcom
41. http://www.biochemistry.org
42. http://www.biochem.ucl.ac.uk
43. http://www.mcb.harvard.edu
44. http://www.gwu.edu
45. http://www.biology.arizona.edu
46. http://www.instructcitcornell.edu
47. http://www.ncbi.nlm.nih.gov
48. http://www.antibodyresource.com
49. http://www.msdssearch.com
50. http:/hazard.com/msds/

❑❑❑

ANNEXURE I

Preparation of Chromic Acid (Cleansing Solution)

Dissolve 63.0 gm of potassium dichromate by heating in 350 ml. distilled water, cool and add concentrated H_2SO_4 into it to make the volume to 1000 ml. H_2SO_4 acid should be added along the slide of the flask slowly. Handle the solution carefully and use rubber glove and apron.

Preparation of Acid and Alkali (1N) Solution

S. No.	*Acid/Alkali*	*Final Volume Made to One Liter with Distilled Water*
1	Concentrated sulfuric acid	30 cc
2	Concentrated hydrochloric acid	100 cc
3	Concentrated Nitric acid	63 cc
4	Oxalic acid dehydrated	62.96 cc
5	Sodium hydroxide	40.00 g
6	Potassium hydroxide	56.00 g

Preparation of Standard Buffer Solutions

Potassium hydrogen phthalate (0.05 M): Dissolve 1.021 g of phthalate in distilled water and make up the volume to 100 ml in a volumetric flask

Phosphate buffer (0.05 M): Dissolve 340 mg of potassium phosphate monobasic (KH_2PO_4) and 355 mg of anhydrous sodium phosphate dibasic (Na_2HPO_4) in distilled water and make up the volume to 100 ml, using a volumetric flask.

Borax (0.01 M): Dissolve 383 mg of borax in distilled water and make up to volume to 100 ml using a volumetric flask.

S. No.	*Chemical*	*Molarity*	*pH*	*Temperature (°C)*
1	Potassium hydrogen phthalate	0.05	4.00 4.01 4.02	25 30 35
2	Phosphate Buffer	0.05	6.86 6.85 6.84	25 30 35
3.	Sodium teteraborate (Borex)	0.01	9.18 9.14 9.10	25 30 35

ANNEXURE II

Molecular Biology Laboratory Waste Disposal

Category	*Description*	*Container*	*Place to Keep*	*Handling*	*Routing*
Non-recyclable waste glass	Plate glass, Pyrex, light bulbs, broken glass chemical containers, Non-contaminated ** plastic serological pipettes	Big cardboard container with heavy plastic liner	Corridor and/or Lab	Removed by Lab staff when full	Solid waste (not recyclable)
Empty chemical containers	Intact, clean triple-rinsed glass and plastic containers; recyclable without caps	Special plastic bucket with half lid	Corridor	Removed by lab staff when full	Recyclable
Lab trash	Non-contaminated** gloves, bench paper, packaging materials, foil, plastic bags, paper towels, weighing boats, bottle caps, fly media, fly embryo plates, culture plates (with or without media); culture and centrifuge tubes (with or without media), filter flasks	Standard waste can	Lab	Removed by custodial staff when full	Solid waste

Chemical Waste	See Lab Safety Manual web site to determine whether hazardous or non-hazardous waste	Hazardous waste in appropriate containers. Non-hazardous waste in trash or drain	Lab	Hazardous waste removed monthly by lab staff.	Solid waste or Hazardous Waste
All sharps	All Pasteur and other glass pipettes, needles, syringes, scalpel blades, razor blades, slides, cover slips	Labeled sharps container	Lab	Placed in medical waste boxes by lab occupants	Solid waste
Other waste	Experimentally cultured stocks, plates or other materials Ethidium bromide gels, poly acryl amide Gels, contaminated animal carcasses, contaminated*** etc.	Standard biohazard waste box with red plastic liner	Lab	Sealed and placed in corridor by lab occupants	Solid waste or Bio Hazardous Waste

** Non-contaminated applies to any material not having been in contact with an infectious agent.

*** Other Waste - Any solid waste generated in the diagnosis, treatment, or immunization of human beings or animals, in research pertaining there to, or in the production or testing of biological, in the following categories: cultures and stocks (of infectious agents)

ANNEXURE III

Hazard Symbols and Precautions

Symbol	*Use For*	*Precaution*
	Harmful-Irritant	Do not breathe vapor and avoid contact with skin and eye
	Highly Flammable	Keep away from combustible material
	Oxidizing	Keep away from combustible material
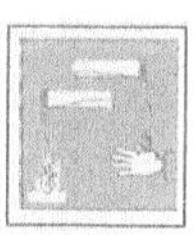	Corrosive	Do not breathe vapor and avoid contact with skin and eye
	Toxic	Avoid contact
	Explosive	Store with extra care keep away from combustible material
	Dangerous for the Environment	Disposal need extra care
	Radioactivity	Avoid contact and use suitable shielding

ANNEXURE IV

Inactivation of Ethidium Bromide

Ethidium bromide is recognized as a potent mutagen and must be disposed of in a responsible manner. We list two alternatives for ethidium bromide waste removal:

1. Ethidium Bromide decontamination from liquid solution can be performed using a filtration/extraction device available at a reasonable price from Schleichjer and Schull, Supelco, Inc. also offer an extraction device for $EtBr_2$ decontamination.
2. A more tedious but effective alternative involves the following. For every 100 ml of $EtBr_2$, add 25 ml of freshly prepared 5% hypo phosphorous acid solution and 12 ml of 0.5 ml fresh sodium nitrite. The final pH of solution should be < 3.0 stir briefly and let stand for 20 hours. Neutralize with an excess of 1M sodium bicarbonate and discard. In contrast to common practice, inactivation by sodium hypochlorite (bleach) is not effective in eliminating mutagenesis (Lunn and Sanditone 1987).

ANNEXURE V

Biological Agents Causing Biohazards

Disease	*Agent*	*Means of Spread*	*Host Animal*
Anthrax	*Bacillus anthraces*	Contact, inhalation, ingestion	Farm animals
Brucellosis	*Brucella spp.*	Contact, ingestion	Swine, dogs, cattle, sheep, goats
Tuberculosis	*Mycobacterium spp.*	Contact, inhalation, ingestion	Primates
Salmonellosis	*Salmonella spp.*	Contact, inhalation, ingestion	Farm animals, rodents, reptiles, amphibians
Tetanus	*Clostridium tetani*	Bite and soil-contaminated puncture wounds	Horses, other equines (also carried by other mammals, and present in soil)
Rabies	*Rabies* virus	Bites, saliva contact	Dogs, bats, other farm animals
Monkey B Virus	*Herpesvirus simiae*	Bite wounds, contact	Old World monkeys
Lymphocytic choriomeningitis (LCM)	*Lymphocytic choriomeningitis* virus	Contact, inhalation	Mice, guinea pigs, hamsters, monkeys
Toxoplasmosis	*Toxoplasma gondii*	Ingestion of oocytes, inhalation	Cats
Ringworm	*Dermatophytes*	Contact	Dogs, cats, guinea pigs, cattle
Histoplasmosis	*Histoplasma capsulatum*	Inhalation of fungi	Dogs, other domestic and wild species

ANNEXURE VI

Risk Group 1 (RG1) Agents

RG1 agents are not associated with disease in healthy adult humans. Examples of RG1 agents include asporogenic *Bacillus subtilis* or *Bacillus licheniformis*, *Bacillus subtilis* or *Bacillus licheniformis* Host-Vector Systems, Exceptions); adeno- associated virus (AAV) types 1 through 4; and recombinant AAV constructs, in which the transgene does not encode either a potentially tumorigenic gene product or a toxin molecule and are produced in the absence of a helper virus. A strain of *Escherichia coli*, *Escherichia coli* K-12 Host Vector Systems, Exceptions) is an RG1 agent if it (1) does not possess a complete lipopolysaccharide (*i.e.*, lacks the O antigen); and (2) does not carry any active virulence factor (*e.g.*, toxins) or colonization factors and does not carry any genes encoding these factors.

Those agents not listed in Risk Groups (RGs) 2, 3 and 4 are not automatically or implicitly classified in RG1; a risk assessment must be conducted based on the known and potential properties of the agents and their relationship to agents that are listed.

Risk Group 2 (RG2) Agents

RG2 agents are associated with human disease which is rarely serious and for which preventive or therapeutic interventions are *often* available.

Bacterial Agents

— *Acinetobacter baumannii* (formerly *Acinetobacter calcoaceticus*)

— *Actinobacillus*

— *Actinomyces pyogenes* (formerly *Corynebacterium pyogenes*)

— *Aeromonas hydrophila*

— *Amycolata autotrophica*

— *Archanobacterium haemolyticum* (formerly *Corynebacterium haemolyticum*)

— *Arizona hinshawii*—all serotypes

— *Bacillus anthracis*

— *Bartonella henselae, B. quintana, B. vinsonii*

— *Bordetella* including *B. pertussis*

— *Borrelia recurrentis, B. burgdorferi*

— *Burkholderia* (formerly *Pseudomonas* species)

— *Campylobacter coli, C. fetus, C. jejuni*

— *Chlamydia psittaci, C. trachomatis, C. pneumoniae*

— *Clostridium botulinum, Cl. chauvoei, Cl. haemolyticum, Cl. histolyticum, Cl. novyi, Cl. septicum, Cl. tetani*
— *Corynebacterium diphtheriae, C. pseudotuberculosis, C. renale*
— Dermatophilus congolensis
— Edwardsiella tarda
— *Erysipelothrix rhusiopathiae*
— *Escherichia coli* - all enteropathogenic, enterotoxigenic, enteroinvasive and strains bearing K1 antigen, including *E. coli* O157:H7
— *Haemophilus ducreyi, H. influenzae*
— *Helicobacter pylori*
— *Klebsiella* - all species except *K. oxytoca* (RG1)
— *Legionella* including *L. pneumophila*
— *Leptospira interrogans* - all serotypes
— *Listeria*
— *Moraxella*
— *Mycobacterium* (except those listed in (RG3)) including *M. avium* complex, *M. asiaticum, M. bovis* BCG vaccine strain, *M. chelonei, M. fortuitum, M. kansasii, M. leprae, M. malmoense, M. marinum, M. paratuberculosis, M. scrofulaceum, M. simiae, M. szulgai, M. ulcerans, M. xenopi*
— *Mycoplasma*, except *M. mycoides* and *M. agalactiae* which are restricted animal pathogens
— *Neisseria gonorrhoeae, N. meningitidis*
— *Nocardia asteroides, N. brasiliensis, N. otitidiscaviarum, N. transvalensis*
— *Rhodococcus equi*
— *Salmonella* including *S. arizonae, S. cholerasuis, S. enteritidis, S. gallinarum-pullorum, S. meleagridis, S. paratyphi, A, B, C, S. typhi, S. typhimurium*
— *Shigella* including *S. boydii, S. dysenteriae, S. flexneri, S. sonnei*
— *Sphaerophorus necrophorus*
— *Staphylococcus aureus*
— *Streptobacillus moniliformis*
— *Streptococcus* including *S. pneumoniae, S. pyogenes*
— *Treponema pallidum, T. carateum*
— *Vibrio cholerae, V. parahemolyticus, V. vulnificus*
— *Yersinia enterocolitica*

Fungal Agents

— *Blastomyces dermatitidis*
— *Cladosporium bantianum,* C. (Xylohypha) *trichoides*
— *Cryptococcus neoformans*
— *Dactylaria galopava* (*Ochroconis gallopavum*)

— Epidermophyton
— *Exophiala* (Wangiella) *dermatitidis*
— *Fonsecaea pedrosoi*
— Microsporum
— *Paracoccidioides braziliensis*
— *Penicillium marneffei*
— *Sporothrix schenckii*
— *Trichophyton*

Parasitic Agents

— *Ancylostoma* human hookworms including *A. duodenale, A. ceylanicum*
— *Ascaris* including *Ascaris lumbricoides suum*
— *Babesia* including *B. divergens, B. microti*
— *Brugia* filaria worms including *B. malayi, B. timori*
— *Coccidia*
— *Cryptosporidium* including *C. parvum*
— *Cysticercus cellulosae* (hydatid cyst, larva of *T. solium*)
— *Echinococcus* including *E. granulosis, E. multilocularis, E. vogeli*
— *Entamoeba histolytica*
— *Enterobius*
— *Fasciola* including *F. gigantica, F. hepatica*
— *Giardia* including *G. lamblia*
— *Heterophyes*
— *Hymenolepis* including *H. diminuta, H. nana*
— *Isospora*
— *Trichinella s—Leishmania* including *L. braziliensis, L. donovani, L. ethiopia, L. major, L. mexicana, L. peruvania, L. tropica*
— *Loa loa* filaria worms
— *Microsporidium*
— *Naegleria fowleri*
— *Necator* human hookworms including *N. americanus*
— *Onchocerca* filaria worms including, *O. volvulus*
— *Plasmodium* including simian species, *P. cynomologi, P. falciparum, P. malariae, P. ovale, P. vivax*
— *Sarcocystis* including *S. sui hominis*
— *Schistosoma* including *S. haematobium, S. intercalatum, S. japonicum, S. mansoni, S. mekongi*
— *Strongyloides* including *S. stercoralis*
— *Taenia solium*

— *Toxocara* including *T. canis*

— *Toxoplasma* including *T. gondii piralis*

— *Trypanosoma* including *T. brucei brucei, T. brucei gambiense, T. brucei rhodesiense, T. cruzi*

— *Wuchereria bancrofti* filaria worms

Viruses

— *Adenoviruses, human—all types*

— *Alphaviruses (Togaviruses)—Group A Arboviruses*

— *Eastern equine encephalomyelitis virus*

— *Venezuelan equine encephalomyelitis vaccine strain TC-83*

— *Western equine encephalomyelitis virus*

Arenaviruses

— *Lymphocytic choriomeningitis* virus (non-neurotropic strains)

— Tacaribe virus complex

— Other viruses

Bunyaviruses

— *Bunyamwera virus*

— *Rift Valley fever virus vaccine strain MP-12*

— *Other viruses*

— *Caliciviruses*

— *Coronaviruses*

— *Flaviviruses* (Togaviruses) - *Group B Arboviruses*

— *Dengue virus serotypes 1, 2, 3, and 4*

— *Yellow fever virus vaccine strain 17D*

— *Other viruses*

— *Hepatitis A, B, C, D, and E viruses*

— *Herpesviruses - except Herpesvirus simiae* (Monkey B virus) —*Cytomegalovirus*

— *Epstein Barr virus*

— *Herpes simplex types 1 and 2*

— *Herpes zoster*

— *Human herpesvirus types 6 and 7*

— *Orthomyxoviruses*

— *Influenza viruses types A, B, and C*

— *Other tick-borne orthomyxoviruses*

— *Papovaviruses*

— *All human papilloma viruses*

— *Paramyxoviruses*

— *Newcastle disease virus*
— *Measles virus*
— *Mumps virus*
— *Parainfluenza viruses types 1, 2, 3, and 4*
— *Respiratory syncytial virus*
— *Parvoviruses*
— *Human parvovirus* (B19)
— *Picornaviruses*
— *Coxsackie viruses types A and B*
— *Echoviruses—all types*
— *Polioviruses—all types, wild and attenuated*
— *Rhinoviruses—all types*
— *Poxviruses—all types except Monkeypox virus (and restricted poxviruses including Alastrim, Smallpox, and Whitepox*
— *Reoviruses—all types including Coltivirus, human Rotavirus, and Orbivirus* (Colorado tick fever virus)

Risk Group 3 (RG3) Agents

RG3 agents are associated with serious or lethal human disease for which preventive or therapeutic interventions may be available.

Bacterial Agents

— *Bartonella*
— *Brucella* including *B. abortus, B. canis, B. suis*
— *Burkholderia* (Pseudomonas) *mallei, B. pseudomallei*
— *Coxiella burnetii*
— *Francisella tularensis*
— *Mycobacterium bovis M. tuberculosis*
— *Pasteurella multocida type B -"buffalo" and other virulent strains*
— *Rickettsia akari, R. australis, R. canada, R. conorii, R. prowazekii, R. rickettsii, R, siberica, R. tsutsugamushi, R. typhi* (R. mooseri)
— *Yersinia pestis*

Fungal Agents

— *Coccidioides immitis* (sporulating cultures; contaminated soil)
— *Histoplasma capsulatum, H. capsulatum var.. duboisii*

Parasitic Agents

None

Viruses and Prions

— *Alphaviruses* (Togaviruses) - *Group A Arboviruses*

— *Semliki Forest virus*

— *St. Louis encephalitis virus*

— *Venezuelan equine encephalomyelitis virus except the vaccine strain TC-83,*

— *Other viruses*

— *Arenaviruses*

— *Flexal*

— *Lymphocytic choriomeningitis virus* (LCM) (neurotropic strains)

— *Bunyaviruses*

— *Hantaviruses including Hantaan virus*

— *Rift Valley fever virus*

— *Flaviviruses* (Togaviruses) - *Group B Arboviruses*

— *Japanese encephalitis virus*

— *Yellow fever virus*

— *Other viruses*

— *Poxviruses*

— *Monkeypox virus*

— *Prions*

— *Transmissible spongioform encephalopathies* (TME) *agents Creutzfeldt-Jacob disease and kuru agents*

— *Retroviruses*

— *Human immunodeficiency virus* (HIV) *types 1 and 2*

— *Human T cell lymphotropic virus* (HTLV) *types 1 and 2*

— *Simian immunodeficiency virus* (SIV)

— *Rhabdoviruses*

— *Vesicular stomatitis virus*

Risk Group 4 (RG4) Agents

RG4 agents are likely to cause serious or lethal human disease for which preventive or therapeutic interventions are not usually available.

Risk Group 4 (RG4) - *Bacterial Agents*

None

Risk Group 4 (RG4) - *Fungal Agents*

None

Risk Group 4 (RG4) - *Parasitic Agents*

None

Risk Group 4 (RG4)–*Viral Agents*

Arenaviruses

— *Guanarito virus*

Lassa virus

Junin virus

Machupo virus

Sabia

Bunyaviruses (Nairovirus)

— *Crimean-Congo hemorrhagic fever virus*

Filoviruses

— *Ebola virus*

— *Marburg virus*

Flaviruses (Togaviruses) - *Group B Arboviruses*

— *Tick-borne encephalitis virus complex including Absetterov, Central European encephalitis, Hanzalova, Hypr, Kumlinge, Kyasanur Forest disease, Omsk hemorrhagic fever, and Russian spring-summer encephalitis viruses*

Herpesviruses (alpha)

Herpesvirus simiae (Herpes B or Monkey B virus)

Paramyxoviruses

Equine morbillivirus

Hemorrhagic fever agents and viruses as yet undefined.

Annexure VII

Biosafety Committees in India

Competent Authority	*Composition*	*Functions*
Recombinant DNA advisory committee (RDAC)	As determined by the department of biotechnology—to consist of experts in their individual capacity.	To review biotechnology developments at national and international levels; to recommend suitable biosafety regulations for India.
Review committee on genetic manipulation ERR	Member secretary, department of biotechnology; Indian Council of Medical Research; Indian Council of Agricultural Research; Council of Scientific and Industrial Research; other experts in their individual capacity.	To issue guidelines for GMO research; to authorise rDNA projects in high risk category III; to authorise controlled field experiments; to permit imports of GMOs for research.
Institutional biosafety committees (IBSC)	Head of the organisation; scientists engaged in rDNA work; biosafety or medical officer; nominee, department of biotechnology.	To seek RCGM approval for category III risk; to ensure adherence with biosafety guidelines; to prepare an emergency plan; to inform DLC, SBCC and GEAC about relevant experiments.
Genetic engineering approval committee (GEAC)	Chair, additional secretary, ministry of environment and forests; co-chair: department of biotechnology representative; representatives from ministry of industrial development, departments of biotechnology and atomic energy; Indian Council of Agricultural Research; Indian Council of Medical Research; Council of Scientific and Industrial Research; Directorate of Plant Protection; Central Pollution Control Board; others in individual capacity.	To authorise commercial use (including import) of GMOs or their products; to authorise large-scale production and release of GMOs and their products into the environment; to mandate restrictions or prohibitions on production, sale, import or use of GMOs, if necessary.
State biotechnology coordination committee (SBCC) ERR	Chief secretary, state government; secretaries, department of environment, health, agriculture, commerce, forests, public works, public health;	To periodically review safety and control measures in institutions handling GMOs; to inspect and take punitive action in case of

	chairman, state pollution control board; state microbiologists and pathologists; other experts in individual capacity	violations through the state pollution control board or the directorate of health; to act as nodal agency at the state level to assess damage, if any, from release of GMOs, and to take on site control measures.
District-level committee (DLC)	District collector; factory inspector; pollution control board representative; chief medical officer; district agricultural officer; public health department representative; district microbiologists/pathologists; municipal corporation commissioner; other experts in individual capacity.	To monitor safety regulations in installations; to investigate compliance with rDNA guidelines and report violations to SBCC or GEAC; to act as nodal agency at district level to assess damage, if any, from release of GMOs and to take on site control measures
Monitoring and evaluation committee (MEC)	Chairman, jointly elected by secretary, department of biotechnology and secretary, department of agricultural research and education. To include plant biotechnologists, plant ecologists, Seed technologists, and plant breeders (nominated by RCGM or ICAR), an NBPGR nominee, an MOEF nominee, and the member-secretary of the RCGM	To undertake field visits at experimental sites; to suggest remedial measures to adjust original trial design; to assist RCGM in collecting and analysing field data; to collect or cause to collect information on comparative agronomic advantages of transgenic plants.

ANNEXURE VIII

Preparation of Buffer Equilibrated Phenol

Material required: Magnetic stir plate, Magnetic stir bar, 2 L glass beaker, Glass container, Aluminium foil, Amber colour bottle, 25 ml glass pipette, pH indicator paper.

Reagents: 8-hydroxiquinoline, Molecular biology grade phenol (Melt crystal in 65°C water bath), 50 mm Tris base (unadjusted pH), 50 mM Tris adjusted to pH 8.0 with HCl.

Procedure

- Place a stir bar in 2 L beaker.
- Add 0.5M of 8-hydroxiquinoline to the glass beaker.
- Gently pour in 50 ml of melted phenol. The phenol will turn yellow.
- Add 500 ml of 500 mM Tris base.
- Cover the beaker with aluminum foil.
- Stir 10 min. at low speed on magnetic stir plate at room temperature.
- Remove from the stir plate
- Let the phase separate
- Aspirate the top phase with glass pipette in to a suitable waste receptacle
- Add 500 ml of 50 mM Tris pH 8.0.
- Repeat steps 5 through 9.
- Check the pH of the phenol phase with indicator paper, if it is not pH 8.0 step 4 through 12 should be repeated.
- Add 250 ml of 50 mM Tris-HCl, pH 8.0 or TE buffer, pH.
- Place in glass container (Amber bottle).
- Store at 4°C (stable for two months) or –20°C

Precautions

- Phenol cause severe burn, gloves, safety glasses and a lab coat should be worn when working with Phenol solutions.
- All manipulations should be carried out in fume hood.

ANNEXURE IX

Miscellaneous Information

1. *Molecular Weight Markers for Agarose Gels and SDS Page*

For DNA Gel:

λDNA Markers (length = 48,502 bp)

λDNA: Hind III digest generates 8 fragments of sizes (in base pairs)
23,130/9,416/6,557/4,361/2,322/2,027/564 and 125.

λDNA: EcoRI digest generates 6 fragments of sizes (in base pairs)
21,226/7,421/5,805/5,643/4,878 and 3,530

pUC-18 plasmid DNA Markers (length= 2,686 bp)

pUC18: MSP I digest generates 13 fragments of sizes (in base pairs)
501/486/404/331/242/190/147/111/110/67/34/34 and 26.

pBR-322 plasmid DNA Markers (length = 4,363 bp)

pBR 322: Alu I digest generates 17 fragments of sizes (in base pairs)
908/659/656/521/403/281/257/226/100/90/63/57/49/46/19/15 and 11

pBR 322: Hinf I digest generates 10 fragments of sizes (in base pairs)
1,631/517/506/396/344/298/221/220/154 and 75.

Molecular Weight Marker of SDS PAGE

S.No.	*Protein*	*Molecular Weight (kDa)*
1.	Aprotinin (bovine lung)	6.5
2.	Cytochrome C	11.7
3.	Lysozyme	14.3
4.	Myoglobin	17
5.	Trypsinogen	24
6.	Carbonic anhydras	29

7.	Egg albumin	45
8.	BSA	66
9.	Haemoglobin (Hb)	68
10.	Immunoglobuline (IgG)	150
11.	β-Galactosidase *(E.coli)*	116
12.	Catalase	240
13.	Myosine	205
14.	Thyroglobulin	670

2. *Physical Constants of the Nucleotide Triphosphates*

Compound	*M.W.*	*λmax (pH 7.0)*	*Σmax × 10^{-3} (pH 7.0)*	*Absorbensy Ratio (280/260)*	*pKa of Base*
ATP	507.2	259	14.7	0.16	4.0
CTP	483.2	280	13.0	2.12	4.8
GTP	523.2	252	13.7	0.67	3.3 & 9.3
UTP	484.2	262	10.0	0.38	9.5
dATP	491.2	259	15.2	0.15	3.6
dCTP	467.2	280	13.1	2.10	4.3
dGTP	507.2	253	13.7	0.67	2.3 & 9.9
dTTP	482.2	267	9.6	0.73	9.3

Conversions: For conversion of base pairs into molecular weight (in Daltons), multiply by 660, as average molecular weight of one base pair = 660 Daltons.

For conversion of nucleotides into molecular weight (in Daltons), multiply by 330 as average molecular 'weight one nucleotide = 330 Daltons.

For conversion of amino acids into molecular weight (in Daltons), multiply by 110 as average molecular weight for an amino acid = 110 Daltons. Therefore, a protein with 300 amino acids will be having a molecular weight of ~33,000 Daltons or 33 kDa.

3. *Isoelectric Points of Some Proteins*

S.No.	*Protein*	*Isoelectric Point*
1	Amyloglucosidase (*A.niger*)	3.6
2	BSA	4.9
3	Carboxipeptidase B (human pancreas)	6.0
4	Casein	4.8
5	Egg albumin	4.6
6	Gelatin	4.8
7	Glucose Oxidase (*A.niger*)	4.2
8	Haemoglobin	6.8
9	Lysozyme	11.0
10	Myoglobin	7.0
11	Ovalbumine (chicken egg)	4.8
12	Pepsin	1.0
13	Ribonuclease	9.6
14	Tubuline	5.5
15	Urease	5.0

4. *Visible Light Absorption and Complimentary Colours*

S.No.	*Wavelength (nm)*	*Filter(Colour absorbed)*	*Complementary (Color Transmitted)*
1	380-435	Violet	Yellowish green
2	436-480	Blue	Yellow
3	481-490	Greenish blue	Orange
4	491-500	Bluish green	Red
5	501-560	Green	Purple
6	561-580	Yellowish green	Violet
7	581-595	Yellow	Blue
8	596-650	Orange	Greenish blue
9	651-780	Red	Bluish green

5. *Properties of Commonly Used Commercial Acids and Bases*

Acid/Base	*MW*	*Specific Gravity*	*Per cent (W/V)*	*Gram/Litre*	*Normality (N)*	*Molarity (M)*
HCl	36.46	1.18	35.5	418.9	11.49	11.49
H_2SO_4	98.08	1.84	98.0	1,803.2	36.77	18.39
HNO_3	63.00	1.42	70.0	910.0	15.80	15.80
H_3PO_4	98.00	1.75	88.0	1,540.0	47.14	15.71
$HClO_4$	100.48	1.54	61.0	939.4	9.35	9.35
CH_3COOH	60.05	1.049	99.7	1,045.9	17.42	17.42
NH_3	17.03	0.910	20.0	227.5	13.36	13.36
NOH	35.0	0.90	28.0	251.0	14.80	14.80

6. *Radioactivity Decay Table for ^{32}P and ^{35}S*

Reference: Short Protocols in Molecular Biology. Edited by F. M. Ausubel *et al*, Wiley, New York.

Decay factors for calculating the amount of radioactivity present at any given time after the reference date For example, to find out the radioactivity present in a vial after 33 days *i.e.,* 4 weeks and 5 days, which was originally containing 50 μCi of an ^{35}S labeled compound on the reference date, see column 4 and row 5 of the table which reads 0.770. So the activity remaining after 33 days = 50 μCi × 0.770 = 38.5 μCi.

^{32}P Half Life: 14.3 days

DAYS	HOURS 0	12	24	36	48	60	72	84
0	1.000	0.976	0.953	0.930	0.908	0.886	0.865	0.844
4	0.824	0.804	0.785	0.766	0.748	0.730	0.712	0.695
8	0.679	0.662	0.646	0.631	0.616	0.601	0.587	0.573
12	0.559	0.546	0.533	0.520	0.507	0.495	0.483	0.472
16	0.460	0.449	0.439	0.428	0.418	0.408	0.398	0.389
20	0.379	0.370	0.361	0.353	0.344	0.336	0.328	0.320
24	0.312	0.305	0.298	0.291	0.284	0.277	0.270	0.264
28	0.257	0.251	0.245	0.239	0.234	0.228	0.223	0.217
32	0.212	0.207	0.202	0.197	0.192	0.188	0.183	0.179
36	0.175	0.170	0.166	0.162	0.159	0.155	0.151	0.147
40	0.144	0.140	0.137	0.134	0.131	0.127	0.124	0.121
44	0.119	0.116	0.113	0.110	0.108	0.105	0.102	0.100
48	0.098	0.095	0.093	0.091	0.089	0.086	0.084	0.082
52	0.080	0.078	0.077	0.075	0.073	0.071	0.070	0.068

^{35}S half Life; 87.4 days

WEEK \ DAYS	0	1	2	3	4	5	6
0	1.000	0.992	0.984	0.976	0.969	0.961	0.954
1	0.946	0.939	0.931	0.924	0.916	0.909	0.902
2	0.895	0.888	0.881	0.874	0.867	0.860	0.854
3	0.847	0.840	0.833	0.827	0.820	0.814	0.807
4	0.801	0.795	0.788	0.782	0.776	0.770	0.764
5	0.758	0.752	0.746	0.740	0.734	0.722	0.722
6	0.717	0.711	0.705	0.700	0.694	0.689	0.683
7	0.678	0.673	0.667	0.662	0.657	0.652	0.646
8	0.641	0.636	0.631	0.626	0.621	0.616	0.612
9	0.607	0.602	0.597	0.592	0.588	0.583	0.579
10	0.574	0.567	0.565	0.560	0.556	0.552	0.547
11	0.543	0.539	0.534	0.530	0.526	0.522	0.518
12	0.514	0.510	0.506	0.502	0.498	0.494	0.490

7. *Storage of Culture*

Numerous methods are used to store strains. A permanent, but time consuming, method for *E. coli* is freeze, drying (Lyopholization), By which strains have been kept alive for more than 30 years. Another permanent and useful method for *E. coli* and *Saccharomyces cerevisiae* is ultra cold storage at – 70°C or in liquid nitrogen. To prevent bursting of the cells during freezing, the culture medium must contain a cryo preservative such as 7% (V/V) dimethyl sulfoxide (DMSO) or 15% (V/V glycerol) To recover cells from frozen cultures the stored cultures is Scraped with a sterile loop or sterile toothpick and ice crystals are allowed to thaw on an agar plate (and then streaked) or placed in sterile growth medium. If the glycerol content in increased to 40-50 %, culture can be stored in a liquid state for several years at – 20°C.

For convenience, stable strains of *E. coli* can be stored at room temperature. As a deep stab culture in airtight tubes containing rich agar. Such slab culture last for several years. They are prepared by autoclaving a screw cap tube containing liquid agar and allowing the agar to harden. Bacteria are picked with a wire loop from an agar plate or a liquid culture and the loop is stabled into the rich agar. The tube is then placed in an incubator and the bacteria are grown for 24 hours, after which the tube is sealed with parafilm and stored at room temp. The lifetime of a deep stab culture is determined primarily by the quality of the airtight seal because it is essential that the agar does not dry.

For routine use in the laboratory, a slant culture is used for *E. coli* or *S. cerevisiae* liquid agar is placed in a small screwcap tube and sterilized the tube is allowed to harden the agar surface is then covered with bacteria or yeast by using either a wire loopful or cells by placing a droplet of culture directly on the surface. The tube is incubated overnight to allow for cell growth and the cap is then tightened. Slant cultures are stored in a refrigerator and will last several months, less if opened frequently.

8. *Glycerol Stocks*

Transfer 0.5 ml of an overnight culture into a labeled 1.5 ml cryo vial and add 0.5 ml of sterile 30% glycerol. Cap the tube and mix gently. Store at – 80°C. Long term viability of stocks depends on the particular strain. After naturally grow bacteria in medium containing 8-10% glycerol in plastic multi well plates and store at – 80°C. This method is typically used for storing cosmid, bacteriophage PI, and BAC and cDNA libraries.

9. *Yeast Glycerol Stock*

Patch out clones on an appropriate plate. Incubate at 30°C for 2 days. Inoculate 6 ml of YPD medium with a "match head full" of cells from the plate (this is a large inoculum; the culture will be turbid before incubation). Add 2 ml of sterile 80% glycerol and mix thoroughly. Transfer 0.5 ml aliquots into 1 ml freezer vial with O-ring seals. Thoroughly shake the freezer vials and freeze at – 60°C or lower (typically – 80°C). Yeast tends to die if frozen at temperature above – 55°C. Yeast strains can be stored at – 80°C indefinitely by using this method. Note that strains grown in YPD medium before freezing have better long term viability those grown in selective medium.

10. *Bacterial Strains and their Phenotypes*

S.No.	*Strain*	*Phenotype*	*Remarks*
1	C600	F, e14⁻ (McrA⁻), *thr⁻ 1, leuB6 ,thi-1, laeYl, supE44, rfb D 1fhu A 21;* The original C600 is *Eco K r* m+* Mcr BC+	A *E. coli* K -12 strain, also known as CR-34, used for transformation of plasmids
2	DH5α	F', *end A* 1, *hsd R* 17 (r_k^- m_k^+) *sup* E44, thi-1, *recA, gyrA (Nal^r), reA* 1, Δ(*lacZYA-argF*) U169, *deoR, (∅80 dlac Δ(laeZ)M15)*	Widely used for transformation and large isolation of plasmids, gives stable transformants with all types of plasmids
3	HB 101	F⁻, *Δ(gpt-proA)62, leu, supE44, ara14, galK2, lacY1, Δ(mcrC-.mrr), rpsL20 (Str^r), xyl-5, mtl-1, recA13.A13.*	A hybrid *between E.coli* K-12 and *E.coli* B, a good host for large scale isolation of plasmids
4	JM83	F⁻, *ara, Δ(lac-proAB), rpsL(Str^r), [∅80 dlac Δ(laeZ)M15]*	Constructed by J. Messing,' a host strain for pUC based vectors, does not require IPTG for induction
5	JM107 (#401-7)	F', *traD36, lac^q, Δ(lacZ)M15, proA⁺B⁺/ e14⁻¹(MeA⁻), Δ(lac-proAB), thi, gyr A96(nal^r),endA1, had R17 ($r^-_k m^-_k$) relA1, supE44.*	As above, but requires IPTG for induction, a host for growth of single stranded phage M13
6	JM 109 (#401-9)	JM 107 *recA 1*	As above but a *rec* mutant.
7	LE 392	F⁻, *e14⁻(mcrA-), hsdR514($r_k^-m_k^+$), supE44, supF58, lac Yl, or Δ(laclZY)6, galK2, galT22, metB1, trpR55*	Constructed by. L. Enquist, a *su+* strain commonly used for propagation of vectors and its recombinants
8	RR1	HB101 *ReeA⁺*	A *rec* derivative of HB101 constructed by R. Rodriguez. This strain can be used to. tansform with high efficiency with plasmid vectors annealed to cDNA by—homopolymeric tailing

11. *DNA Storage*

DNA can be conveniently stored in caped and parafilm sealed micro centrifuge tubes with the DNA dissolved in sterile water, TE suffer, or 100% ethanol. It is important to remove potentially damaging agents such as phenol, nucleases, and ethidium bromide from the DNA before dissolving the DNA in water buffer or ethanol. In the short term (several days) storage of DNA in TE buffer, pH 8.0 at 4°C is acceptable and is preferred if the DNA stock is to be used repeatedly over the course of a few weeks. Freezing thawing storage of DNA at –20°C in now frost free freezers is acceptable for months DNA may be stored for longer period at –70°C.

ANNEXURE X

Periodic Table of the Elements

1 H																	2 He
3 Li	4 Be											5 B	6 C	7 N	8 O	9 F	10 Ne
11 Na	12 Mg											13 Al	14 Si	15 P	16 S	17 Cl	18 Ar
19 K	20 Ca	21 Sc	22 Ti	23 V	24 Cr	25 Mn	26 Fe	27 Co	28 Ni	29 Cu	30 Zn	31 Ga	32 Ge	33 As	34 Se	35 Br	36 Kr
37 Rb	38 Sr	39 Y	40 Zr	41 Nb	42 Mo	43 Tc	44 Ru	45 Rh	46 Pd	47 Ag	48 Cd	49 In	50 Sn	51 Sb	52 Te	53 I	54 Xe
55 Cs	56 Ba	57 La	72 Hf	73 Ta	74 W	75 Re	76 Os	77 Ir	78 Pt	79 Au	80 Hg	81 Tl	82 Pb	83 Bi	84 Po	85 At	86 Rn
87 Fr	88 Ra	89 Ac	104 Rf	105 Db	106 Sg	107 Bh	108 Hs	109 Mt	110 110	111 111	112 112						

58 Ce	59 Pr	60 Nd	61 Pm	62 Sm	63 Eu	64 Gd	65 Tb	66 Dy	67 Ho	68 Er	69 Tm	70 Yb	71 Lu
90 Th	91 Pa	92 U	93 Np	94 Pu	95 Am	96 Cm	97 Bk	98 Cf	99 Es	100 Fm	101 Md	102 No	103 Lr

Alkali Metal	Alkaline Metal	Basic Metal
Halogen	Noble Gas	Non Metal
Rare Earth	Semi-Metal	Transition Metal